Cândida Ferreira

Gene Expression Programming

Studies in Computational Intelligence, Volume 21

Editor-in-chief
Prof. Janusz Kacprzyk
Systems Research Institute
Polish Academy of Sciences
ul. Newelska 6
01-447 Warsaw
Poland
E-mail: kacprzyk@ibspan.waw.pl

Further volumes of this series can be found on our homepage:
springer.com

Vol. 5. Da Ruan, Guoqing Chen, Etienne E. Kerre, Geert Wets (Eds.)
Intelligent Data Mining, 2005
ISBN 3-540-26256-3

Vol. 6. Tsau Young Lin, Setsuo Ohsuga, Churn-Jung Liau, Xiaohua Hu, Shusaku Tsumoto (Eds.)
Foundations of Data Mining and Knowledge Discovery, 2005
ISBN 3-540-26257-1

Vol. 7. Bruno Apolloni, Ashish Ghosh, Ferda, Alpaslan, Lakhmi C. Jain, Srikanta Patnaik (Eds.)
Machine Learning and Robot Perception, 2005
ISBN 3-540-26549-X

Vol. 8. Srikanta Patnaik, Lakhmi C. Jain, Spyros G. Tzafestas, Germano Resconi, Amit Konar (Eds.)
Innovations in Robot Mobility and Control, 2005
ISBN 3-540-26892-8

Vol. 9. Tsau Young Lin, Setsuo Ohsuga, Churn-Jung Liau, Xiaohua Hu (Eds.)
Foundations and Novel Approaches in Data Mining, 2005
ISBN 3-540-28315-3

Vol. 10. Andrzej P. Wierzbicki, Yoshiteru Nakamori
Creative Space, 2005
ISBN 3-540-28458-3

Vol. 11. Antoni Ligęza
Logical Foundations for Rule-Based Systems, 2006
ISBN 3-540-29117-2

Vol. 13. Nadia Nedjah, Ajith Abraham, Luiza de Macedo Mourelle (Eds.)
Genetic Systems Programming, 2006
ISBN 3-540-29849-5

Vol. 14. Spiros Sirmakessis (Ed.)
Adaptive and Personalized Semantic Web, 2006
ISBN 3-540-30605-6

Vol. 15. Lei Zhi Chen, Sing Kiong Nguang, Xiao Dong Chen
Modelling and Optimization of Biotechnological Processes, 2006
ISBN 3-540-30634-X

Vol. 16. Yaochu Jin (Ed.)
Multi-Objective Machine Learning, 2006
ISBN 3-540-30676-5

Vol. 17. Te-Ming Huang, Vojislav Kecman, Ivica Kopriva
Kernel Based Algorithms for Mining Huge Data Sets, 2006
ISBN 3-540-31681-7

Vol. 18. Chang Wook Ahn
Advances in Evolutionary Algorithms, 2006
ISBN 3-540-31758-9

Vol. 19. Ajita Ichalkaranje, Nikhil Ichalkaranje, Lakhmi C. Jain (Eds.)
Intelligent Paradigms for Assistive and Preventive Healthcare, 2006
ISBN 3-540-31762-7

Vol. 20. Wojciech Pecznek, Agata Pó³rola
Advances in Verification of Time Petri Nets and Timed Automata, 2006
ISBN 3-540-32869-6

Vol. 21 Cândida Ferreira
Gene Expression Programming: Mathematical Modeling by an Artificial Intelligence, 2006
ISBN 3-540-32796-7

Cândida Ferreira

Gene Expression Programming

Mathematical Modeling by an Artificial Intelligence

Second, revised and extended edition

 Springer

Dr. Cândida Ferreira
Chief Scientist
Gepsoft Ltd.
73 Elmtree Drive
Bristol BS13 8NA
United Kingdom
E-mail : candidaf@gepsoft.com

ISSN print edition: 1860-949X
ISSN electronic edition: 1860-9503

ISBN 978-3-642-069321-1 Springer Berlin Heidelberg New York

Springer is a part of Springer Science+Business Media
springer.com
© Springer-Verlag Berlin Heidelberg 2006
Softcover reprint of the hardcover 2nd edition 2006

Cover design: deblik, Berlin
Typesetting: by the author and TechBooks
Printed on acid-free paper SPIN: 11506591   89/Strasser   5 4 3 2 1 0

To José Simas
For All the Dreams

and

To my Grandfather, Domingos de Carvalho
For His Vision

Preface to the Second Edition

The idea for this second edition came from Janusz Kacprzyk on April 29, 2005, who kindly invited me to his new Springer series, Studies in Computational Intelligence. The initial plan was to correct the usual typos and mistakes but leave the book unchanged, as Janusz thought (and I agreed with him) that it was the proper moment for a second edition. But then there was the problem of the new format and I had to reformat and proofread everything again. And I just thought that while I was at it, I might as well change some things in the book to make it more enjoyable and interesting. Foremost in my thoughts was the restructuring of chapter 4, The Basic GEA in Problem Solving. In that chapter, buried together with a wide variety of problems, were several important new algorithms that I wanted to bring to the forefront. These algorithms include: the GEP-RNC algorithm (the cornerstone of several new other algorithms); automatically defined functions; polynomial induction; and parameter optimization. So I removed all these materials from chapter 4 and gave them the deserved attention by writing four new chapters (chapter 5 Numerical Constants and the GEP-RNC Algorithm, chapter 6 Automatically Defined Functions in Problem Solving, chapter 7 Polynomial Induction and Time Series Prediction, and chapter 8 Parameter Optimization). Then this new structure just begged for me to include one of the last additions to the GEP technique – decision trees – that I was regretfully unable to include in the first edition (I implemented decision trees in August of 2002, two months before sending the manuscript to the printer). So, chapter 9, Decision Tree Induction, is totally new to this second edition and an interesting addition both to the book and to GEP. The last three chapters, chapter 10 Design of Neural Networks, chapter 11 Combinatorial Optimization, and chapter 12 Evolutionary Studies, remain basically unchanged.

With all this restructuring in chapter 4, I was able to develop several new topics, including solving problems with multiple outputs in one go and designing parsimonious solutions both with parsimony pressure and user defined functions, also interesting new extensions to the GEP technique. Furthermore, the section on Logic Synthesis was totally restructured and an interesting analysis of the most common universal logical systems is presented.

Chapter 3, The Basic Gene Expression Algorithm, with the exception of section 2, Fitness Functions and the Selection Environment, remains practically unchanged. In section 2, however, I introduce several new fitness functions that are then used to explore more efficiently the solution landscapes of the problems solved in the book.

The inversion operator is one of the latest additions to the GEP technique, and you will notice an extra entry for it in chapters 3 and 10. Furthermore, you'll also notice that both IS and RIS transposition were slightly modified so that the transposon sizes were automatically chosen rather than a priori set. But unbeknownst to me, this slightly different implementation had consequences in the performance of these operators, and the new implementation is slightly more unpredictable in terms of performance. This is the reason why in the Evolutionary Studies of chapter 12, the old implementation of these operators is still used. And to be fair, the comparison of the inversion operator with these transposition operators should also use a fixed set of sizes for the inverted sequences. And this is the reason why inversion is not analyzed in chapter 12.

Cândida Ferreira

December 20, 2005

Preface to the First Edition

I developed the basic ideas of gene expression programming (GEP) in September and October of 1999 almost unaware of their uniqueness. I was reading Mitchell's book *An Introduction to Genetic Algorithms* (Mitchell 1996) and meticulously solving all the computer exercises provided at the end of each chapter. Therefore, I implemented my first genetic algorithm and I also implemented what I thought was a genetic programming (GP) system. Like a GP system, this new system could also evolve computer programs of different sizes and shapes but, surprisingly, it surpassed the old GP system by a factor of 100-60,000. So, what happened here? What was responsible for this astounding difference in performance? For an evolutionary biologist, the answer is quite straightforward: this new system – gene expression programming – simply crossed the phenotype threshold. This means that the complex computer programs (the phenotype) evolved by GEP are totally encoded in simple strings of fixed length (the chromosomes or genotype). The separation of the genotype from the phenotype is comparable to opening a Pandora box full of good things or possibilities. Of these good things, perhaps the most important is that there are virtually no restrictions concerning the number or type of genetic operators used. Another important thing is that the creation of higher levels of complexity becomes practically a trivial task. Indeed, it was trivial to create a multigenic system from a unigenic one and a multicellular system from a unicellular one. And each new system creates its own box of new possibilities, which enlarges considerably the scope of this new technique.

In this first book on gene expression programming I describe thoroughly the basic gene expression algorithm and numerous modifications to this new algorithm, providing all the implementation details so that anyone with elementary programming skills (or willing to learn them) will be able to implement it themselves. The first chapter briefly introduces the main players of biological gene expression in order to show how they relate to the main players of artificial evolutionary systems in general and GEP in particular. The second chapter introduces the players of gene expression programming, showing their structural and functional organization in detail. The language especially created to express the genetic information of GEP chromosomes is also described in this chapter. Chapter 3 gives a detailed description of the basic gene expression algorithm and the basic genetic operators. In addition, a very simple problem is exhaustively dissected, showing all the individual programs created during the discovery process in order to demystify the workings of adaptation and evolution. Chapter 4 describes some of the applications of the basic gene expression algorithm, including a large body of unpublished materials, namely, parameter optimization, evolution of Kolmogorov-Gabor polynomials, time series prediction, classifier systems, evolution of linking functions, multicellularity, automatically defined functions, user defined functions and so forth. The materials of Chapter 5 are also new and show how to simulate complete neural networks with gene expression programming. Two benchmark problems are solved with these GEP-nets, providing an effective measure of their performance. Chapter 6 shows how to do combinatorial optimization with gene expression programming. Multigene families and several combinatorial-specific operators are introduced and their performance evaluated on two scheduling problems. The last chapter discusses some important and controversial evolutionary topics that might be refreshing to both evolutionary computists and evolutionary biologists.

Acknowledgments

The invention of a new paradigm can often create strong resistance, especially if it seems to endanger long established technologies and enterprises. The publication of my work on scientific journals and conferences, which should be forums for discussing and sharing new ideas, became a nightmare and both my work and myself were outright dismissed and treated with scorn. Despite the initial opposition and due to a set of happy circumstances and resources, I was finally able to make my work known and available to all. I am deeply indebted to José Simas, an accomplished graphic and web designer and software developer, for believing in me and in GEP from the beginning and for helping its expansion and promotion on the World Wide

Web. Together we founded Gepsoft and developed software based on gene expression programming which is already helping numerous scientists and engineers worldwide. And thanks to Gepsoft it was possible for me to concentrate fully on the writing of this book and on the development of several new algorithms. Indeed, my work at Gepsoft benefited tremendously from my writing and vice versa.

I am also very grateful to Pedro Carneiro, a talented musician with an avid mind, for reading and editing the first three chapters of the manuscript. José Simas also read several drafts of the manuscript, accompanying the process from the beginning and contributing with valuable discussions and suggestions. He is, in fact, my first reader and I always write with him in my mind.

Finally, I would like to thank José Gabriel, a talented printer and skilled craftsman, for his involvement in the making of this book from the start and for handling its printing with special care.

Cândida Ferreira
October 15, 2002

List of Symbols

ADF	Automatically defined function
ADF-RNC	ADFs with random numerical constants
AND	AND or Boolean function of two arguments #8
APS	Gepsoft Automatic Problem Solver
CA	Cellular automata
Dc	Gene domain for encoding random numerical constants
DT	Decision tree
Dt	Gene domain for encoding the thresholds of neural networks
Dw	Gene domain for encoding the weights of neural networks
EDT	Evolvable decision trees
EDT-RNC	Evolvable decision trees with numeric attributes
ET	Expression tree
FN	False negatives
FP	False positives
GA	Genetic algorithm
GEA	Gene expression algorithm
GEP	Gene expression programming
GEP-ADF	GEP with automatically defined functions
GEP-EDT	GEP for inducing evolvable decision trees
GEP-KGP	GEP for inducing Kolmogorov-Gabor polynomials
GEP-MO	GEP for solving problems with multiple outputs
GEP-NC	GEP with numerical constants
GEP-nets	GEP for inducing neural networks
GEP-NN	GEP for inducing neural networks
GEP-PO	GEP for parameter optimization
GEP-RNC	GEP with random numerical constants
GKL	Gacs-Kurdyumov-Levin rule
GOE	Greater Or Equal or Boolean function of two arguments #13

GP	Genetic programming
GT	Greater Than or Boolean function of two arguments #4
HZero	Parameter optimization algorithm with a head size of zero
IC	Initial configuration
IF	Rule of three arguments #202, If $a = 1$, then b, else c
IS	Insertion sequence elements
LOE	Less Or Equal or Boolean function of two arguments #11
LT	Less Than or Boolean function of two arguments #2
MAJ	Majority function of three arguments or rule #232
MGF	Multigene family
MIN	Minority function of three arguments or rule #23
MUX	3-Multiplexer or rule #172, If $a = 0$, then b, else c
NAND	NAND or Boolean function of two arguments #7
NC	Numerical constants
NLM	NAND-like module
NOR	NOR or Boolean function of two arguments #1
NPV	Negative predictive value
NXOR	NXOR or Boolean function of two arguments #9
OR	OR or Boolean function of two arguments #14
ORF	Open reading frame
PPV	Positive predictive value
RIS	Root insertion sequence elements
RNC	Random numerical constants
TAP	Task assignment problem
TN	True negatives
TP	True positives
TSP	Traveling salesperson problem
UDF	User defined function
ULM	Universal logical module
XOR	Exclusive-OR or Boolean function of two arguments #6

Contents

1   Introduction:
The Biological Perspective

The aim of this chapter is to bring into focus the basic differences between gene expression programming (GEP) and its predecessors, genetic algorithms (GAs) and genetic programming (GP). All three algorithms belong to the wider class of Genetic Algorithms (the use of capitals here is meant to distinguish this wider class from the canonical GA) as all of them use populations of individuals, select the individuals according to fitness, and introduce genetic variation using one or more genetic operators. The fundamental difference between the three algorithms resides in the nature of the individuals: in GAs the individuals are symbolic strings of fixed length (chromosomes); in GP the individuals are nonlinear entities of different sizes and shapes (parse trees); and in GEP the individuals are also nonlinear entities of different sizes and shapes (expression trees), but these complex entities are encoded as simple strings of fixed length (chromosomes).

If we have in mind the history of life on Earth (see, for instance, Dawkins 1995 or Maynard Smith and Szathmáry 1995), we can see that the difference between GAs and GP is only superficial: both systems use only one kind of entity that works both as genotype and body (phenotype). These kinds of systems are condemned to have one of two limitations: on the one hand, if they are easy to manipulate genetically, they lose in functional complexity (the case of GAs); on the other hand, if they exhibit a certain amount of functional complexity, they are extremely difficult to reproduce with modification (the case of GP).

In his book, *River Out of Eden*, Richard Dawkins (1995) gives a list of thresholds of any life explosion. The first is the "replicator threshold" which consists of a self-copying system in which there is hereditary variation. Also important is that replicators survive by virtue of their own properties. The second threshold is the "phenotype threshold" in which replicators survive by virtue of causal effects on something else. This "something else" is what

Cândida Ferreira: *Gene Expression Programming*, Studies in Computational Intelligence (SCI) **21**, 1–27 (2006)
www.springerlink.com

is called the phenotype or body, that is, the entity that faces the environment and does all the work. A simple example of a replicator/phenotype system is the DNA/protein system of life on Earth. It is believed that for life to move beyond a very rudimentary stage, the phenotype threshold must be crossed (e.g., Dawkins 1995; Maynard Smith and Szathmáry 1995).

Similarly, the entities of both GAs and GP (simple replicators) survive by virtue of their own properties. Understandingly, there has been an effort in the last years in the scientific community to cross the phenotype threshold in evolutionary computation. The most outstanding effort is developmental genetic programming or DGP (Banzhaf 1994) where binary strings are used to encode mathematical expressions. The expressions are decoded using a five-bit binary code, called genetic code. However, contrary to its analogous natural genetic code, this "genetic code", when applied to binary strings, frequently produces invalid expressions (in nature there is no such thing as a structurally incorrect protein). Therefore, a huge amount of computational resources goes into editing these illegal structures, which limits this system considerably. Not surprisingly, the gain in performance of DGP over GP is minimal (Banzhaf 1994; Keller and Banzhaf 1996).

Gene expression programming is an example of a full-fledged replicator/ phenotype system where the chromosomes/expression trees form a truly functional, indivisible whole (Ferreira 2001). Indeed, in GEP there is no such thing as an invalid expression tree or program. Obviously, the interplay of GEP chromosomes and expression trees requires an unambiguous translation system to transfer the language of chromosomes into the language of expression trees. Furthermore, we will see that the structural organization of GEP chromosomes allows the unconstrained modification of the genome, creating the perfect conditions for evolution to occur. Indeed, the varied set of genetic operators developed to introduce genetic modification in GEP populations always produce valid expression trees, making GEP a simple artificial life system, well established beyond the replicator threshold.

To help the non-biologist understand the fundamental difference between gene expression programming and the other genetic algorithms and why GEP is such a leap forward in evolutionary computation, it is useful to know a little more about the structure and function of the main players of biological gene expression and how they work together. What follows is a very brief introduction to the structure and function of the main molecules of information metabolism and how mutation in proteins relates to evolution. If you wish to pursue these questions further, any textbook on biochemistry will do, although I recommend the third edition of *Biochemistry* by Mathews et al. (1999) for its clarity and elegant presentation.

1.1 The Entities of Biological Gene Expression

In the cell, the expression of the genetic information is a very complex process involving hundreds of molecules. For our purposes, though, it is enough to know the basics about the structure and function of the main players: DNA, RNA, and proteins.

DNA is the carrier of the genetic information and the proteins read and express that information. RNA is a working copy of DNA and, although its existence makes sense in the environment of the cell, its equivalent is of little use in a computer system like GEP. Nonetheless, it is important to know the structure and properties of this molecule in order to understand the fundamental difference between GEP and the other genetic algorithms.

1.1.1 DNA

DNA molecules are long, linear strings of four nucleotides (represented by A, T, C, and G). Each DNA molecule is, in fact, a double helix in which one of the strings is the complementary of the other and, thus, adds nothing to the information contained in a single string. In the structure of the double helix, A pairs with T, and C with G (Figure 1.1). The double-stranded, complementary nature of DNA is fundamental for the replication of the genetic information in the cell, but is of little importance in a computer system like GEP or GAs. Indeed, the chromosomes of both GEP and GAs are single-stranded and their replication is done by simple program instructions.

The information stored in DNA consists of the sequence of the four nucleotides, which is called the primary structure of DNA. The secondary structure of DNA consists of the different kinds of double helixes it can form and, most important to us, DNA lacks a tertiary structure, which consists of a unique three-dimensional arrangement of the molecule. Indeed, DNA molecules fold, forming random coils. And because complex functionality such as catalytic activity is closely related to tertiary structure, DNA molecules are useless for doing much of the work that needs to be done in a cell.

However, the simple DNA molecule is excellent to store information. In the structure of the double helix, the complementary nucleotides face each

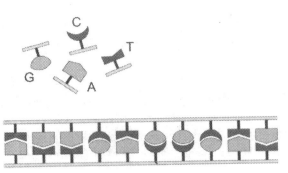

Figure 1.1. Base pairing in the double stranded DNA molecule. Note that the bulkier G and A pair, respectively, with the smaller C and T, putting the two strands exactly the same distance apart. Note also that the information contained in one strand is basically the same contained in the other. In fact, DNA strands are said to be complementary.

other and are locked in the interior of the double helix. This makes DNA chemically inert and stable, which are desirable qualities of information keepers. In fact, in the cell, DNA is further protected in the protected environment of the nucleus in eukaryotes or the nucleoid in prokaryotes. But most important to us, is that DNA is incapable of both catalytic activity and structural diversity: first, the potential functional groups (the bases A, T, C and G) are locked up in the interior of the helix and, second, the molecule lacks tertiary structure, another prerequisite for catalytic activity and structural diversity.

Simplifying, DNA may be seen as a long string composed of four different letters (A, T, C, and G) in which the sequence of the letters (or primary structure) consists of the genetic information. The genetic information or the blueprints of all organisms on Earth are written on the four letter language of DNA. For instance, the average mammal genome contains about 5×10^9 base pairs (or letters, if only one strand is considered) of DNA and codes for approximately 300,000 protein and RNA genes which are the immediate products of expression of the genome. Below we will see how the genetic information is expressed as proteins.

1.1.2 RNA

As I said earlier, in the global picture of information metabolism, RNA might be seen as a working copy of a particular sequence of DNA. When a protein gene is expressed, a copy of the gene in the form of messenger RNA (mRNA) is made and used to direct the synthesis of the protein. Thus, in terms of information, the RNA copy contains exactly the same information as the original DNA.

The messenger RNA is not the only kind of RNA molecule working in a cell. Despite its central role in information decoding, structurally, mRNAs are very dull. The structural diversity discussed below is typical of the other classes of RNA, namely transfer RNA (tRNA) and ribosomal RNA (rRNA).

Like DNA, RNA molecules are also long, linear strings of four nucleotides (ribonucleotides, in this case: A, U, C, and G). In contrast to DNA, RNA molecules are single-stranded and some of them are capable of folding in a unique three-dimensional structure. One of the reasons for the folding of RNA molecules resides in the existence of short sequences which are complementary to other sequences within the same molecule. Obviously, if these complementary sequences were to stumble upon each other, short double helixes would be formed. These intramolecular double helixes are indeed fundamental for the unique three-dimensional structure of some RNA molecules.

Thus, like proteins, some RNA molecules can have a unique three-dimensional structure (tertiary structure) and therefore can exhibit some degree of structural and functional diversity. The rules of complementarity in RNA double helixes are much the same as in DNA, with A pairing with U, and C with G. In RNA molecules with tertiary structure, some nucleotides are involved in helix formation and therefore are not chemically available, but other functional groups are free and exposed and thus can engage in different kinds of interactions and even participate in biological catalysis. Indeed, this, together with a unique three-dimensional structure, allows RNA molecules to function as real biological catalysts (ribozymes).

So, despite its reduced chemical vocabulary, RNA is the kind of molecule that can simultaneously function as genotype and phenotype, that is, as a simple replicator. Note, however, how in such cases the genotype and the phenotype are tied up together: any modification on the replicator is immediately reflected in its performance. That is, there is no room for subtle or neutral changes in these systems. And subtle and neutral changes are fundamental to an efficient evolution (Kimura 1983; see also Ferreira 2002c for a discussion of the role of neutrality in artificial evolutionary systems).

Another important constraint of these simple replicator systems can be very well illustrated using the artificial system of GP, also a simple replicator system. In GP, if one were to introduce genetic variation as freely as it is done in nature, most modifications made on the parse trees would have resulted in invalid structures. Indeed, only a very limited number of modifications can be made on GP parse trees in order to guarantee the creation of valid structures. The problem with this kind of system is that extremely efficient search operators such as point mutation cannot be used. Instead, an inefficient sub-tree swapping is used so that valid parse trees are always produced. Nevertheless, no matter how carefully genetic operators are implemented, there are obviously limits to what grafting and pruning can do, and the search space in such systems can never be thoroughly explored.

1.1.3 Proteins

Proteins are linear, long strings of 20 different amino acids and they consist of the immediate expression of the genetic information stored in DNA. This means that the four-letter language of DNA is translated into the more complex 20-letter language of proteins. Obviously, there must be some kind of code (genetic code) to translate the language of the four nucleotides into the language of 20 amino acids. In order to specify each of the 20 amino acids there should be at least 20 DNA "words". By using triplets of nucleotides (codons) for each amino acid, $4^3 = 64$ different three-letter "words" are possible. This is more than adequate to code for the 20 amino acids and, in fact, most amino acids have multiple codons, as only three of the 64 codons code for the instruction "stop synthesis". There is also a codon for a "start synthesis" instruction, but this codon also codes for methionine, one of the 20 amino acids found in proteins. The genetic code is virtually universal, meaning that all organisms on Earth with very few exceptions use the same codons to translate the language of their genes into proteins (the genetic code is shown in section 1.2.4, Figure 1.6).

Thus, the information for proteins is decoded triplet by triplet at a time and expressed as linear sequences of amino acids. Although the amino acid sequence of the protein reflects the sequence of the corresponding DNA molecule, the protein has a unique three-dimensional structure and exhibits unique properties. Because of the richer chemical alphabet of proteins, the linear strings of amino acids fold in special ways giving each protein its individual three-dimensional structure. This unique three-dimensional structure or tertiary organization of proteins, together with the vast chemical repertoire of amino acids, allows proteins to play numerous roles, amongst them the role of biological catalysts or enzymes. In fact, proteins are the real workers of the cell.

Note again that, like RNA, proteins can function simultaneously as genotype and phenotype. Note, however, that despite the richer functional diversity, such systems are also equally constrained: any modification in the replicator is immediately reflected in its performance.

In theoretical terms, the differences between DNA, RNA, and proteins are most useful to help understand the fundamental differences between GAs, GP and GEP. Both GAs and GP are simple replicator systems, using only one kind of entity: linear strings of 0's and 1's in the case of GAs, and complex ramified structures composed of several different elements in the case of GP. Many believe that a simple "RNA world" existed in the early history of life, perhaps contemporary to a simple "protein world". RNA and proteins somehow started working together, recruiting also DNA. The complex DNA/protein system of life on Earth is the descendant of this evolutionary process.

It is surprising that computer scientists some 4 billion years after these events took pretty much the same steps of life on Earth, first inventing simple replicator systems and only later inventing sophisticated replicator/phenotype systems. The genetic algorithm invented by Holland in the 60's (Holland 1975) is analogous to a simple RNA replicator with its linear chromosomes and limited functionality, whereas the algorithm popularized by Koza (1992) is analogous to a simple protein replicator with its richer functionality. Curiously enough, the conscious attempts to create a genotype/phenotype system, despite trying very hard to emulate the DNA/protein system, are far from being the desired leap forward (Banzhaf 1994, Ryan et al. 1998).

On the other hand, the full-fledged genotype/phenotype system of gene expression programming was invented in 1999 by myself (Ferreira 2001), totally unaware of all the hard work done by other researchers to create a genotype/phenotype system. In fact, I first heard of GP in Mitchell's book (Mitchell 1996) and was so impressed that I tried to make a GP on my own. I suppose I just applied what I knew from biochemistry and evolution and, therefore, it never crossed my mind to make a system without an autonomous genome. Obviously, the complicated things of information metabolism were discarded as they are irrelevant to a computer system where the rules are not dictated by chemistry. Consequently, double-stranded chromosomes, RNA-like intermediates, and complicated genetic codes with complicated translation mechanisms did not make their way into gene expression programming. Furthermore, I also knew that for a genotype/phenotype machine to run smoothly, the genetic operators could not be constrained and they should always produce valid structures. And the result was the first truly functional genotype/phenotype system that can be easily implemented using any programming language, as nothing in this algorithm depends on the workings of a particular language.

1.2 Biological Gene Expression

Of the information stored in DNA molecules, what most interests us are the genes coding for proteins. In eukaryotes, an average of only 2% of the genome codes for proteins, whereas the rest is mainly repetitive sequences and introns (noncoding sequences that interrupt the coding sequences of genes). The genomes of prokaryotes and virus are, however, much more compact and they are essentially organized in genes that code for proteins. For our purpose it is enough to have in mind the simplest genomes, since they are already extremely complex life machines.

1.2.1 Genome Replication

When in 1953 James D. Watson and Francis Crick proposed the model for the double helix, the mechanism of DNA replication became obvious to all, and it was only a matter of time to isolate and study the essential participants and understand the details of the mechanism. The complementary, double-stranded DNA molecule opens itself and each strand serves as template for the synthesis of the respective complementary strand (Figure 1.2). For instance, when a virus replicates within a cell, it copies its genome hundreds of times, forming hundreds of new virions in a few minutes.

As fundamental as it is in biology (not only in DNA replication but also in DNA repair), the double-stranded nature of DNA molecules is of little use to replicate the genomes in a computer system such as GEP or GAs, where single-stranded chromosomes are copied using simple computer instructions. More important to us are the modifications that allow the creation of genetic diversity. In the next section, I will briefly introduce the fundamental mechanisms of genetic variation.

1.2.2 Genome Restructuring

In the cell, the processes of information restructuring are related to different functions, like, for instance, response to environmental stresses, repair of

Figure 1.2. Replication of DNA molecules. Each strand acts as a template for a new, complementary strand. When copying is complete, there will be two daughter DNA molecules, each identical in sequence to the mother molecule.

DNA, control of gene expression, and creation of genetic diversity. For our purposes, though, it is only necessary to understand the role they play in evolution, especially how genetic modification is created and how this genetic diversity reflects itself in the numerous protein variants available in the genetic pool of a species.

In nature, the incredible diversity of all organisms, living and extinct, is largely a consequence of the effects of the restructuring processes that take place on the genomes of the organisms, creating a diversity of protein functions. Likewise, in GEP, populations of individuals (computer programs) evolve by developing new abilities and becoming better adapted to the environment due to the genetic modifications accumulated over a certain number of generations. However, thanks to simple mechanisms especially created to prevent mass extinction, in computer systems it is possible to hasten considerably the rate of evolution and make it even faster than the evolution of new strains of virus.

Mutation

When a particular genome replicates itself and passes on the genetic information to the next generation, the sequence of the daughter molecule sometimes differs from that of the mother in one or more points. In spite of the virtual perfection of the replication machinery, sometimes a mismatched nucleotide is introduced in the newly synthesized strand. Although cells have mechanisms for correcting most mismatches (and for this, the complementary double-stranded DNA is extremely useful), some of them are not repaired and are passed on to the next generation (Figure 1.3).

In nature, the rate at which mutation occurs is tightly controlled and different groups of organisms have different mutation rates, with virus and bacteria having higher mutation rates than eukaryotes. Of those, virus of course have the highest mutation rates as a single virion can leave hundreds or even thousands of progeny per infected cell, testing several new genomes in one generation.

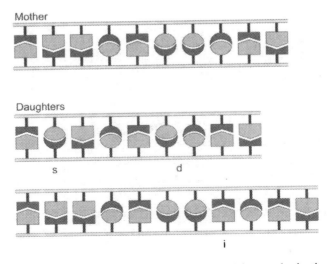

Figure 1.3. Mutations in the DNA sequence of genes. A base substitution (**s**), a small deletion (**d**) and a small insertion (**i**) are shown here.

It is important to analyze more closely the effects of point mutation on the protein itself. In a gene, the replacement of one nucleotide by another can have different effects: (1) the new codon might code for a new amino acid (these are called missense mutations); (2) the new codon might code for a "stop" codon, truncating the protein, or else a "stop" codon mutates into an amino acid codon, elongating the chain (nonsense mutations); (3) fairly frequently, though, point mutations in a gene have no effect at all in the protein sequence as the new codon might code for the same amino acid (neutral mutations); (4) in addition, in eukaryotes, where the sequence of most genes is interrupted by noncoding regions (introns), if a mutation occurs in an intron it has no effect whatsoever in protein sequence (also an example of a neutral mutation).

In another kind of protein mutation, large or small fragments may be inserted or deleted in the coding region of a gene. Large insertions or deletions almost invariably result in the production of a defective protein. The effect of short insertions or deletions depends on whether or not these modifications cause a shift in the reading frame of the gene (if they do, they are called frameshift mutations). If the fragment deleted or inserted is a multiple of three, then one or more codons are removed or added, resulting in the deletion or insertion of one or more amino acids in the protein. The consequences of these non-frameshift deletions/insertions are similar to the ones caused by missense mutations.

The effects of these kinds of mutations on the structure and functionality of a protein can be quite different. Point mutations may be neutral in effect, either not changing the amino acid at all or changing it by another that functions equally well in that position. The deletion/insertion of codons may also be of little consequence, changing only slightly the protein function. Occasionally, such mutations increase the efficiency of a protein, conferring some selective advantage for the organism itself. On the other hand, nonsense mutations and frameshift mutations have, almost every time, a lethal effect, especially if the new protein is fundamental to the survival of the organism. Nonetheless, very occasionally, such mutations might give rise to new, revolutionary traits.

We will see that, in gene expression programming, most mutations, including point mutations and small insertions, have a profound effect in the structure and function of expression trees, more resembling the nonsense and frameshift mutations that occur in nature. Nonetheless, this type of mutation is extremely important for GEP evolvability and several new traits are introduced in this manner. However, less drastic mutations can also be found in gene expression programming. As a matter of fact, some mutations change expression trees very smoothly and they might slightly or significantly increase the efficiency of the expression tree. Furthermore, in gene expression programming, some mutations also have a clear neutral effect. For instance, mutations in the noncoding regions of genes have no effect whatsoever in the structure of expression trees. Other neutral mutations are not so easy to spot because they result in structurally different expression trees. In this case, the new expression tree is equivalent (in mathematical terms) to the parental expression tree. We will see that all kinds of mutation, from the most conservative to the most radical, are important to the evolution of good computer programs.

Recombination

Proteins gradually evolve by accumulating different kinds of mutations over eons of time. But mutation is not the only source of genetic diversity. In the remainder of this section other important genetic operators are presented. One such operator is recombination. In nature there are varied kinds of recombinational processes, involved in different processes and playing different functions. However, during all recombinational processes some fragments of genetic material are exchanged between two distinct donor molecules, as such that genetic information from each donor is present in the offspring (Figure 1.4). For example, during sexual reproduction two paired homologous chromosomes exchange DNA fragments.

The simple image of homologous recombination where homologous chromosomes (chromosomes with extensive sequence homology) are paired, can be useful to help understand the recombinational processes used to create genetic diversity in GEP populations, although in the latter case no sequence homology is required. This simple image is nonetheless very handy, because, in GEP, recombining chromosomes share a structural homology and also because two new daughter chromosomes are created in the process. Thus, during recombination, two chromosomes (not necessarily homologous) are paired and exchange some material between them, forming two new daughter chromosomes. Note, however, that due to the structural homology, a fragment of a particular gene occupying a particular position in the chromosome is never exchanged for a fragment of a gene in a different position; or a fragment of the gene tail is never exchanged for a fragment of the head. In

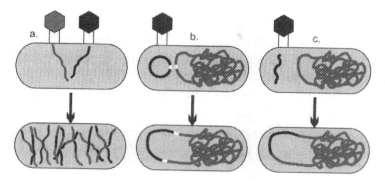

Figure 1.4. Three recombinational processes: **a)** homologous recombination; **b)** site-specific recombination; and **c)** non-homologous recombination. Note that the transforming power of site-specific recombination and non-homologous recombination is much more profound than homologous recombination.

this sense, note that GEP recombination has no counterpart in biology despite its apparent resemblance to homologous recombination.

Transposition

Transposable genetic elements consist of genes that can move from place to place within the genome. There are three classes of transposable elements with different structures and mechanisms of transposition. They exist both in eukaryotes and prokaryotes and cause different effects on the chromosome they move to. For instance, they can inactivate genes by interrupting the coding sequence of a gene; they can activate an adjacent gene by providing a promoter or transcriptional activator; or they can restructure a chromosome by producing homologous sequences that might afterwards be used in homologous recombination. The existence of these "jumping genes" which move from one chromosome to another without respect for boundaries, including species boundaries, greatly altered our views of evolution, as something smoothly driven by neutral mutations alone. The effects of transposition are too drastic to be apparently of any use. And so are the effects of non-homologous recombination or the effects of nonsense or frameshift mutations. Nonetheless, transposition (and non-homologous recombination) is not only fairly frequent but also widely distributed throughout the living world, and many are the marks it left behind in living organisms. Notwithstanding, most of the times, the effect of transposition and some kinds of non-homologous recombination is deleterious. Very occasionally, though, some "promising monsters" are formed and a completely new protein is created. In nature, in fact, very different proteins – both in structure and functionality – share the same construction motifs (domains) probably due to these kinds of genetic modification.

Both the structural details and the varied mechanisms of the different classes of transposons are of little importance in an artificial evolutionary system such as gene expression programming. The transposable elements used in gene expression programming are an oversimplification of those found in nature. First, in GEP, transposable elements were designed to transpose only within the same chromosome. Second, GEP transposable elements might be entire genes or fragments of a gene. Third, any gene or fragment is eligible to become a transposable element, without requirements for particular identifying sequences. Fourth, the transposable element is copied in its entirety at the target site. Finally, in gene transposition, the donor sequence is deleted in the place of origin, whereas in fragment transposition the donor sequence stays unchanged, usually producing two homologous sequences resident in the same chromosome. We will see that, in gene expression programming, simple and repetitive sequences might be created using transposition.

Gene Duplications

Gene duplication plays an important role in protein evolution. Although the mechanism of gene duplication is unknown, occasionally a gene is copied twice during replication. This kind of transformation is potentially harmless and may be advantageous if the protein produced is needed in large amounts. On the other hand, with time, the two copies of the gene might start to evolve independently. One copy may continue to express the original protein, but the other may evolve into an entirely different protein.

In gene expression programming, occasionally, genes are also duplicated. Although there is no special operator for gene duplication, a gene might get duplicated through the combined effects of gene transposition and recombination. Interestingly, chromosomes with duplicated genes are commonly found among the best individuals of GEP populations.

1.2.3 Transcription

The expression of the genetic information into proteins does not proceed directly from DNA. For the language of DNA to be transferred into the language of proteins, an intermediate molecule is necessary. This molecule is a special class of RNA, called messenger RNA, and the process of its synthesis is called transcription.

During transcription the sequence of a gene is copied into an mRNA, using one strand of the DNA molecule as template (Figure 1.5). The rules of complementarity in DNA/RNA double-strands are much the same as in DNA, with rA pairing with dT ("r" stands for RNA and "d" for DNA), rU pairing with dA, rG with dC, and rC with dG. According to the complementary rules, an exact copy of the gene is made in the form of an mRNA. This messenger RNA, containing all the necessary signals for protein synthesis, namely the "start" and "stop" signals, is carried to the appropriate place in the cell and used as template for protein synthesis. For example, in eukaryotes, the mRNAs synthesized in the nucleus must go to the cytosol where the machinery of translation is located.

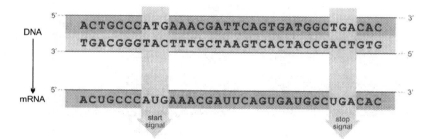

Figure 1.5. Relationship of the DNA to the mRNA. Note that the mRNA is complementary to the DNA strand from which it is transcribed. Note also that there are sequences upstream of the start signal and sequences downstream of the stop signal that will not be translated into amino acids. The sequence shown here is unrealistically short to illustrate both the start and the stop signals.

As I said earlier, the need for an mRNA intermediate makes sense in the environment of the cell, but is of little use in simple computer systems such as gene expression programming, in which small and simple genomes are currently devoid of sophisticated mechanisms of regulation of gene expression. Perhaps in the near future we will see sophisticated GEP-like computer systems with more complex genomes and capable of complex somatic differentiation. Maybe then an mRNA-like intermediate will be necessary to ensure a differential pattern of gene expression.

1.2.4 Translation and Posttranslational Modifications

The process of protein synthesis in the cell involves a very complex machinery, with hundreds of molecules. The main players of translation consist of: (1) the mRNA molecules that work as template for protein synthesis; (2) the ribosomes where the actual decoding takes place; and (3) a special class of RNAs (transfer RNA) that carries the appropriate amino acid to the complex mRNA/ribosome.

Translation

The logistics of the synthesis of both DNA and RNA molecules, based on the complementarity of nucleotides, is fairly simple when compared with the synthesis of proteins. Chemically, there is no simple way of pairing directly the triplet codons with the appropriate amino acids. Indeed, most of the sophisticated machinery of translation evolved to solve exactly this problem. The amino acid must be correctly attached to special molecules which are then coupled to the correct codons in the mRNA molecule. This special class of molecules are also RNA molecules (transfer RNAs), but they are structurally and functionally very different from mRNA. As we have seen, tRNAs have tertiary structure and can therefore have varied functionalities. Although the identification of the correct amino acids is not made by the tRNAs themselves, their unique three-dimensional structure is fundamental to their correct identification by particular enzymes. Each such enzyme recognizes both a particular amino acid and the appropriate tRNA, further attaching the amino acid to the tRNA. Furthermore, each tRNA contains also a nucleotide sequence (the anticodon) that is complementary to the appropriate codon, through which the coupling of the correct amino acid carrier to the mRNA is made.

The set of rules to translate the triplet codons of mRNA into amino acids is the genetic code. Figure 1.6 shows the 64 codons and the amino acids or instructions each codes for. The amino acids are represented by a three-letter abbreviation and by a one-letter abbreviation often used in describing the primary structure in proteins. (I'll stick to this one-letter abbreviation to represent protein chains in all the figures of this chapter.)

The message in the mRNA molecule is read one codon at a time, each codon being correctly paired with the appropriate tRNA through the anticodon, and the transported amino acids are linked one after another, forming a long, linear protein chain whose sequence exactly reflects the sequence of

	U	C	A	G	
U	UUU UUC } Phe - F UUA UUG } Leu - L	UCU UCC UCA UCG } Ser - S	UAU UAC } Tyr - Y UAA stop UAG stop	UGU UGC } Cys - C UGA stop UGG } Trp - W	U C A G
C	CUU CUC CUA CUG } Leu - L	CCU CCC CCA CCG } Pro - P	CAU CAC } His - H CAA CAG } Gln - Q	CGU CGC CGA CGG } Arg - R	U C A G
A	AUU AUC AUA } Ile - I AUG Met - M start	ACU ACC ACA ACG } Thr - T	AAU AAC } Asn - N AAA AAG } Lys - K	AGU AGC } Ser - S AGA AGG } Arg - R	U C A G
G	GUU GUC GUA GUG } Val - V	GCU GCC GCA GCG } Ala - A	GAU GAC } Asp - D GAA GAG } Glu - E	GGU GGC GGA GGG } Gly - G	U C A G

Figure 1.6. The genetic code as expressed in mRNA. Three of the 64 codons are stop signals. The start codon also codes for metionine. Note that the code is redundant with many codons coding for the same amino acid. The amino acid corresponding to each codon is given by the three-letter and one-letter abbreviation often used in describing amino acid sequences in proteins.

the gene (Figure 1.7). It is worth mentioning that ribosomes – key particles in this complex process of protein synthesis – are huge macromolecular structures composed of numerous proteins and another class of RNA molecules, ribosomal RNAs. Like tRNAs, rRNAs also have unique three-dimensional structures. Not surprisingly, rRNAs also participate as real enzymes in the myriad of chemical reactions that occur in the ribosomal machine.

Fortunately for us, the chemical intricacies of translation are of limited interest in a computer system like gene expression programming. In a

Figure 1.7. An oversimplification of translation, showing the essential elements: the mRNA template, the start and stop codons, a charged tRNA, and the emergent protein chain. The sequence shown here is unrealistically short to display both the start and stop codons.

computer system as such, the rules of translation (in the broader sense) are simply defined and simply applied: we don't have to deal with chemistry. Indeed, we don't have to deal with either intermediate transcription processes or with complicated genetic codes requiring complicated translation mechanisms. The genetic code of gene expression programming is a simple one-to-one relationship between the symbols of the genome and the functions and variables (also called terminals or leaves in the jargon of evolutionary computation) they represent.

Posttranslational Modifications

When the machinery of translation reaches a stop signal, a non-functional protein chain is released. Immediately after their release, protein chains are subjected to a variety of modifications. The first of these so called posttranslational modifications is common to all proteins and consists in the folding of the protein chain in its unique three-dimensional structure. Some proteins are further subjected to other posttranslational modifications like, for instance, the chemical modification of some amino acids which greatly enriches the language of proteins; the formation of covalent bonds between particular amino acids; and the removal of some fragments to shorten the chain length. Finally, some folded protein chains (subunits) must aggregate with other subunits to form a multi-subunit protein. Such multi-subunit proteins include many of the most important enzymes and transport proteins in the cell. These proteins are said to have a quaternary structure, the highest level of protein organization.

What is important to understand here, is that each level of protein organization is built on the lower levels, and everything is dictated by the primary structure of the protein, which obviously is ultimately dictated by the gene. Another important thing is that when a protein chain folds itself, amino acids separated by long stretches in the linear chain might be brought together in the final three-dimensional structure; the opposite might also happen and neighbor amino acids in the protein chain may face completely different moieties and be involved in distinct aspects of the structure or functionalities of the protein. Finally, the most important fact about protein structure and function is that the function of a particular protein is dictated by its unique three-dimensional structure, which is ultimately dictated by the sequence of the gene. Note, however, how far a protein is from its DNA sequence! Such is the power of simple transformations and the beauty of emergence.

The expression trees of gene expression programming are also the products of simple transformations and the results are equally overwhelming. We will also see that, in GEP, the expression trees also fold in particular ways, bringing together elements distant in the gene, and separating others that were close. We will also see that some expression trees have a quaternary structure, being composed of smaller subunits (sub-expression trees) that are linked together by different kinds of posttranslational interactions.

1.3 Adaptation and Evolution

Much of the diversity we see in the living world, results from the accumulation of mutations (in the broad sense) in proteins. If we take any protein, for instance, hemoglobin, and analyze its sequence among the individuals of one population, we will see that there are numerous protein variants, differing in one or several amino acids. Most of these variants work with equal efficiency, but some of them may exhibit slight differences in function. In certain environments some of these variants are better adapted than others and may confer some advantage to the individuals expressing them. If we continue this investigation further and analyze the hemoglobin molecules from different species, we will see that there are, in this case, considerable differences between their hemoglobins. Although these different hemoglobins play exactly the same function, they seem wonderfully adapted to the natural environment of the particular species. Indeed, the modifications that occur at the molecular level in proteins enable populations of organisms to develop new abilities, adapt to new environments, and ultimately become new species.

For populations to adapt in the long run, the individual organisms must be selected to reproduce. In terms of evolution, the survival of a particular organism is only important if this organism leaves progeny. It is the individual's progeny that might exhibit new traits and thus be better adapted to the natural environment. And the better adapted an organism becomes, the higher the probability of being selected and leaving more offspring. The variation or genetic diversity we find in nature among organisms is, in fact, the raw material for selection as any organism in the struggle for existence exploits any advantage it may have upon others to guarantee its survival. The more successful individuals leave more progeny and these better adapted organisms (and the criterion for better adapted is that they survived) may increase in frequency in the population, altering its character with time. But, in nature, the process of adaptation never comes to a rest due to the fact that organisms not only change the same environment in which selection occurs but also because more individuals are produced than can survive. Even in stable ecosystems evolution is under way.

In evolutionary computation the term "fitness" is widely used but its meaning differs from the current meaning in evolution theory today. In evolutionary computation the term fitness has the meaning it had in Darwin's day: a quality of organisms likely to be favored by selection. In fact, in all artificial genetic algorithms individuals are selected according to this fitness. In evolution theory, though, fitness is a measure that incorporates both survival and reproductive success.

This shows a very important difference between adaptive computer systems and natural systems. In nature, organisms are selected against a multitude of factors and why or how a new trait is selected is not always clear. Therefore the fitness of an individual can only be measured by the progeny it leaves. But in computer systems the fitness of an individual in a certain environment is easily evaluated, and this measure can be rigorously used to determine selection. However, some scientists like to introduce a random factor in selection to mimic natural selection, and a simple way of implementing this kind of selection is by roulette-wheel sampling (see, e.g., Goldberg 1989). Each individual receives a slice of a circular roulette-wheel proportional to its fitness. The roulette is spun, and the bigger the slice the higher the probability of being selected. And, as it happens with all non-deterministic phenomena, sometimes the improbable happens whereas the highly probable does not happen. Nevertheless, I prefer this kind of selection because it mimics nature more faithfully and works very well in all populations (different selection schemes will be discussed in chapter 12). Indeed, this kind of selection, together with the cloning of the best individual of each generation (simple elitism), works very well, allowing a very efficient search through the fitness landscape.

Finally, another important difference between natural systems and computer systems is that in computer systems it is possible to measure rigorously the fitness as we know exactly what lies ahead and what we want and, therefore, only individuals more or less fit to do a predetermined job are selected. Consequently, it is fundamental the way we analyze the task at hand and choose the conditions (selection environment or fitness cases) under which individuals breed and are selected because, for once, we are probably going to get what we asked for.

1.4 Genetic Algorithms

Genetic Algorithms were invented by John Holland in the 1960s and they apply biological evolution theory to computer systems (Holland 1975). And like all evolutionary computer systems, GAs are an oversimplification of biological evolution. In this case, solutions to a problem are usually encoded in fixed length strings of 0's and 1's (chromosomes), and populations of such strings (individuals or candidate solutions) are manipulated in order to evolve a good solution to a particular problem. From generation to generation individuals are reproduced with modification and selected according to fitness. Modification in the original genetic algorithm was introduced by the search operators of mutation, crossover, and inversion, but more recent applications started favoring mutation and crossover, dropping inversion in the process.

It is worth pointing out that GAs' individuals consist of naked chromosomes or, in other words, GAs' individuals are simple replicators. And like all simple replicators, the chromosomes of GAs work both as genotype and phenotype. This means that they are simultaneously the objects of selection and the guardians of the genetic information that must be replicated and passed on with modification to the next generation. Consequently, whatever is done in the genome will affect fitness and selection. To make this important feature of GAs clearer, compare this situation with the current state of nature where individuals are selected by virtue of the properties of their bodies alone: only the body of the individual and the abilities it can perform are important to the selection process; the state of its genome is irrelevant.

The variety of functions GAs' chromosomes are able to play is severely limited by this dual function they have (genotype and phenotype) and by their structural organization, especially the simple language of the chromosomes and their fixed length. Indeed, GAs' chromosomes very much resemble simple RNA replicators, in which the linear RNA genome is also capable of exhibiting limited structural and functional diversity. In both cases, the whole structure of the replicator determines the functionality and, therefore, the fitness of the individual. For instance, in such systems it is not possible to use only a particular region of the replicator as a solution to a problem; the whole replicator is always the solution: nothing more, nothing less. As a result, these systems are highly constrained.

1.5 Genetic Programming

Genetic Programming, invented by Cramer in 1985 (Cramer 1985) and further developed by Koza (1992), finds an alternative to fixed length solutions through the introduction of nonlinear structures (parse trees) with different sizes and shapes. The alphabet used to create these structures is also more varied than the 0's and 1's of GAs' individuals, creating a richer, more versatile system of representation. Notwithstanding, GP individuals also lack a simple, autonomous genome: like the linear chromosomes of GAs, the nonlinear structures of GP are also naked replicators cursed with the dual role of genotype/phenotype.

The parse trees of GP resemble protein molecules in their use of a richer alphabet and in their complex and unique hierarchical representation. Indeed, parse trees are capable of exhibiting a great variety of functionalities. The problem with these complex replicators is that their reproduction with modification is highly constrained in evolutionary terms, simply because the modifications must take place on the parse tree itself and, consequently, only a limited range of modification is possible. Indeed, the genetic operators of GP operate at the tree level, modifying or exchanging particular branches between trees. For instance, the simple, yet high-performing, point mutation cannot be used as, most of the times, it would generate structural impossibilities. As a comparison, it is worth emphasizing that, in nature, the expression of any protein gene always results in a valid protein structure.

Despite its lack of linear chromosomes, GP is also a genetic algorithm as it uses populations of individuals, selects them according to fitness, and introduces genetic variation using genetic operators. Thus, the fundamental difference between GP and GAs resides in the nature of the individuals and, consequently, in the way in which they are reproduced with modification to allow adaptation.

The individuals of genetic programming are usually LISP programs represented as parse trees (Figure 1.8). What is particularly interesting is that parse trees may assume different sizes and shapes in the process of evolution. As such, populations may evolve and discover solutions of greater complexity.

As stated previously, in GP, the genetic operators act directly on the parse tree and, although at first sight this might appear advantageous, it greatly limits this technique (it is impossible to make an orange tree produce mangos only by grafting and pruning). The genetic operators must be very carefully applied so that only valid structures are formed. Consider, for instance, crossover, the most used and often the only search operator used in GP (Figure 1.9). In this case, selected branches are exchanged between two parent trees to create offspring. The idea behind its implementation was to exchange smaller, mathematically concise blocks in order to evolve more complex, hierarchical solutions composed of smaller building blocks. Effectively, GP crossover very much resembles the pruning and

a.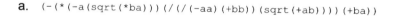

b.

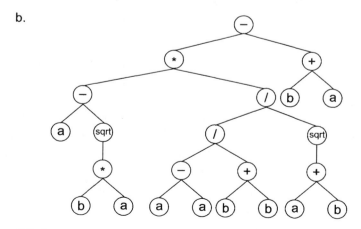

Figure 1.8. A computer program in LISP **(a)** and its tree representation **(b)**. Note that LISP operators precede their arguments, e.g., $(a + b)$ is written as $(+\ a\ b)$.

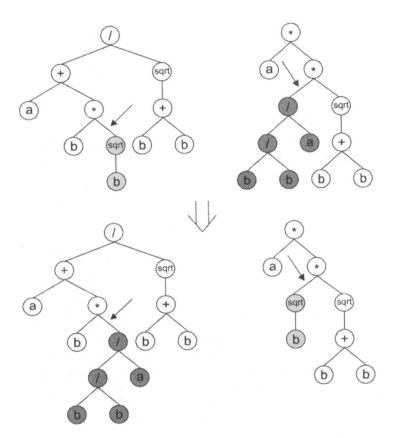

Figure 1.9. Tree crossover in genetic programming. The arrows indicate the crossover points.

grafting of trees and, like these, has a very limited power. This kind of tree crossover was also devised because its implementation in LISP is trivial and the parse trees it creates are always legal LISP programs.

The mutation operator in GP also differs from biological point mutation in order to guarantee the creation of syntactically correct LISP programs. The mutation operator selects a node in the parse tree and replaces the branch underneath by a new randomly generated branch (Figure 1.10). Notice that the overall shape of the tree is not greatly changed by this kind of mutation, especially if lower nodes are preferentially chosen as mutation targets.

Permutation is the third operator used in GP and the most conservative of the three. During permutation, the arguments of a randomly chosen function

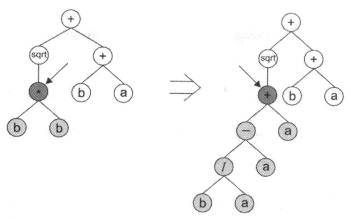

Figure 1.10. Tree mutation in genetic programming. The arrow indicates the mutation point. Note that the mutation operator creates a randomly generated branch at the mutation point.

are randomly permuted (Figure 1.11). In this case the overall shape of the tree remains unchanged.

In summary, in GP the operators resemble more of a conscious mathematician than the blind way of nature. But in adaptive systems the blind way of nature is much more efficient and systems such as GP are highly limited in evolutionary terms. For instance, the implementation of other operators in GP, such as the simple yet high-performing point mutation (Ferreira 2002a), is unproductive as most mutations would have resulted in syntactically incorrect structures (Figure 1.12). Obviously, the implementation of other

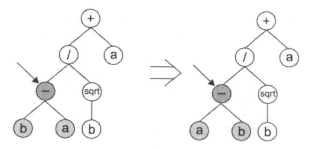

Figure 1.11. Permutation in genetic programming. The arrow indicates the point of permutation. Note that the arguments of the permuted function changed places in the daughter tree. After permutation the shape of the tree remains unchanged.

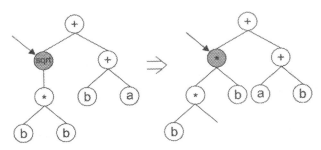

Figure 1.12. Illustration of a hypothetical event of point mutation in genetic programming. Note that the daughter tree is an invalid structure.

operators such as transposition or inversion raises similar difficulties and the search space in GP remains vastly unexplored.

Although Koza described these three operators as the basic GP operators, tree crossover is practically the only genetic operator used in most GP applications (Koza 1992, 1994; Koza et al. 1999). Consequently, no new genetic material is introduced in the genetic pool of GP populations. Not surprisingly, huge populations of parse trees must be used in order to prime the initial population with all the necessary building blocks so that good solutions could be discovered by just moving these initial building blocks around.

Finally, due to the dual role of the parse trees (genotype and phenotype), genetic programming is also incapable of a simple, rudimentary expression; in all cases, the entire parse tree is the solution: nothing more, nothing less.

1.6 Gene Expression Programming

Gene expression programming was invented by myself in 1999 (Ferreira 2001), and incorporates both the simple, linear chromosomes of fixed length similar to the ones used in GAs and the ramified structures of different sizes and shapes similar to the parse trees of GP. And since the ramified structures of different sizes and shapes are totally encoded in the linear chromosomes of fixed length, this is equivalent to say that, in GEP, the genotype and phenotype are finally separated from one another and the system can now benefit from all the evolutionary advantages this brings about.

Thus, the phenotype of gene expression programming consists of the same kind of ramified structure used in GP. But the ramified structures evolved by GEP (called expression trees) are the expression of a totally autonomous genome. Therefore, with GEP, a remarkable thing happened: the second evolutionary threshold – the phenotype threshold – was crossed (Dawkins 1995). And this means that only the genome (slightly modified) is passed on to the next generation. Consequently, one no longer needs to replicate and mutate rather cumbersome structures as all the modifications take place in a simple linear structure which only later will grow into an expression tree.

The pivotal insight of gene expression programming consisted in the invention of chromosomes capable of representing any parse tree. For that purpose a new language – *Karva* language – was created in order to read and express the information encoded in the chromosomes.

Furthermore, the chromosome structure was designed to allow the creation of multiple genes, each coding for a smaller program or sub-expression tree. It is worth emphasizing that gene expression programming is the only genetic algorithm with multiple genes. Indeed, the creation of more complex individuals composed of multiple genes is extremely simplified in truly functional genotype/phenotype systems. In fact, after their inception, these systems seem to catapult themselves into higher levels of complexity and countless new ideas are waiting to be explored. In this book we will encounter some of these rather complex entities like, for instance, the uni- and multicellular systems, where different cells put together different combinations of genes and evolvable genotype/phenotype artificial neural networks and decision trees that not only learn but also adapt their structures to solve a wide variety of problems.

The basis for all this novelty resides on the simple, yet revolutionary structure of GEP genes. This structure not only allows the encoding of any conceivable program but also allows an efficient evolution. This versatile structural organization also allows the implementation of a very powerful set of genetic operators which can then very efficiently search the solution space. As in nature, the search operators of gene expression programming always generate valid structures (be they complex mathematical expressions, or complex artificial neural networks, or sophisticated decision trees) and therefore are remarkably suited to creating genetic diversity.

In the next chapter we are going to learn all the details about the structural and functional organization of GEP chromosomes; how the language of the chromosomes is translated into the language of the expression trees; how the chromosomes work as genotype and the expression trees as phenotype; and how an individual program is created, matured, and reproduced, leaving offspring with new properties, thus, capable of adaptation.

2 The Entities of Gene Expression Programming

In contrast to its analogous cellular gene expression, GEP gene expression is rather simple. The main players in gene expression programming are only two: the chromosomes and the expression trees, the latter consisting of the expression of the genetic information encoded in the former. The process of information decoding (from the chromosomes to the expression trees) is called translation. And this translation implies obviously a kind of code and a set of rules. The genetic code is very simple: a one-to-one relationship between the symbols of the chromosome and the functions and terminals they represent. The rules are also quite simple: they determine the spatial organization of the functions and terminals in the expression trees and the type of interaction between sub-expression trees in multigenic systems.

Therefore, there are two languages in gene expression programming: the language of the genes and the language of the expression trees, and we will see that the sequence or structure of one of these languages is more than sufficient to infer exactly the other. In nature, although the inference of the sequence of proteins given the sequence of genes and vice versa is possible, very little is known about the rules that determine the folding of the protein. And the expression of a protein gene is not complete before the folding of the protein, that is, strings of amino acids only become proteins when they are correctly folded into their native three-dimensional structure. The only thing we know for sure about protein folding is that the sequence of the amino acids determines the folding. However, the rules that orchestrate the folding are still unknown. Fortunately for us, in GEP, thanks to the simple rules that determine the structure of expression trees and their interactions, it is possible to infer immediately the phenotype (the final structure, which is equivalent to the folded protein molecule) given the sequence of a gene, and vice versa. This bilingual and unequivocal system is called *Karva* language. The details of this language are explored in this chapter.

Cândida Ferreira: *Gene Expression Programming*, Studies in Computational Intelligence (SCI) **21**, 29–54 (2006)
www.springerlink.com

2.1 The Genome

In gene expression programming, the genome or chromosome consists of a linear, symbolic string of fixed length composed of one or more genes. Despite their fixed length, we will see that GEP chromosomes code for expression trees with different sizes and shapes.

2.1.1 Open Reading Frames and Genes

The structural organization of GEP genes is better understood in terms of open reading frames (ORFs). In biology, an ORF, or coding sequence of a gene, begins with the start codon, continues with the amino acid codons, and ends at a termination codon. However, a gene is more than the respective ORF, with sequences upstream of the start codon and sequences downstream of the stop codon. Although in GEP the start site is always the first position of a gene, the termination point does not always coincide with the last position of a gene. It is common for GEP genes to have noncoding regions downstream of the termination point. (For now we will not consider these noncoding regions, as they do not interfere with the product of expression.)

Consider, for example, the algebraic expression:

$$\sqrt{(a-b)\times(c+d)} \tag{2.1}$$

It can also be represented as a diagram or expression tree (ET):

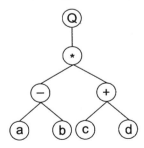

where "Q" represents the square root function.

This kind of diagram representation is in fact the phenotype of GEP genes, the genotype being easily inferred from the phenotype as follows:

```
01234567
Q*-+abcd
```
(2.2)

which is the straightforward reading of the ET from left to right and from top to bottom (exactly as we read a page of text). The expression (2.2) is an ORF, strarting at "Q" (position 0) and terminating at "d" (position 7). I named these ORFs *K-expressions* (from Karva language).

Note that this notation differs from both the postfix and prefix representations used in different GP implementations with arrays or stacks (Keith and Martin 1994). Figure 2.1 compares Karva notation both with postfix and prefix expressions.

K-expression:
```
01234567
Q*-+abcd
```

Postfix:
```
01234567
ab-cd+*Q
```

Prefix:
```
01234567
Q*+dc-ba
```

Figure 2.1. Comparison of Karva notation with both postfix and prefix representations. In all cases, the expression (2.1) is represented.

Consider another ORF, the following K-expression:

```
01234567890
Q*b**+baQba
```
(2.3)

Its expression as an ET is also very simple and straightforward. For its complete expression, the rules governing the spatial distribution of functions and terminals must be followed. First, the start of the ORF corresponds to the root of the ET (this root, though, is at the top of the tree), forming this node the first line of the ET. Second, depending on the number of arguments of each element (functions may have a different number of arguments, whereas terminals have an arity of zero), in the next line are placed as many nodes as there are arguments to the elements in the previous line. Third, from left to right, the new nodes are filled consecutively with the elements of the ORF. This process is repeated until a line containing only terminals is formed. So, for the K-expression (2.3) above, the root of the ET will be formed by "Q", the symbol at position 0:

The square root function has only one argument, so the next line requires only one node, which is filled with the next symbol at position 1, that is, "*":

The multiplication function takes two arguments and, therefore, in the next line are placed two more nodes. These nodes are then filled with the symbols at positions 2 and 3, that is, "b" and "*", obtaining:

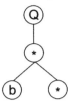

In this line we have a leaf node and a bud node representing a function of two arguments (multiplication). Therefore two more nodes are required to build the next line. And in this case, they are filled with the elements at positions 4 and 5, namely "*" and "+", giving:

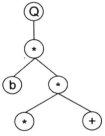

Now we have a line with two function buds, each expanding into two more branches. Thus, four new nodes are required in the next line. And they are filled with the elements "b", "a", "Q", and "b", occupying respectively positions 6, 7, 8, and 9 in the ORF, obtaining:

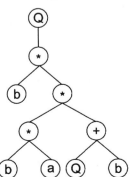

In this new line, although there are four nodes, just one of them is a bud node, whereas the remaining three are leaf nodes. From the leaf nodes obviously no more growth will be sprouting, but from the bud node another branch will be formed. Thus, the required branch is placed below the bud node and filled with the next element in the ORF, the symbol "a" at position 10:

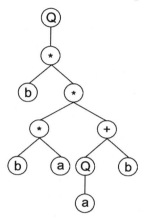

With this step, the expression of the K-expression (2.3) is complete as the last line contains only nodes with terminals. We will see that, thanks to the structural organization of GEP genes, the last line of all ETs generated by this technique will contain only terminals. And this is equivalent to say that all programs evolved by GEP are syntactically correct. Indeed, in GEP, there is no such thing as an invalid expression or computer program.

Looking at the structure of ORFs only, it is difficult or even impossible to see the advantages of such a representation, except perhaps for its simplicity and elegance. However, when open reading frames are analyzed in the context of a gene, the advantages of this representation become obvious. As I said before, the chromosomes of gene expression programming have fixed length, and they are composed of one or more genes of equal length. Therefore the length of GEP genes is also fixed. Thus, in gene expression programming, what varies is not the length of genes, which is constant, but the length of the ORFs. Indeed, the length of an ORF may be equal to or less than the length of the gene. In the first case, the termination point coincides with the end of the gene, and in the latter the termination point is somewhere upstream of the end of the gene. And this obviously means that GEP genes have, most of the times, noncoding regions at their ends.

And what is the function of these noncoding regions at the end of GEP genes? We will see that they are the essence of gene expression programming and evolvability, for they allow the modification of the genome through the use of virtually any kind of genetic operator without any kind of restriction, always producing syntactically correct programs. And this opens up new grounds for the exploration of the search space as all the recesses of the fitness landscape are now accessible. Thus, in gene expression programming, the fundamental property of genotype/phenotype systems – syntactic closure – is intrinsic, allowing the totally unconstrained manipulation of the genotype and, consequently, an efficient evolution. Indeed, this is the paramount difference between gene expression programming and previous GP implementations, with or without linear genomes, all of them either limiting themselves to inefficient genetic operators or checking exhaustively all the newly created programs for syntactic errors (for a review on genetic programming with linear genomes see Banzhaf et al. 1998).

Let us now analyze the structural organization of GEP genes in order to understand how they invariably code for syntactically correct computer programs and why their revolutionary structure allows the unconstrained application of virtually any search operator and, therefore, guarantees the generation of the quintessential genetic variation fundamental for the evolution of good solutions.

2.1.2 Structural and Functional Organization of Genes

So, what is so special about the structure of GEP genes? Well, they are composed of two different domains – a head and a tail domain – each with different properties and functions. The head domain is used mainly to encode the functions chosen for the problem at hand, whereas the tail works as a buffer or reservoir of terminals in order to guarantee the formation of only valid structures. Thus, the head domain contains symbols that represent both functions and terminals, whereas the tail is composed of only terminals.

For each problem, the length of the head h is chosen, whereas the length of the tail t is a function of h and the number of arguments of the function with more arguments n_{max} (also called maximum arity) and is evaluated by the equation:

$$t = h \cdot (n_{max} - 1) + 1 \qquad (2.4)$$

Consider a gene for which the set of functions F = {Q, *, /,-, +} and the set of terminals T = {a, b}, thus giving n_{max} = 2. And if we chose an h = 15, then $t = 15 \cdot (2 - 1) + 1 = 16$ and the length of the gene g is 15 + 16 = 31. One such gene is shown below (the tail is shown in bold):

```
01234567890123456789012345678 90
*b+a-aQab+//+b+babbabbbababbaaa
```
(2.5)

It codes for the following ET:

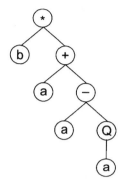

Note that in this case the ORF ends at position 7, whereas the gene ends at position 30 and, therefore, a noncoding region of 23 elements, apparently doing nothing, exists downstream of the termination point.

Suppose now a mutation occurred in the coding region of the gene, specifically at position 6, changing the "Q" into "*". Then the following gene is obtained:

```
01234567890123456789012345678 90
*b+a-a*ab+//+b+babbabbbababbaaa
```
(2.6)

And its expression gives:

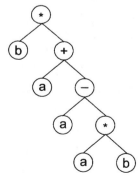

In this case, the termination point shifts just one position to the right (position 8), with the new expression tree managing to keep most of the characteristics of the original one.

Consider another mutation in chromosome (2.5) above, the substitution of "a" at position 5 by "+". In this case, the following daughter chromosome is obtained:

```
0123456789012345678901234567890
*b+a-+Qab+//+b+babbabbbababbaaa
```
(2.7)

And its expression results in the following ET:

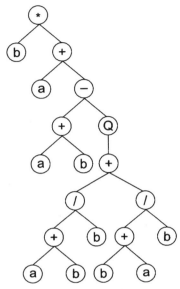

In this case, the termination point shifts 12 positions to the right (position 19), enlarging significantly the new expression tree.

Obviously the opposite might also happen, and the expression tree might shrink. For instance, suppose that a mutation occurred at position 2 in chromosome (2.5) above, changing the "+" into "Q":

```
0123456789012345678901234567890
*bQa-aQab+//+b+babbabbbababbaaa
```
(2.8)

Now its expression results in the following ET:

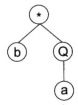

In this case, the ORF ends at position 3, shortening the original ET in four nodes and creating a new daughter tree very different from the original one.

So, despite their fixed length, each gene has the potential to code for expression trees of different sizes and shapes, being the simplest composed of only one node (when the first element of a gene is a terminal) and the largest composed of as many nodes as the length of the gene (when all the elements of the head are functions with maximum arity).

It is evident from the examples above, that any modification made in the genome, no matter how profound, always results in a structurally correct expression tree. Obviously, the structural organization of genes must be preserved, always maintaining the boundaries between head and tail and not allowing symbols from the function set on the tail. We will pursue these matters further in the next chapter (section 3.3) where the mechanisms and the effects of different genetic operators are thoroughly analyzed.

2.1.3 Multigenic Chromosomes

In nature, chromosomes usually code for more than one gene, as complex individuals require complex genomes. Indeed, the evolution of more complex entities is only possible through the creation of multigenic genomes. Not surprisingly, gene expression programming also explores the advantages of multigenic systems.

The chromosomes of gene expression programming are usually composed of more than one gene of equal length. For each problem, the number of genes, as well as the length of the head, are chosen a priori. Each gene codes for a sub-ET and the sub-ETs interact with one another forming a more complex entity. The details of such interactions will be fully explained in section 2.2, Expression Trees and the Phenotype. For now we will focus exclusively on the construction of sub-ETs from their respective genes.

Consider, for example, the following chromosome with length 39, composed of three genes, each with $h = 6$ and a length of 13 (the tails are shown in bold):

```
0123456789012012345678901201234 56789012
*Qb+*/bbbabab-a+QbQbbbababa/ba-/*bbaaaaa          (2.9)
```

It has three open reading frames, and each ORF codes for a particular sub-ET (Figure 2.2). We know already that the start of each ORF coincides with the first element of the gene and, for the sake of clarity, for each gene it is always indicated by position zero; the end of each ORF, though, is only evident upon construction of the respective sub-ET. As shown in Figure 2.2, the first ORF ends at position 9 (sub-ET$_1$); the second ORF ends at position 6 (sub-ET$_2$); and the last ORF ends at position 2 (sub-ET$_3$).

a.   012345678901201234567890120123456789012
     Qb+/**bbbabab**-a+QbQ**bbababa**/ba-/***bbaaaaa**

b.   Sub-ET₁   Sub-ET₂   Sub-ET₃

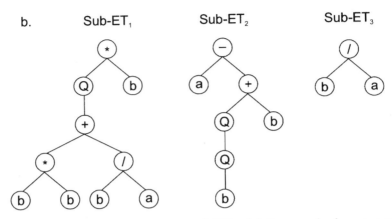

Figure 2.2. Expression of GEP genes as sub-ETs. **a)** A three-genic chromosome with the gene tails shown in bold. Position 0 marks the start of each gene. **b)** The sub-ETs codified by each gene.

In summary, the chromosomes of gene expression programming contain several ORFs, each ORF coding for a structurally and functionally unique sub-ET. We will see that, depending on the problem at hand, these sub-ETs may be selected individually according to their respective fitness (for example, in problems with multiple outputs), or they may form a more complex, multi-subunit ET and be selected according to the fitness of the whole, multi-subunit ET. The patterns of expression and the details of selection will be often discussed in this book. However, keep in mind that each sub-ET is both a separate entity and part of a more complex, hierarchical structure, and, as in all complex systems, the whole is more than the sum of its parts.

2.2 Expression Trees and the Phenotype

In nature, the phenotype has multiple levels of complexity, being the most complex the organism itself. But tRNAs, proteins, ribosomes, cells, and so forth, are also products of expression and all of them are ultimately encoded in the genome.

In contrast to nature, in gene expression programming, the expression of the genetic information is very simple. Nonetheless, as we have seen in section 2.1, GEP chromosomes can have different structural organizations and, therefore, the individuals encoded in those structures have obviously different degrees of complexity. The simplest individuals we have encountered so far are encoded in a single gene, and the "organism" is, in this case, the product of one gene – an ET composed of only one subunit. In other cases, the "organism" is a multi-subunit ET in which the different subunits are linked together by a particular linking function. And in other cases, the "organism" is composed of different sub-ETs in which the different sub-ETs are responsible for a particular facet of the problem at hand. In this section we will discuss different aspects of the expression of the genetic information in gene expression programming, drawing attention to the different levels of phenotypic complexity.

2.2.1 Information Decoding: Translation

From the simplest individual to the most complex, the expression of the genetic information in gene expression programming starts with translation or, in other words, with the construction of all the sub-ETs.

Consider the following chromosome composed of just one gene (the tail is shown in bold):

```
0123456789012345
NIAbObbaaaabaabb
```
(2.10)

The symbols {A, O, N, I} represent, respectively, the Boolean functions AND, OR, NOT, and IF (if $a = 1$, then b; else c), where the first two functions take two arguments, NOT takes one argument, and the IF function takes three arguments. In this case, the product of translation is the following expression tree with nine elements:

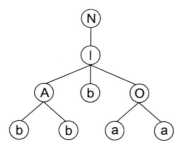

which, as you can easily check, is a perfect, albeit rather verbose, solution to the NOR function.

Thus, for this simple individual, a single unit ET is the "organism" and, in this case, the expression of the genetic information ends with translation.

When the genome codes for more than one gene, each gene is independently translated as a sub-ET, and two different kinds of "organism" might be formed: in the first kind, the sub-ETs are physically connected to one another by a particular linking function; in the second kind, the sub-ETs work together to solve the problem at hand but there are no direct connections between them. Let's now make this clearer with two examples.

Consider, for instance, the following chromosome composed of three different genes (the tails are shown in bold):

```
01234567890123456789012345678 9
AOaabaaaabNabaaaaaabINNbababaa
```
(2.11)

It codes for three different sub-ETs (Figure 2.3), each one representing a particular Boolean expression. Usually, these sub-ETs or sub-programs are part of a bigger program in which the sub-ETs are linked by a particular linking function. For instance, if we linked the sub-ETs one after the other by the Boolean function OR or AND, two different programs would be represented by chromosome (2.11) above. Thus, for this individual, the expression of the genetic information starts with the translation of the sub-ETs, but

a.
```
01234567890123456789012345678 9
AOaabaaaabNabaaaaaabINNbababaa
```

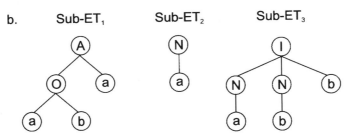

Figure 2.3. Translation of GEP genes as Boolean sub-ETs. **a)** A three-genic chromosome with the gene tails shown in bold. **b)** The sub-ETs codified by each gene. Note that the full expression of the chromosome will require some kind of interaction between the sub-ETs. Indeed, the program encoded in the chromosome only makes sense if the interactions between sub-programs are specified. For instance, three different programs would be obtained if the linking were done by OR, AND, or IF.

ends only after the linking of the sub-ETs by a particular linking function. And therefore the "organism" in this case will consist of a multi-subunit expression tree composed of three smaller subunits.

For problems with multiple outputs, however, the different sub-ETs encoded in the genome are engaged in the identification of just one kind of output and, therefore, they are not physically connected to one another: they remain more or less autonomous agents working together to solve the problem at hand. For instance, in classification problems a particular sub-ET is responsible for the identification of a particular class.

Consider, for instance, the chromosome below composed of three different genes created to solve a classification task with three distinct classes:

```
01234567890120123456789012012345678901 2
-/dac/dacaccd//-aacbbbabcd-d/+c*dbdbacd                    (2.12)
```

It codes for three different sub-ETs, each one representing a rather complex algebraic expression (Figure 2.4).

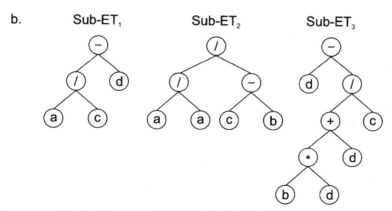

a.   01234567890120123456789012012345678901 2
     -/dac/dacaccd//-aacbbbabcd-d/+c*dbdbacd

b.   Sub-ET₁   Sub-ET₂   Sub-ET₃

Figure 2.4. Translation of GEP genes as algebraic sub-ETs. **a)** A three-genic chromosome with the tails shown in bold. **b)** The sub-ETs codified by each gene. Note that, after translation, the sub-ETs might either form multi-subunit expression trees composed of smaller sub-ETs or remain isolated as a single-unit expression tree. For instance, in problems with just one output, the sub-ETs might be linked by a particular linking function or, in problems with multiple outputs, each sub-ET is responsible for the identification of a particular output.

In this classification task, for gene expression to be complete, the output of each sub-ET must first be converted into 0 or 1 using a rounding threshold, below which the output is converted into 0, 1 otherwise. Then the sub-ETs must be subjected to the following rules in order to determine the class:

IF (Sub-ET$_1$ = 1 AND Sub-ET$_2$ = 0 AND Sub-ET$_3$ = 0), THEN Class 1;
IF (Sub-ET$_1$ = 0 AND Sub-ET$_2$ = 1 AND Sub-ET$_3$ = 0), THEN Class 2;
IF (Sub-ET$_1$ = 0 AND Sub-ET$_2$ = 0 AND Sub-ET$_3$ = 1), THEN Class 3.

Let's make this more concrete with a simple example, a small sub-set of the iris dataset (Fisher 1936) where the first five samples of each type of iris in the original dataset are used (Table 2.1). Indeed, the program (2.12) above was created using the training samples shown in Table 2.1. And, as you can easily confirm with the help of Figure 2.4, this model classifies all the samples correctly using a rounding threshold of 0.5.

So, for problems with multiple outputs, multiple sub-programs are encoded in the chromosome and the "organism" is the result of an intricate collaboration between all sub-programs, in which each sub-program is engaged in the discovery of a particular facet of the global problem.

Table 2.1. The iris sub-set.

Sepal length (a)	Sepal width (b)	Petal length (c)	Petal width (d)	Type
5.1	3.5	1.4	0.2	class 1 (seto.)
4.9	3.0	1.4	0.2	class 1 (seto.)
4.7	3.2	1.3	0.2	class 1 (seto.)
4.6	3.1	1.5	0.2	class 1 (seto.)
5.0	3.6	1.4	0.2	class 1 (seto.)
7.0	3.2	4.7	1.4	class 2 (vers.)
6.4	3.2	4.5	1.5	class 2 (vers.)
6.9	3.1	4.9	1.5	class 2 (vers.)
5.5	2.3	4.0	1.3	class 2 (vers.)
6.5	2.8	4.6	1.5	class 2 (vers.)
6.3	3.3	6.0	2.5	class 3 (virg.)
5.8	2.7	5.1	1.9	class 3 (virg.)
7.1	3.0	5.9	2.1	class 3 (virg.)
6.3	2.9	5.6	1.8	class 3 (virg.)
6.5	3.0	5.8	2.2	class 3 (virg.)

2.2.2 Posttranslational Interactions and Linking Functions

We have already seen that translation results in the formation of sub-ETs with different sizes and shapes, and that the complete expression of the genetic information requires the interaction of these sub-ETs with one another. Only then will the individual be fully expressed. A very common and useful strategy consists in the linking of sub-ETs by a particular linking function. Indeed, most mathematical and Boolean applications are problems of just one output and, therefore, can benefit from this strategy, in which more complex programs are designed by linking together smaller sub-programs.

When the sub-ETs are algebraic expressions or Boolean expressions, any mathematical or Boolean function with more than one argument can be used to link the sub-ETs in a final, multi-subunit ET. For algebraic expressions, the most frequently chosen functions to link the sub-ETs are addition, subtraction, multiplication, or division. For Boolean expressions, the most frequently chosen linking functions are all the interesting functions of two arguments (functions 1, 2, 4, 6, 7, 8, 9, 11, 13, and 14, or, more intelligibly, NOR, LT, GT, XOR, NAND, AND, NXOR, LOE, GOE, and OR, respectively), or functions of three arguments such as the already familiar IF function (if $a = 1$, then b; else c) or the 3-multiplexer (also easily described as an IF THEN ELSE statement: if $a = 0$, then b; else c, which, as you can see, is very similar to the IF function).

However, the linking of sub-ETs with functions of two arguments is much simpler, as any number of sub-ETs can be linked together one after the other (see Figure 2.5). On the other hand, the linking of sub-ETs with linking functions of more than two arguments, say n arguments, is more problematic as it requires n^n sub-ETs for a correct linking (see Figure 2.6). Let's now make this clearer with two examples.

For instance, consider the following chromosome, encoding three algebraic sub-ETs linked by addition (the tails are shown in bold):

```
012345678901201234567890120123456789012
QaQ+-Qbbaaaba+Q+ab+abababa*-**b+aabbaba
```
(2.13)

As you can see in Figure 2.5, the sub-ETs are linked together one after the other in an orderly fashion. Note that the multi-subunit ET encoded in chromosome (2.13) could be linearly encoded as the following K-expression:

```
0123456789012345678901
++*Q+-*aQ+*b+aab+abbaab
```
(2.14)

a. `0123456789012012345678901201234567890123456789012`
   `QaQ+-Q`**`bbaaaba`**`+Q+ab+`**`abababa`**`*-**b+`**`aabbaba`**

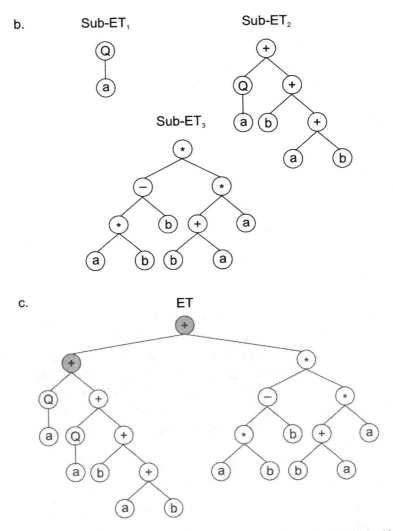

Figure 2.5. Expression of multigenic chromosomes encoding sub-ETs linked by a two-argument function. **a)** A three-genic chromosome with the tails shown in bold. **b)** The sub-ETs codified by each gene. **c)** The result of posttranslational linking with addition. The linking functions are shown in gray.

However, the use of multigenic chromosomes is more appropriate to evolve solutions to complex problems, for they permit the modular construction of more complex, hierarchical structures, where each gene codes for a smaller building block (see the section Testing the Building Blocks Hypothesis in chapter 12). These small building blocks are separated from each other and, therefore, can evolve with a certain degree of independence.

And now consider another chromosome, this time encoding three Boolean sub-ETs, linked by a three-argument function (the tails are shown in bold):

```
012345678901234501234567890123450123456789012345
IOaIAcbaaacaacacAOcaIccabcbccbacIONAAbbbbacbcbbc          (2.15)
```

As you can see in Figure 2.6, at least three genes are required to link the sub-ETs with the IF(a,b,c) function. Note also that if more sub-ETs were needed, the simplest organization would require at least nine sub-ETs so that they could be linked properly by the three-argument function. Again, the multi-subunit ET encoded in chromosome (2.15) could be linearized, forming the following K-expression:

```
012345678901234567890123456789
IIAIOaIOcONAAcbaaaIAbbbbacccaac                          (2.16)
```

In summary, to express fully a chromosome, the information concerning the kind of interaction between the sub-ETs must also be provided. Therefore, for each problem, the type of linking function or type of interaction between sub-ETs is chosen a priori. We can start with addition for algebraic expressions or OR for Boolean rules but, in some cases, another linking function might be more appropriate (like multiplication or AND, for instance). The idea, of course, is to find a good solution, and different linking functions can be used to explore different recesses of the fitness landscape, increasing the odds of finding Mount Everest. Notwithstanding, the basic gene expression algorithm can be easily modified to enable the evolution of linking functions. And an elegant and interesting way of solving this problem consists in the creation of homeotic genes that encode a developmental program or cell in which different combinations of sub-ETs are brought together by following the linking interactions operating in that particular cell (see how this is achieved in the next section).

a.   012345678901234501234567890123450123456789012345
     IOaIAcbaaacaacacAOcaIccabcbccbacIONAAbbbbacbcbbc

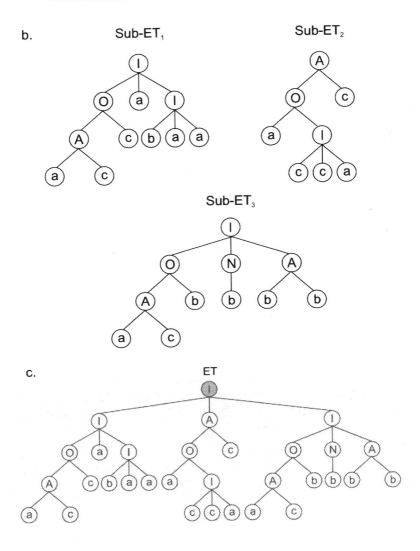

b.   Sub-ET₁   Sub-ET₂

     Sub-ET₃

c.   ET

Figure 2.6. Expression of multigenic chromosomes encoding sub-ETs linked by a three-argument function. **a)** A three-genic chromosome with the gene tails shown in bold. **b)** The sub-ETs codified by each gene. **c)** The result of posttranslational linking with IF. The linking function is shown in gray.

2.3 Cells and the Evolution of Linking Functions

The linking of sub-ETs by a particular linking function is simple and highly efficient in evolutionary terms. Indeed, from unigenic to multigenic systems, efficiency increases considerably (see, for instance, section 5 of chapter 12 for a discussion of The Higher Organization of Multigenic Systems). Despite this artificial increase in complexity ("artificial" in the sense that it was not evolved by the system itself), evolution in multigenic systems still occurs efficiently and therefore they can be efficiently used to evolve good solutions to virtually all kinds of problems.

In principle, it is possible to impose from outside higher levels of complexity, but this is no guarantee that an increase in performance will be achieved. Natural evolutionary systems are not created this way: higher levels of complexity are created above lower levels and the evolution of complexity occurs more or less continuously.

Notwithstanding, the linking of sub-ETs in gene expression programming can be implemented by using a higher level of complexity. For that purpose a special class of genes was created – homeotic genes – that control the development of the individual. The expression of such genes results in different main programs or cells, that is, they determine which genes are expressed in each cell and how the sub-ETs of each cell interact with one another. Or stated differently, homeotic genes determine which sub-ETs are called upon (and how often) in which main program or cell and what kind of connections they establish with one another. How this is done is explained in the next section.

2.3.1 Homeotic Genes and the Cellular System

Homeotic genes have exactly the same kind of structural organization as conventional genes and they are built using an identical process. This means that they also contain a head and tail domain, with the heads containing, in this case, linking functions (so called because they are in fact used to link the different sub-ETs encoded in the conventional genes) and a special class of terminals – genic terminals – representing conventional genes, which, in the cellular system, encode different sub-ETs or sub-programs; the tails contain obviously only genic terminals.

Consider, for instance, the following chromosome:

```
012345601012345601012345601012345601234560123
/+a/abbba+*-abbabb/b*+abaab+Q/0*210212
```
(2.17)

It codes for three conventional genes and one homeotic gene (shown in bold). The three conventional genes code, as usual, for three different sub-ETs, with the difference that now these sub-ETs may be invoked multiple times from different places, or, stated differently, they will act as automatically defined functions (ADFs). And the homeotic gene controls the interactions between the different sub-ETs or, in other words, determines which functions are used to link the sub-ETs or ADFs and how the linking is established (Figure 2.7).

a.   `012345601012345601012345601`**`01234560123`**
      `/+a/abbba+*-abbabb/b*+abaab`**`+Q/0*210212`**

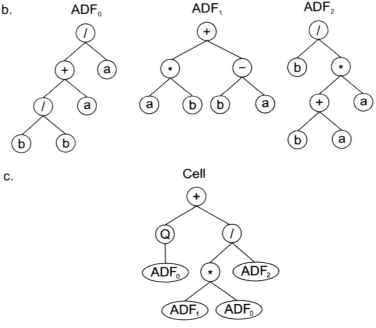

Figure 2.7. Expression of chromosomes with a single homeotic gene encoding a main program or cell. **a)** The chromosome composed of three conventional genes and one homeotic gene (shown in bold). **b)** The sub-ETs or ADFs codified by each conventional gene. **c)** The final main program or cell. Note that the cellular system allows code reuse as each ADF might be called several times by the main program.

It is worth emphasizing how flexible and dynamic the linking of sub-ETs became with this new method, allowing not only the use of any kind of linking function (even functions of just one argument can now be used as linkers) but also the use of different linking functions at a time as opposed to just one static linker. And to top it off, this new system is totally unsupervised, being the linking functions totally sorted out by the evolutionary process itself.

As Figure 2.7 clearly shows, this kind of representation not only allows the evolution of linking functions but also allows code reuse. Thus, this is also an extremely elegant form of implementing ADFs in gene expression programming. Indeed, any ADF in this cellular representation can not only be used as many times as necessary but also establish different interactions with the other ADFs in the main program or cell. For instance, in the particular case of Figure 2.7, ADF_0 is used twice in the main program, whereas ADF_1 and ADF_2 are both used just once.

It is worth pointing out that homeotic genes have their specific length and their specific set of functions. And these functions can take any number of arguments (functions with one, two, three, ..., n arguments). For instance, in the particular case of chromosome (2.17), the head length of the homeotic gene h_H is equal to five, whereas for the conventional genes $h = 4$; and the function set of the homeotic gene consists of $F_H = \{+, *, /, Q\}$, whereas for the conventional genes the function set consists of $F = \{+, -, *, /\}$.

In summary, as Figure 2.7 emphasizes, the cellular system of gene expression programming is not only a form of elegantly allowing the totally unconstrained evolution of linking functions in multigenic systems, but also an extremely elegant and flexible way of encoding automatically defined functions that can be called an arbitrary number of times from an arbitrary number of different places.

2.3.2 Multicellular Systems with Multiple Main Programs

The use of more than one homeotic gene results obviously in a multicellular system where each homeotic gene puts together a different combination of sub-expression trees.

Consider, for instance, the following chromosome:

```
0123456010123456010123456010123456 0123456
/Q+*babab/+a/abbab*Q-bbaaab*1+1020*Q*1202        (2.18)
```

It codes for three conventional genes and two homeotic genes (shown in bold). And its expression results in two different cells or main programs, each expressing different genes in different ways (Figure 2.8). And as you can see in Figure 2.8, in this particular case, ADF_0 is called both from $Cell_0$ and $Cell_1$; ADF_1 is called twice from $Cell_0$ and just once from $Cell_1$; and ADF_2 is only called from $Cell_1$.

The applications of these multicellular systems are multiple and varied and, like the multigenic systems, they can be used both in problems with just one output and in problems with multiple outputs. In the former case, the best program or cell accounts for the fitness of the individual; in the latter, each cell is responsible for a particular facet of a multiple output task such as a classification task with multiple classes. We will pursue these questions further in chapter 6, where automatically defined functions are discussed in more detail.

a. 0123456010123456010123456010**1234560123456**
    /Q+*babab/+a/abbab*Q-bbaaab**1+1020*Q*1202**

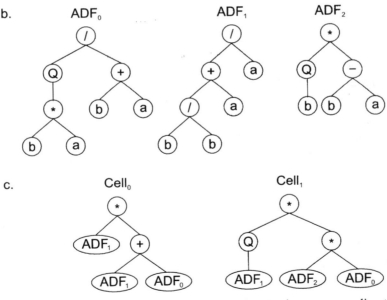

Figure 2.8. Expression of chromosomes with two homeotic genes encoding two different main programs or cells. **a)** The chromosome composed of three conventional genes and two homeotic genes (shown in bold). **b)** The sub-ETs or ADFs codified by each conventional gene. **c)** Two different main programs expressed in two different cells. Note how different cells put together different consortiums of genes.

2.4 Other Levels of Complexity

As we have seen in the examples above, although a very simple system, gene expression programming exhibits already a complex development and also uses different chromosomal organizations. So far, we have been dealing with genes (both conventional and homeotic) containing only head/tail domains. But gene expression programming regularly uses other chromosomal organizations that are more complex than the basic head/tail domain. These complex chromosomal structures consist of clusters of functional units composed of conventional head/tail domains plus one or more extra domains. The extra domains usually code for several one-element sub-ETs. And all the sub-ETs encoded in the different domains interact with one another, forming a more complex entity with a complex network of interactions.

One such architecture was developed to manipulate random numerical constants for function finding problems (Ferreira 2001, 2003). For instance, the following chromosome contains an extra domain Dc (shown in bold) encoding random numerical constants:

```
0123456789012345**6789012**
+*?+?*+a??aaa??**09081345**
```
                                                                  (2.19)

As you can see in Figure 2.9, the translation of the head/tail domain is done in the usual fashion, but, after translation, additional processing is needed in order to replace the question marks in the tree by the numerical constants they represent. In chapter 5, Numerical Constants and the GEP-RNC Algorithm, we will learn how these sub-ETs interact with one another so that the individual is fully expressed.

Multiple domains are also used to design neural networks totally encoded in a linear genome. These neural networks are one of the most complex individuals evolved by GEP. In this case, the neural network architecture is encoded in a conventional head/tail domain, whereas the weights and thresholds are encoded in two extra domains, Dw and Dt, each encoding several one-element sub-ETs. For instance, the chromosome below contains two extra domains encoding the weights and the thresholds of the neural network encoded in the head/tail domain (the domains are shown in different shades):

```
0123456789012345678**9012345678901234567**89010
DUDTUDcdabdcabacbad**4299840979148240926**75841
```
                                                                  (2.20)

As you can see in Figure 2.10, the translation of the head/tail domain encoding the neural network architecture is also done in the usual fashion, but the

a.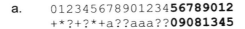

b.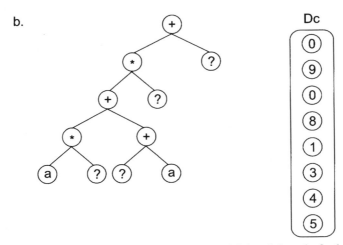

Figure 2.9. Translation of chromosomes with an additional domain for handling random numerical constants. **a)** The chromosome composed of a conventional head/tail domain and an extra domain (Dc) encoding random numerical constants represented by the numerals 0-9 (shown in bold). **b)** The sub-ETs codified by each domain. The one-element sub-ETs encoded in Dc are placed apart together. "?" represents the random numerical constants encoded in the numerals of Dc. How all these sub-ETs interact will be explained in chapter 5.

weights of the connections and the thresholds of the neurons must be assigned posttranslationally. In chapter 10, Design of Neural Networks, we will learn the rules of their complete development and how populations of these complex individuals evolve, finding solutions to problems in the form of adaptive neural networks totally encoded in linear genomes.

2.5 Karva Language: The Language of GEP

We have already seen that each gene codes for a particular sub-ET, and that each sub-ET corresponds to a specific K-expression or open reading frame. Due to the simplicity and elegance of this correspondence, K-expressions are, per se, extremely compact, intelligible computer programs. We have already seen how multi-subunit expression trees can be easily converted into linear K-expressions, and this can be easily done for any algebraic or Boolean

a. 01234567890123456789**90123456789012345**6789010
    DUDTUDçdabdcabacbad**42998409791482409**2675841

b.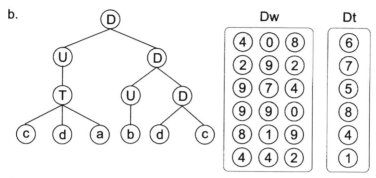

Figure 2.10. Translation of chromosomes with two extra domains for handling the weights and thresholds of neural networks. **a)** A multi-domain chromosome composed of a conventional head/tail domain encoding the neural network architecture, and two extra domains – one encoding the weights (Dw) of the neural network encoded in the head/tail domain and another the thresholds (Dt). Dw and Dt are shown in different shades. **b)** The sub-ETs codified by each domain. The one-element sub-ETs encoded in Dw and Dt are placed apart together. "U", "D", and "T" represent, respectively, neurons or functions with connectivity one, two, and three. How all these sub-ETs interact will be shown in chapter 10.

expression. Indeed, the language of gene expression programming – Karva language – is a versatile representation that can be used to evolve relatively complex programs as simple, extremely compact, symbolic strings. In fact, there is already commercially available software such as Automatic Problem Solver by Gepsoft that automatically converts K-expressions and GEP chromosomes into several programming languages, such as C, C++, C#, Visual Basic, VB.NET, Java, Fortran, VHDL, Verilog, and others.

Another advantage of Karva notation is that it can be used to evolve highly sophisticated programs using any programming language. Indeed, the original GEP implementation was written in C++, but it can be done in virtually any programming language, as it does not rely on any quirks of a particular programming language. As a comparison, it is worth pointing out that early GP implementations relied greatly on LISP because the sub-tree swapping that occurs during reproduction in that system is very simple to implement in that programming language.

In the next chapter, the implementation details of the gene expression algorithm will be fully analyzed, starting with the creation of the initial population and finishing with selection and reproduction to generate the new individuals of the new generation.

3 The Basic Gene Expression Algorithm

The fundamental steps of the gene expression algorithm (GEA) are schematically represented in Figure 3.1. The process begins with the random generation of the chromosomes of a certain number of individuals (the initial population). Then these chromosomes are expressed and the fitness of each individual is evaluated against a set of fitness cases (also called selection environment which, in fact, is the input to a problem). The individuals are then selected according to their fitness (their performance in that particular environment) to reproduce with modification, leaving progeny with new traits. These new individuals are, in their turn, subjected to the same developmental process: expression of the genomes, confrontation of the selection environment, selection, and reproduction with modification. The process is repeated for a certain number of generations or until a good solution has been found.

In this chapter, we will analyze with great detail all the fundamental steps of this evolutionary algorithm, starting with the random generation of the chromosomes of all the individuals of the initial population and finishing with their selection and reproduction with modification, which obviously leads to the creation of the new individuals of the next generation. The goal consists not only in studying the logistics of the gene expression algorithm but also in understanding why and how populations of computer programs evolve from generation to generation, becoming better and better solutions to the problem at hand.

We have already seen that populations of entities, be they organisms or computer programs, evolve because individuals are reproduced with modification, giving rise to genetic diversity, which is the raw material of evolution. This genetic diversity is the basis for a differential selection and, therefore, plays a central role in evolution. Thus, we are going to analyze thoroughly the mechanisms and effects of all the agents of genetic diversity – the genetic operators. Each genetic operator is going to be used to solve the

Cândida Ferreira: *Gene Expression Programming*, Studies in Computational Intelligence (SCI) **21**, 55–120 (2006)
www.springerlink.com

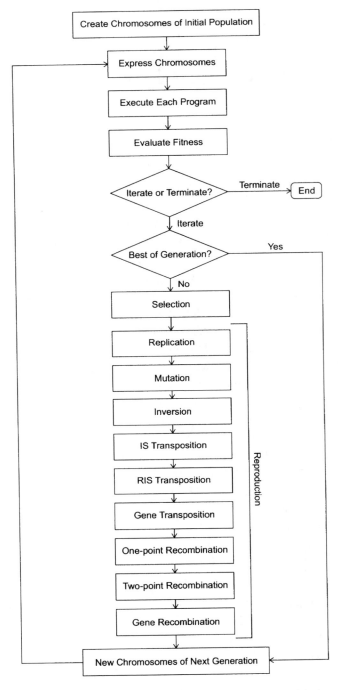

Figure 3.1. The flowchart of a gene expression algorithm.

same simple problem so that entire populations could be fully analyzed and the power of each genetic operator could be completely understood. The problem used throughout this chapter to illustrate the internal workings of all the genetic operators is very simple and easy to understand and consists of the Boolean majority function of three arguments, which will be fully explained below.

Furthermore, the chapter closes with the exhaustive analysis of a simple symbolic regression problem, taking us through all the phases in the design of good models with gene expression programming.

3.1 Populations of Individuals

Like all genetic algorithms, gene expression programming works with populations of individuals and, therefore, some kind of initial population must be created in order to get things started. Subsequent populations are descendants, via genetic modification, of this initial or founder population.

We have already seen that, in the genotype/phenotype system of gene expression programming, we only need to create the simple, linear chromosomes of the individuals without having to bother about the structural soundness of the programs they code for, as we know for sure that their complete development will always result in syntactically correct programs. Thus, in GEP, it is only necessary to generate randomly the simple chromosomal structures of the individuals of the initial population to get things going. And this is indeed a trivial task and one of the great advantages of this technique.

Thus, for each problem, we must choose the symbols used to create the chromosomes, that is, we must choose the set of terminals (the variables and constants used in a problem) and the set of functions we believe to be appropriate to solve the problem at hand. We must also choose the length of each gene, the number of genes per chromosome and how the products of their expression interact with one another. And, finally, we must provide for a selection environment (the set of fitness cases) against which the fitness of each individual is evaluated. Then, according to fitness in that particular environment, individuals are selected to reproduce with modification, giving birth to the new members of the next generation. This new population is then subjected to the same developmental process, giving rise to yet another new population. This process is repeated for a certain number of generations or until a good solution has been found.

3.1.1 Creation of the Initial Population

The chromosomes of the individuals of the initial population are randomly generated using the symbols representing the functions and terminals thought appropriate to solve the problem at hand. These initial individuals are the first set of candidate solutions to the problem at hand. Because they are totally random and not yet toughened up by the environment, these founder individuals are almost always not very good solutions. Notwithstanding, they are everything that is necessary to get things started, as evolution takes care of the rest and, soon enough, very good solutions will start to appear in the population. Let us illustrate this with a concrete example.

Suppose, for instance, that we wanted to know how to express the Boolean Majority(a, b, c) function in terms of ANDs, ORs, and NOTs. In this case, the choice of the function set is not problematic and consists of F = {A, O, N}, representing, respectively, the Boolean functions AND, OR, and NOT. The choice of the terminal set is also simple and consists of T = {a, b, c}, representing the three arguments to the majority function. Therefore, for this problem, the heads of the genes will be randomly generated using six different symbols {A, O, N, a, b, c}, whereas the tails will be randomly generated using a smaller alphabet of just three symbols {a, b, c}.

The truth table for the majority function is shown in Table 3.1. For this problem, the complete set of transition states is used as the selection environment. So, the fitness of each candidate solution will be evaluated against this set of fitness cases (also called training set, as it is used to evaluate the performance during the adaptive process). A good fitness function for this simple problem is also not very difficult to design and will correspond to the number of fitness cases correctly evaluated by a particular individual.

Table 3.1
Majority function.

a	b	c	y
0	0	0	0
0	0	1	0
0	1	0	0
0	1	1	1
1	0	0	0
1	0	1	1
1	1	0	1
1	1	1	1

Thus, with the symbols drawn from the function and terminal sets, the chromosomes of the initial population are randomly generated. Obviously, the heads of the genes are created using elements from both **F** and **T**, whereas the tails of the genes are created using only elements drawn from **T**. Figure 3.2 shows a small initial population (also called generation 0) of 10 individuals randomly created to solve the majority function problem. The chromosomes are composed of two genes, each with a length of 7 ($h = 3$ and $t = 4$), and they encode sub-ETs which are posttranslationally linked by the Boolean function OR.

```
Generation N:  0
01234560123456
NaObaacOAbbcca-[0]
AaNcbbaNcOaacc-[1]
OONcbbbNcbcbca-[2]
ANNcaacNcObaab-[3]
AbObcbcOAacaac-[4]
AcNbcbbAONbbcc-[5]
NAcbcacNbOaaba-[6]
NbNbbaaAacbacb-[7]
NAAaccaONacbbb-[8]
AAaccacNcaabab-[9]
```

Figure 3.2. The chromosomes of a small initial population created to solve the Majority(a, b, c) function problem. The randomly generated chromosomes are composed of two genes, encoding sub-ETs linked by OR.

It is worth emphasizing how easily the initial populations of gene expression programming are created. The simple structures of the chromosomes are randomly created using the symbols of the function and terminal sets and nothing whatsoever needs to be done to monitor their structural soundness: we know for sure that, without exception, all of them encode valid programs. As a comparison, the random generation of parse trees in genetic programming must be done with extreme care in order to guarantee the formation of syntactically correct programs. Not surprisingly, the generation of initial populations in GP is a complicated, time consuming process.

After the random generation of the chromosomes of the individuals of the initial population, the chromosomes are expressed, and the fitness of each individual is evaluated. Figure 3.3 shows the fitness of each individual at the majority task (for simplicity, only the chromosomes of the individuals are

```
Generation N:  0
01234560123456
NaObaacOAbbcca-[0]  =  4
AaNcbbaNcOaacc-[1]  =  2
OONcbbbNcbcbca-[2]  =  4
ANNcaacNcObaab-[3]  =  2
AbObcbcOAacaac-[4]  =  6
AcNbcbbAONbbcc-[5]  =  4
NAcbcacNbOaaba-[6]  =  2
NbNbbaaAacbacb-[7]  =  3
NAAaccaONacbbb-[8]  =  4
AAaccacNcaabab-[9]  =  4
```

Figure 3.3. The chromosomes of the individuals of the initial population and their respective fitness (the value after the equal sign). The fitness corresponds to the number of fitness cases (see Table 3.1) correctly solved by the program encoded in each chromosome. The chromosomes encode sub-ETs linked by OR.

shown, but you shouldn't have any problems in drawing the programs they encode and checking their fitnesses). Then, according to fitness, the individuals are selected to reproduce with modification. For instance, all the 10 individuals of the initial population shown in Figure 3.2 are viable (that is, have non-zero fitness) and are therefore eligible to be selected to reproduce. These viable individuals, among themselves, create as many new individuals as there are individuals in the population. As such, throughout a run, the population size is kept unchanged. The descendants of the selected individuals consist of the new members of the next generation.

The expression of the best individuals of generation 0 (chromosome 4) is shown in Figure 3.4. As you can easily check, this individual is able to solve correctly six out of eight fitness cases, and thus has a fitness equal to six.

You have probably noticed that all the genes created randomly in the initial population have a function at position 0, which, as you already know, corresponds to the root of the sub-ETs. Although I have chosen to have all the genes in the initial population starting with a function, this is not very important in multigenic systems. This feature of initial populations is a remnant of the evolutionary process of GEP itself, for the first GEP implementation used chromosomes composed of only one gene. In those systems, genes coding for one-element ETs were of little use and therefore were kept to a minimum within populations. However, when multigenic chromosomes were introduced, I did not change this feature of initial populations because

a.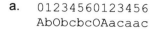

b.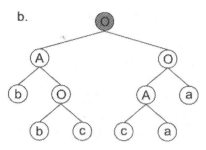

Figure 3.4. Expression of the best individual of generation 0 (chromosome 4).
a) The chromosome composed of two genes. **b)** The program encoded in the
chromosome (the linking function is shown in gray). This Boolean function solves
correctly six out of eight fitness cases (see Table 3.1).

terminals can be inserted at the start position of a gene in future generations
by the genetic operators. And if this happens to be advantageous to solve a
certain task, this kind of gene can and will certainly evolve. We will see later
on in this chapter that point mutation and inversion allow the evolution of
genes with a terminal at the start position or, in other words, allow the evolution
of simple sub-ETs composed of only one node.

The problem with the random generation of initial populations is that,
sometimes, especially if the population size is small or the set of fitness
cases is not broad enough or the fitness function is excessively tight, it just
might happen that none of the initial chromosomes encode viable individuals and the run is aborted (this is highly improbable for Boolean problems,
though, unless a very restrictive fitness function is used). A simple form of
circumventing this problem consists of introducing a start up control. In gene
expression programming, this control is by default set to one viable individual, and was used in all the problems presented in this book. We will see
that this does not hinder the evolutionary process, as GEP populations can
efficiently evolve even when all the individuals are descendants of a sole
viable founder (Ferreira 2002d; see also the discussion of The Founder Effect in section 2 of chapter 12). Thus, this control mechanism, together with
the cloning of the best individual (see section 3.1.2 below), prevents the
existence of failed runs in gene expression programming. So, if all the individuals of the initial population happened to have zero fitness, another initial
population would have been randomly generated. This process is repeated
until at least one viable individual is created and the evolutionary process
can resume.

3.1.2 Subsequent Generations and Elitism

The descendants of the viable individuals of the initial population shown in Figure 3.3, are shown in Figure 3.5. In order to simplify the analysis of the evolutionary history of the population, only mutation was used as source of genetic modification at a rate equivalent to two point-mutations per chromosome, which in this case corresponds to a mutation rate of 0.143.

Note that chromosome 0 of generation 1 is an exact copy of the best individual of the initial population (chromosome 4 in Figure 3.3). Indeed, from generation to generation, the best individual (or one of the best) is replicated unchanged into the next generation. If the best fitness is shared by more than one individual in the population, the last one is chosen to become a clone. Thus, in the case of the population of generation 1, chromosome 6 will be replicated without modification and will occupy the first place in the next generation (see Figure 3.7 below). When this happens, the individual occupying that position dies. This is important to keep in mind when we will be analyzing the outcome of certain operators, like recombination, where one of the daughter chromosomes might not be present in the next generation. The cloning of the best, also called simple elitism, guarantees that at least one descendant will be viable (obviously, this only happens in problems where the selection environment is kept the same from generation to generation), keeping at the same time the best trait during the process of adaptation. Elitism plays also another role: it allows the use of several modification operators at relatively high rates without the risk of causing a mass extinction.

```
Generation N: 1
01234560123456
AbObcbcOAacaac-[0]  = 6
NaObabcOAbbcca-[1]  = 4
NAAacccONOcbbb-[2]  = 3
aaNcbcaNcOaacc-[3]  = 4
AcNbcbbOONbbcc-[4]  = 4
AcNbcbbAONbbcc-[5]  = 4
AbObcbcOAAcaac-[6]  = 7
NAAaccaONacbbb-[7]  = 4
AaNcbbaNcNaacc-[8]  = 2
cNNcaabNcObaab-[9]  = 4
```

Figure 3.5. The descendants of the initial population shown in Figure 3.3. As you can see, one of them is better than the best of the previous generation.

In nature, at least in periods of stasis, the rates of mutation are tightly controlled and usually very small, and adaptation and evolution occur very smoothly. In GEP, populations are always evolving at high creative rates in what Eldredge and Gould called creative periods in their theory of punctuated equilibrium (Eldredge and Gould 1972). So, in artificial evolutionary systems, evolution can be much faster and kept in constant turmoil, allowing the discovery of very good solutions in record time.

Let's now take a closer look at the structure of the best individual of generation 1 (chromosome 6). Its expression is shown in Figure 3.6. As you can see, this program is slightly better than the best of the previous generation, solving correctly a total of seven fitness cases. As for its structure, we can easily guess that it is a descendant of the best of the previous generation (chromosome 4), as they only differ at position 2 in the second gene:

```
01234560123456
AbObcbcOAacaac-[0,4] = 6
AbObcbcOAAcaac-[1,6] = 7
```

As you can see, this new individual suffered just one point-mutation during reproduction, changing the "a" at position 2 in gene 2 into "A". This resulted in a bulkier sub-ET_2 encoding a different Boolean expression (compare Figures 3.4 and 3.6), conferring to this new individual a slightly better performance than its mother.

Figure 3.7 shows the next generation (generation 2) created in this evolutionary process. And as you can see, the best individual of this generation (chromosome 1) has maximum fitness and therefore codes for a perfect

a.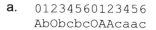
```
01234560123456
AbObcbcOAAcaac
```

b.

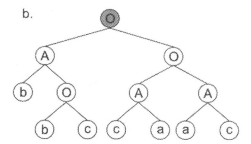

Figure 3.6. Expression of the best individual of generation 1 (chromosome 6). **a)** The chromosome composed of two genes. **b)** The program encoded in the chromosome (the linking function is shown in gray). This Boolean function solves correctly seven out of eight fitness cases (see Table 3.1).

```
Generation N: 2
01234560123456
AbObcbcOAAcaac-[0]  = 7
AbcbcbcOAAcaab-[1]  = 8
AaNbbbaNcNaacc-[2]  = 3
ANNbabbOONbbcc-[3]  = 3
AcNbcbbAONbbcc-[4]  = 4
AbAccbcOAAcaac-[5]  = 7
NaObaccOAbbcca-[6]  = 4
AbObcbcOAacaac-[7]  = 6
AbabbbcOAacaab-[8]  = 6
AbObcbbOAacaac-[9]  = 6
```

Figure 3.7. The second generation of computer programs evolved to solve the Majority(a, b, c) function problem. The individuals of this generation are the immediate descendants of the selected individuals of the previous generation. Note that a perfect program with maximum fitness was discovered (chromosome 1), thus better than all its ancestors.

solution to the majority function. As it is shown by its expression in Figure 3.8, this perfect solution involves a total of 11 nodes, two more nodes than the most parsimonious solutions to the majority function problem that can be designed using the primitives AND, OR, and NOT.

Let's now take a closer look at the event that led to the creation of this perfect solution. By comparing its genome with the genomes of all the individuals of the previous generation, there is no doubt that this new individual is a direct descendant of the best individual of generation 1, chromosome 6:

a. 01234560123456
AbcbcbcOAAcaab

b.

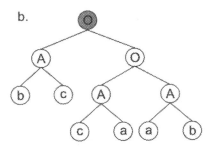

Figure 3.8. A perfect solution to the majority function. **a)** The chromosome encoding two sub-ETs linked by OR. **b)** The perfect solution encoded in the chromosome (the linking function is shown in gray).

```
01234560123456
AbObcbcOAAcaac-[1,6] = 7
AbcbcbcOAAcaab-[2,1] = 8
```

As you can see, two point-mutations had to occur in order to create this perfect solution: the first one changed the "O" at position 2 in gene 1 into "c", shortening the parental sub-ET in two nodes; and the second, although a less drastic change in terms of structure, managed to change the expression encoded in sub-ET$_2$ by changing the "c" at position 6 in gene 2 into "b".

It is worth noticing that, in this run, some individuals contain genes encoding one-element sub-ETs (namely, chromosomes 3 and 9 of generation 1) as they have a terminal at the start position of one of their genes. Obviously, all these genes were created by point mutation, as only mutation was used to create genetic modification in this problem. But we will see that, in gene expression programming, not only mutation but also inversion are capable of introducing a terminal at the root of sub-ETs (see section 3.3 for a detailed analysis of the genetic operators used in gene expression programming).

3.2 Fitness Functions and the Selection Environment

Fitness functions and selection environments are the two faces of fitness and are, therefore, intricately connected. When we speak of the fitness of an individual, on the one hand, it is always relative to a particular environment and, in the other, it is also relative to the measure (the fitness function) we are using to evaluate them. Consequently, the success of a problem not only depends on the way the fitness function is designed but also on the quality of the selection environment.

3.2.1 The Selection Environment

We know already that the selection environment is the input to the evolutionary system and, as such, the usefulness of the output, that is, the models designed by the system, will depend greatly on the quality of the environment we choose for them to blossom.

Thus, the first requirement is a set of fitness cases representative of the problem at hand. And the second consists of a well-balanced set in order to avoid the creation of models that can only solve a partial, and often marginal, aspect of the overall problem. Indeed, if these models are allowed to lurk into populations, they can seriously hinder evolution for they have usually very simple structures that are not only easily discovered but also extremely resilient. And because they can have a relatively high fitness, they will compete fiercely with any generalist that might appear, trapping the system in a local optimum for an indefinite length of time.

Irrespective of unbalanced or unrepresentative datasets, populations do sometimes get stuck in local optima for long periods of time and might need a little help to get out of there. A way of doing this consists of using a different measure to evaluate fitness. Indeed, by changing the fitness function we are changing the fitness landscape, and what was a peak on the previous landscape might now be a valley, and the population might be able to find a new path to the elusive global optimum. This is the reason why it is handy to have more than one fitness function at our disposal so that we can try a few of them on a problem. With this in mind, some of the most common fitness functions, used not only in symbolic regression but also in classification and logic synthesis, are described below.

3.2.2 Fitness Functions for Symbolic Regression

One important application of evolutionary computation is symbolic regression, where the goal is to find a symbolic expression that performs well for all fitness cases or, in other words, to find a model that is good at numeric prediction. And there are several measures that can be used to evaluate how good these models are. Some of them are based on the absolute error between predicted and target values. Others are based on the relative rather than absolute error. Others still are based on statistical indexes that measure the correlation between the values predicted by the model and the target values. Which of these fitness functions is more appropriate to search the fitness landscape in any given problem is a matter that can only be determined by studying the problem itself. But fitness functions based on standard measures such as the mean squared error or R-square are virtually universal and can be used to evolve very good models for all kinds of problems.

Number of Hits

The *number of hits* fitness function favors models that perform well for all fitness cases within a certain error (that is, the precision that is chosen for the evolved models) of the correct value. This error can be either the absolute or relative error.

More formally, the fitness $f_{(ij)}$ of an individual program i for fitness case j is evaluated by the formula:

$$\text{If } E_{(ij)} \leq p, \text{ then } f_{(ij)} = 1; \text{ else } f_{(ij)} = 0 \tag{3.1}$$

where p is the precision and $E_{(ij)}$ is the error of an individual program i for fitness case j, and is expressed by equation (3.2a) if the error chosen is the absolute error, and by equation (3.2b) if the error chosen is the relative error:

$$E_{(ij)} = \left| P_{(ij)} - T_j \right| \tag{3.2a}$$

$$E_{(ij)} = \left| \frac{P_{(ij)} - T_j}{T_j} \cdot 100 \right| \tag{3.2b}$$

where $P_{(ij)}$ is the value predicted by the individual program i for fitness case j and T_j is the target value for fitness case j. So, for these fitness functions, maximum fitness $f_{\max} = n$, where n is the number of fitness cases.

Precision and Selection Range

The *precision and selection range* fitness function explores the idea of a selection range and a precision. The selection range is used as a limit for selection to operate, above which the performance of a program on a particular fitness case contributes nothing to its fitness. And the precision is the limit for improvement as it allows the fine-tuning of the evolved solutions as accurately as possible.

More formally, the fitness f_i of an individual program i is expressed by equation (3.3a) if the error chosen is the absolute error, and by equation (3.3b) if the error chosen is the relative:

$$f_i = \sum_{j=1}^{n} \left(R - \left| P_{(ij)} - T_j \right| \right) \tag{3.3a}$$

$$f_i = \sum_{j=1}^{n} \left(R - \left| \frac{P_{(ij)} - T_j}{T_j} \cdot 100 \right| \right) \tag{3.3b}$$

where R is the selection range, $P_{(ij)}$ the value predicted by the individual program i for fitness case j (out of n fitness cases) and T_j is the target value for fitness case j. Note that the absolute value term corresponds, in both equations, to the error (absolute in the former, relative in the latter). This term is what is called the precision p. Thus, for a perfect fit, $P_{(ij)} = T_j$ for all fitness cases and $f_i = f_{max} = n \cdot R$.

Mean Squared Error

The *mean squared error* fitness function is based on the standard mean squared error, which, on its turn, is based on the absolute error. But obviously the relative error can also be used and two different fitness functions can be designed: one based on the absolute error and the other on the relative.

More formally, the mean squared error E_i of an individual program i is expressed by equation (3.4a) if the error chosen is the absolute error, and by equation (3.4b) if the error chosen is the relative:

$$E_i = \frac{1}{n}\sum_{j=1}^{n}\left(P_{(ij)} - T_j\right)^2 \tag{3.4a}$$

$$E_i = \frac{1}{n}\sum_{j=1}^{n}\left(\frac{P_{(ij)} - T_j}{T_j}\right)^2 \tag{3.4b}$$

where $P_{(ij)}$ is the value predicted by the individual program i for fitness case j (out of n fitness cases) and T_j is the target value for fitness case j. Thus, for a perfect fit, $P_{(ij)} = T_j$ for all fitness cases and $E_i = 0$. So, the mean squared error ranges from 0 to infinity, with 0 corresponding to the ideal.

As it stands, E_i cannot be used directly to measure the fitness of the evolved models since, for fitness proportionate selection, the fitness must increase with efficiency. Thus, to evaluate the fitness f_i of an individual program i, the following equation is used:

$$f_i = 1000 \cdot \frac{1}{1+E_i} \tag{3.5}$$

which obviously ranges from 0 to 1000, with 1000 corresponding to the ideal.

The fact that both these fitness functions are not only based on standard indicators but also very easy to implement, makes them very attractive; they can indeed be used to find good solutions to virtually all problems. Furthermore, they can also be used as a basis for designing other fitness functions. For instance, a slightly different kind of fitness function can be designed by taking the square root of the mean squared error. By doing this we are giving it the same dimensions as the quantity being predicted, so it might be interesting to use this measure instead of the straight mean squared error.

R-square

The *R-square* fitness function is based on the standard R-square, which returns the square of the Pearson product moment correlation coefficient. This coefficient is a dimensionless index that ranges from -1 to 1 and reflects the extent of a linear relationship between the predicted values and the target values. When the Pearson correlation coefficient R_i equals 1, there is a perfect positive linear correlation between target T and predicted P values, that is, they vary by the same amount. When $R = -1$, there is a perfect negative linear correlation between T and P, that is, they vary in opposite ways (when T increases, P decreases by the same amount). When $R = 0$, there is no correlation between T and P. Intermediate values describe partial correlations and the closer to -1 or 1 the better the model.

The Pearson product moment correlation coefficient R_i of an individual program i is evaluated by the equation:

$$R_i = \frac{n\sum_{j=1}^{n}(T_j P_{(ij)}) - \left(\sum_{j=1}^{n}T_j\right)\left(\sum_{j=1}^{n}P_{(ij)}\right)}{\sqrt{\left[n\sum_{j=1}^{n}T_j^2 - \left(\sum_{j=1}^{n}T_j\right)^2\right]\left[n\sum_{j=1}^{n}P_{(ij)}^2 - \left(\sum_{j=1}^{n}P_{(ij)}\right)^2\right]}} \tag{3.6}$$

where $P_{(ij)}$ is the value predicted by the individual program i for fitness case j (out of n fitness cases); and T_j is the target value for fitness case j.

The fitness f_i of an individual program i is a function of the squared correlation coefficient (the so called R-square) and is expressed by the equation:

$$f_i = 1000 \cdot R_i^2 \tag{3.7}$$

and therefore ranges from 0 to 1000, with 1000 corresponding to the ideal.

3.2.3 Fitness Functions for Classification and Logic Synthesis

Although very different, classification and logic synthesis share one similarity: their predictables or dependent variables are both binary and, consequently, both these problems can use the same kind of fitness function to evaluate the fitness of the evolved models. However, the vast majority of fitness functions (and the most colorful, I might add) were originally designed for classification problems, where it is usually not enough to just guess right or wrong. There are indeed different kinds of wrong decisions and different kinds of right decisions, and all of them have different costs and benefits. And all these different costs and benefits can be explored to design the right fitness function to evolve the right kind of model. For instance, there is a difference between classifying correctly a malignant tumor and classifying correctly a benign one.

More formally, there are four different possible outcomes of a single prediction for a two-class problem with classes "1" ("yes") and "0" ("no"). A false positive FP is when the outcome is incorrectly classified as "yes" (or "positive"), when it is in fact "no" (or "negative"). A false negative FN is when the outcome is incorrectly classified as negative when it is in fact positive. True positives TP and true negatives TN are obviously correct classifications. In the first case, the outcome is correctly classified as positive; and, in the latter, the outcome is correctly classified as negative.

Keeping track of all these possible outcomes is such an error-prone activity, that they are usually shown in what is appropriately called a confusion matrix (Figure 3.9). Good results correspond to large numbers down the main diagonal (TP plus TN) and to small numbers (ideally zero) off main diagonal positions (FP plus FN).

These two kinds of error, false positives and false negatives, will generally have different costs. And obviously the two kinds of correct classifications, true positives and true negatives, will have different benefits. Thus, for each problem, one can explore these differences between different costs and benefits to evolve the right kind of model by using different kinds of fitness functions. Let's take a look at some examples below.

Predicted Class

	Yes	No
Actual Class Yes	TP	FN
No	FP	TN

Figure 3.9. Confusion matrix for a two-class prediction.

Number of Hits

The *number of hits* fitness function is very simple and corresponds to the number of samples correctly classified. More formally, the fitness f_i of an individual program i is evaluated by the formula:

$$f_i = h \tag{3.8}$$

where h is the number of fitness cases correctly evaluated (number of hits). So, for this fitness function, maximum fitness f_{max} is given by the formula:

$$f_{max} = n \tag{3.9}$$

where n is the total number of fitness cases.

As you would recall, this fitness function was efficiently used to solve the Majority(a, b, c) function problem of section 3.1. But sometimes a more sophisticated fitness function is needed because, both in classification and logic synthesis, it is indeed very easy to discover lots of simple programs with relatively high fitness that are not particularly good predictors. On the one hand, the fitness of most programs will only reflect the 50% likelihood of correctly solving a binary function; and on the other, such simple programs as the constant zero $f(x_1, ..., x_n) = 0$ or constant one $f(x_1, ..., x_n) = 1$ functions are very easily discovered and will most probably also score a high fitness. For more complex applications, all these simple programs might hinder the evolutionary process and therefore it is advisable to reduce them to a minimum. A simple way of doing this is shown below.

Hits with Penalty

The *hits with penalty* fitness function was designed in order to avoid the dissemination of useless programs that could indefinitely trap the system in local optima. For that the following simple strategy can be used: for each program i, both the number of true positives TP_i and true negatives TN_i is evaluated and as long as either TP_i or TN_i remain equal to zero, this model will be considered unviable and therefore it won't be selected to reproduce.

More formally, the fitness f_i of an individual program i is evaluated by the formula:

$$\text{IF } (TP_i = 0 \text{ OR } TN_i = 0), \text{ THEN } f_i = 0; \text{ ELSE } f_i = h \tag{3.10}$$

where h is the number of fitness cases correctly evaluated (or number of hits, which obviously corresponds to $h = TP_i + TN_i$). So, for a perfect solution, maximum fitness corresponds to the total number of fitness cases and is also evaluated by equation (3.9).

Sensitivity / Specificity

The *sensitivity/specificity* fitness function is based both on the sensitivity and specificity indicators, both commonly used in the medical field. The sensitivity reflects the probability of the diagnostic test finding disease among those who have the disease or the proportion of people with disease who have a positive test result. And the specificity reflects the probability of the diagnostic test finding no disease among those who do not have the disease or the proportion of people free of a disease who have a negative test.

More formally, and perhaps more clearly, the sensitivity SE_i of an individual program i is evaluated by the equation:

$$SE_i = \frac{TP_i}{TP_i + FN_i} \tag{3.11}$$

and the specificity SP_i is evaluated by:

$$SP_i = \frac{TN_i}{TN_i + FP_i} \tag{3.12}$$

By multiplying both these indicators and using this new index as basis to measure the fitness of the evolved models, one forces the discovery of models that have both high sensitivity and specificity, since it would be relatively simple to maximize the sensitivity by minimizing the specificity and vice versa. Indeed, this kind of fitness function is extremely valuable in situations where highly unbalanced training sets are being used, that is, datasets with an excess of positive or negative instances. So, the sensitivity/specificity SS_i of an individual program i is evaluated by the equation:

$$\tag{3.13}$$

And for evaluating the fitness f_i of an individual program i, the following equation is used:

$$f_i = 1000 \cdot SS_i \tag{3.14}$$

which obviously ranges from 0 to 1000, with 1000 corresponding to the ideal.

Positive Predictive Value / Negative Predictive Value

The *positive predictive value / negative predictive value* fitness function is based both on the positive predictive value (PPV) and negative predictive value (NPV) indicators. Both these indicators are also commonly used in medicine, with the PPV reflecting the percentage of people with a positive diagnostic test result who actually have the disease, and the NPV reflecting the percentage of people with a negative diagnostic test who do not have the disease.

More formally, the positive predictive value PPV_i of an individual program i is evaluated by the equation:

$$PPV_i = \frac{TP_i}{TP_i + FP_i}, \text{ where } TP_i + FP_i \neq 0 \tag{3.15}$$

and the negative predictive value NPV_i is evaluated by:

$$NPV_i = \frac{TN_i}{TN_i + FN_i}, \text{ where } TN_i + FN_i \neq 0 \tag{3.16}$$

And again, by multiplying both these indicators and using this new index as basis to measure the fitness of the evolved models, one forces the discovery of models that have both high PPV and NPV. Thus, the PPV/NPV PN_i of an individual program i is evaluated by the equation:

$$PN_i = PPV_i \cdot NPV_i \tag{3.17}$$

And for evaluating the fitness f_i of an individual program i, the following equation is used:

$$f_i = 1000 \cdot PN_i \tag{3.18}$$

which obviously ranges from 0 to 1000, with 1000 corresponding to the ideal.

And now that we know how to measure the fitness of different kinds of evolving programs, let's see how they are selected to reproduce.

3.2.4 Selection Mechanism

In gene expression programming, individuals are selected according to fitness by roulette-wheel sampling (see, for instance, Goldberg 1989). This means that each individual receives a slice of the roulette-wheel proportional to its fitness. Then the roulette is spun as many times as there are individuals in the population so that the population size is maintained from generation to generation. And, obviously, the bigger the slice the higher the probability of being selected.

This kind of selection, together with the simple elitism presented in section 3.1.2, was used in all the problems of this book. It is true that with this selection scheme, sometimes some of the best individuals might be lost, while the worse or mediocre ones are passed on to the next generation. But this is not necessarily bad: for one thing, these good individuals might be created again and be more fortunate in the next round and, for another, the descendants of the mediocre individuals won't necessarily be mediocre and adaptation just goes on, perhaps through an unexpected path that might lead to the global optimum. Nonetheless, due to the cloning of the best individual of each generation, the survival and reproduction (without modification) of the best is guaranteed, what, in practical terms, means that at least the best trait is never lost and a continuous improvement is also accomplished.

Other selection schemes can be found in the literature but the most popular – roulette-wheel, deterministic, and tournament selection – will be compared in chapter 12 (section 7, Analysis of Different Selection Schemes). And we will see that, if elitism is present, no appreciable difference exists between all these methods. Indeed, evolutionarily speaking, all of them are very good if the rest is also good. And the "rest" is the essence of each artificial evolutionary algorithm. And, at least for GEP, this essence does not reside in the kind of selection mechanism used but in the power of the genetic operators operating on a fully functional genotype/phenotype system. As we will see in the next section, the genetic operators are the real "eagles" of the fitness landscape.

3.3 Reproduction with Modification

According to both the fitness and the luck of the draw, individuals are selected to reproduce with modification. And this creates the fundamental genetic diversity that will allow adaptation in the long run.

In nature, several modifications are introduced during the replication of the genomes (e.g., mutations, small deletions/insertions, etc.); others, like homologous recombination and yet more mutations, occur after replication. Thus, in nature, it is not always possible to know when a modification took place. In gene expression programming, though, we know exactly when modifications in the genome occur. On the one hand, all of them occur after the replication of the genome, which in itself is flawless. On the other, the genetic operators perform in an orderly fashion, starting with replication and continuing with mutation, inversion, transposition, and recombination. It is worth pointing out, however, that the order in which the latter modify the genome is not important to the final outcome.

Except for replication, which copies exactly the genomes of all the selected individuals, all the remaining operators randomly pick up the chromosomes to be subjected to a certain modification. And, except for mutation, each operator is not allowed to modify a chromosome more than once. For instance, for a population of 10 individuals, a crossover rate of 0.8 means that eight different chromosomes are randomly chosen to recombine.

Thus, in GEP, a chromosome might be randomly chosen to be modified by more than one modification operator during its reproduction. Therefore, during reproduction, the modifications carried out by the different genetic operators accumulate in the chromosomes, to the extent that the new population is usually very different from the old one.

Presented below, in the order in which they are usually applied, are the most commonly used genetic operators, obviously starting with replication and selection and finishing with the modification operators of mutation, inversion, transposition, and recombination.

3.3.1 Replication and Selection

Although vital, replication in artificial evolutionary systems is the most uninteresting operator of all: by itself, it contributes nothing to genetic variation. But together with selection, it can already create genetic drift, changing the proportions of the particular kinds of individuals with time. And most importantly, together with selection and genetic modification, it allows adaptation and evolution.

So, the selection operator chooses which individuals are going to be reproduced according to their fitnesses and the luck of the roulette. And this means that the fitter the individual the higher the probability of leaving more offspring. When selection is done by roulette-wheel sampling, the roulette is spun as many times as there are individuals in the population, thus maintaining the same population size from generation to generation. As for the replication operator, it copies exactly the chromosomes of the individuals picked up by the selection operator. And the copied chromosomes are none other than the chromosomes of the individuals of the next generation. But reproduction is not yet completed. Before that, the replicated chromosomes will undergo genetic modification. This modification is done by the genetic operators of mutation, inversion, transposition, and recombination (called modification operators in order to distinguish them from the genetic operators of replication and selection). But for now let's concentrate on selection and replication.

Figure 3.10 gives an example of how the selected individuals are replicated (the modification operators and elitism were switched off so that replication and selection could be better understood). For instance, chromosome 0, one of the best of generation 0, left two daughters (chromosomes 1 and 3 of generation 1); chromosome 1, also one of the best of this generation, left only one descendant (chromosome 9 of generation 1); chromosome 7, also one of the best, left two daughters (chromosomes 2 and 4 of generation 1); but chromosome 8, also one of the best, died without leaving offspring; the second best (chromosome 3) was more lucky and left two daughters (chromosomes 7 and 8 of generation 1); and although the most unfit of generation 0 (chromosomes 2, 4, and 5) did not reproduce, chromosome 6, a mediocre individual, left the biggest progeny (chromosomes 0, 5, and 6 of generation 1). The outcome of such an "evolutionary" process with just replication and

```
Generation N: 0                    Generation N: 1
01234560123456                     01234560123456
OOOaaabAAAcabb-[0]  = 6            OONbcaaAaAaacc-[0]  = 4
AAcbbabNONaaac-[1]  = 6            OOOaaabAAAcabb-[1]  = 6
ANaccbcNAAcbbc-[2]  = 2            AcOaccbAbNbabc-[2]  = 6
OAOccbaAOAbcab-[3]  = 5            OOOaaabAAAcabb-[3]  = 6
AAAbbabNcObcca-[4]  = 3            AcOaccbAbNbabc-[4]  = 6
NbacabbNbccbbc-[5]  = 2            OONbcaaAaAaacc-[5]  = 4
OONbcaaAaAaacc-[6]  = 4            OONbcaaAaAaacc-[6]  = 4
AcOaccbAbNbabc-[7]  = 6            OAOccbaAOAbcab-[7]  = 5
AaOacccAbbbaca-[8]  = 6            OAOccbaAOAbcab-[8]  = 5
AOAcaaaNaNbaab-[9]  = 4            AAcbbabNONaaac-[9]  = 6
```

Figure 3.10. Illustration of replication and selection. Only replication and roulette-wheel selection are switched on so that these operators could be better understood. Note, for instance, that chromosome 8 of generation 0 (one of the best of this generation) did not leave descendants, whereas chromosome 6 (a mediocre individual) left the biggest progeny (chromosomes 0, 5, and 6 of generation 1).

selection operating, is shown in Figure 3.11, where we can see that by generation 13 all the individuals are descendants of only one individual: in this case, chromosome 0 of generation 0. Indeed, replication, together with just selection, is only capable of causing genetic drift. And, although useful for searching the fitness landscape, genetic drift by itself cannot create genetic diversity. Only the modification operators have that power.

```
Generation N: 13
01234560123456
000aaabAAAcabb-[0] = 6
000aaabAAAcabb-[1] = 6
000aaabAAAcabb-[2] = 6
000aaabAAAcabb-[3] = 6
000aaabAAAcabb-[4] = 6
000aaabAAAcabb-[5] = 6
000aaabAAAcabb-[6] = 6
000aaabAAAcabb-[7] = 6
000aaabAAAcabb-[8] = 6
000aaabAAAcabb-[9] = 6
```

Figure 3.11. Illustration of genetic drift. In this extreme case, after 13 generations the population lost all genetic diversity, and all its members are descendants of one chromosome, in this case, chromosome 0 of generation 0 (see Figure 3.10).

3.3.2 Mutation

Of the operators with intrinsic modification power, mutation is the most efficient (see a discussion of the Genetic Operators and Their Power in chapter 12). With mutation, populations of individuals adapt very efficiently, allowing the evolution of good solutions to virtually all problems. Typically, I use a mutation rate p_m equivalent to two one-point mutations per chromosome. Considering the relatively small length of GEP genomes, this mutation rate is much higher than the mutation rates found in nature (see, e.g., Futuyma 1998). Indeed, thanks to elitism, we can have GEP populations subjected to very high mutation rates and, nevertheless, evolving very efficiently. As a comparison, the human genome is about 6×10^9 base pairs long and only about 120 new mutations are introduced per genome per generation.

In gene expression programming, mutations are allowed to occur anywhere in the chromosome. However, the structural organization of chromosomes must be obviously preserved. Thus, in the heads of genes, any symbol can change into another (function or terminal); and in the tails, terminals can only change into terminals. This way, the structural organization of chromosomes is maintained and all the new individuals produced by mutation are structurally correct programs.

It is worth emphasizing that GEP point mutation is totally unconstrained. This means that, in the heads, functions can be replaced by other functions without concern for the number of arguments each one takes; functions can also be replaced by terminals and vice versa; and obviously terminals can also be replaced by other terminals. Indeed, no restrictions whatsoever exist and, therefore, mutation can be completely exploited to roam thoroughly the fitness landscape.

The workings of mutation can be analyzed in the evolutionary history shown in Figure 3.12. For this analysis, mutation was the only source of genetic variation. The mutation rate was, as usual, equivalent to two one-point mutations per chromosome, which, for these chromosomes of length 14, corresponds to a mutation rate $p_m = 0.143$. The populations shown in Figure 3.12 were obtained in a run created to solve the already familiar majority function problem using the number of hits fitness function of section 3.2.3. Note that chromosome 1 of generation 6 encodes a perfect solution to the majority function and, therefore, has maximum fitness.

By analyzing the sequence of this perfect solution, one can easily guess that its most probable ancestors are chromosomes 0, 1, and 3 of generation 5. In the first case, only one point mutation would have occurred during reproduction, whereas two point mutations would have been required in the last two cases. Figure 3.13 compares the sub-ETs of one of these putative ancestors (chromosome 3) with the daughter sub-ETs, i.e., before and after mutation. As you can see, in this case, two point mutations occurred during reproduction of chromosome 3: one changed the "N" at position 1 in gene 1 into "a"; and another changed the "a" at position 3 in gene 2 into "b". Note that the first mutation changed significantly the sub-ET$_1$, shortening the original sub-ET in one node. Note also that, although the second mutation did not change the shape of sub-ET$_2$, the expression encoded in this new sub-ET is no longer the same.

Let's now analyze more closely the populations shown in Figure 3.12. On the one hand, we can see that several mutations have a neutral effect. For instance, chromosome 7 of generation 6 is a descendant of chromosome 4 of generation 5. These chromosomes differ only at positions 1 and 4 in gene 2. As you can see in Figure 3.14, the expression of these chromosomes results in

```
Generation N: 0
01234560123456
NabbabbAAccbcb-[0] = 3
NAabbcaNbbbcca-[1] = 2
OcOcaaaNaOabaa-[2] = 4
AaAcccbAbccbbc-[3] = 7
AObbabaAOcaabc-[4] = 7
AAAbaacONOaabc-[5] = 4
AAccbcaNNcbbac-[6] = 6
NOccabaOcbabcc-[7] = 4
NOAcbbbAaNabca-[8] = 2
NacbbacAbccbbc-[9] = 3
```

. . .

```
Generation N: 5                      Generation N: 6
01234560123456                       01234560123456
AabbabcAOcaabc-[0] = 7               AOAbabacbcaaba-[0] = 7
babbabcAOcaabc-[1] = 7               AabbabcAOcbabc-[1] = 8
AOAacbcOAOcaac-[2] = 6               AabbabcAccaabc-[2] = 7
ANbbabcAOcaabc-[3] = 6               NAAbaacONaaacc-[3] = 4
AOAbabacbcaaba-[4] = 7               AOAbabacbcaaba-[4] = 7
AabcaccAONaabc-[5] = 6               AabbbbcAONaabb-[5] = 6
AOAccbaAbaabbc-[6] = 6               AOAbabacNcabba-[7] = 7
AObcabaAbNcaba-[7] = 6               AOAccbaAbaabbc-[6] = 6
NAAbbacONOacca-[8] = 3               NAAbbacONOacaa-[8] = 4
AONbabacbcaaba-[9] = 5               AObbabaAAcaabc-[9] = 7
```

Figure 3.12. An initial population and its later descendants created via mutation in order to solve the Majority(a, b, c) function problem. The chromosomes encode sub-ETs linked by OR. Note that none of the later descendants are identical to their ancestors of the initial population. The perfect solution found in generation 6 (chromosome 1) and one of its putative ancestors (chromosome 0 of generation 5) are shown in bold. Note that chromosomes 1 and 3 of generation 5 are also good candidates to be the predecessors of this perfect solution; in both cases, two point mutations would have occurred during reproduction.

identical sub-ETs, due to the fact that both mutations occurred downstream of the termination point of ORF_2. As we have seen, mutations occurring in the noncoding region of a gene have a neutral effect, as they have no expression in the "organism" itself. But let's not forget that they play nonetheless an important role in evolution, because these noncoding sequences might become active in the blink of an eye due to the actions of the modification operators (see a discussion of The Role of Neutrality in Evolution in chapter 12).

a.   01234560123456
      ANbbabcAOc**aa**bc-[m] = 6
      A**a**bbabcAOc**b**abc-[d] = 8

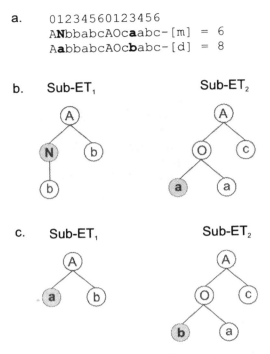

b.   Sub-ET₁   Sub-ET₂

c.   Sub-ET₁   Sub-ET₂

Figure 3.13. Illustration of point mutation. **a)** The mother and daughter chromosomes with the mutation points in bold. **b)** The sub-ETs encoded in the mother chromosome (before mutation). **c)** The sub-ETs codified by the daughter chromosome (after mutation). The mutated nodes are shown in gray. Note that the first mutation changed significantly the sub-ET₁, shortening the original sub-ET in one node. Note also that with these two point mutations, a perfect solution to the Majority(a, b, c) function was discovered (the sub-ETs are linked by OR).

On the other hand, we can see that, in gene expression programming, mutations in the coding sequence of a gene have most of the times a very profound effect, reshaping drastically the sub-ETs . However, this capability to reshape profoundly expression trees is fundamental for an efficient evolution (see chapter 12 for a discussion of the Genetic Operators and Their Power). Indeed, the results presented throughout this book clearly show that our too human wish to keep intact the small functional building blocks and recombine them carefully without disrupting them (as is done in genetic programming through tree crossover) is conservative and works poorly in evolutionary terms. In fact, genotype/phenotype systems can find much more efficient ways of creating their own building blocks and manipulating them

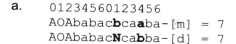

a.   01234560123456
        AOAbabac**b**caa**b**a-[m] = 7
        AOAbabac**N**ca**b**ba-[d] = 7

b.   Sub-ET₁   Sub-ET₂

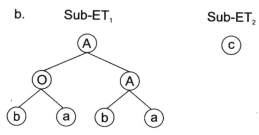

Figure 3.14. Illustration of neutral mutations. **a)** The mother and daughter chromosomes with the mutation points shown in bold. **b)** The sub-ETs encoded by both chromosomes (they are exactly the same and, therefore, are shown just once). Note that the point mutations shown in (a) have no expression in sub-ET₂ as they occurred downstream of the termination point.

to create new ones. These building blocks are very different from the ones a mathematician would have chosen but, nonetheless, they work much more efficiently.

And finally, it is worth emphasizing that in gene expression programming there are no constraints both in the type of mutation and the number of mutations in a chromosome because, in all cases, the newly created individuals will always be syntactically correct programs. This important feature distinguishes gene expression programming from all GP-style systems, where straight point mutations would have resulted most of the times in invalid programs. And an evolutionary algorithm unable to use freely this powerful operator is severely restricted, since mutation is the most important agent of genetic diversification (Ferreira 2002a; see also the discussion of the Genetic Operators and Their Power in chapter 12).

3.3.3 Inversion

We know already that the modifications bound to make a big impact occur usually in the heads of genes. Therefore, the inversion operator was restricted to these regions. Here any sequence might be randomly selected and inverted. It is worth pointing out that, since the inversion operator was restricted to the heads of genes, there is no danger of a function ending up in the tails and, consequently, all the new individuals created by inversion are syntactically correct programs.

In gene expression programming, the inversion operator randomly chooses the chromosome, the gene within the chromosome to be modified, and the start and termination points of the sequence to be inverted. Typically, a small inversion rate p_i of 0.1 is used as this operator is seldom used as the only source of genetic variation.

The workings of inversion can be analyzed in the evolutionary history shown in Figure 3.15. For this analysis, a much higher inversion rate of 1.0 was used as only inversion was used to create genetic diversity. Again, the problem we are trying to solve is the majority function and the chromosomes encode in this case two sub-ETs linked by OR. Note that chromosome 6 of generation 19 has maximum fitness and therefore encodes a perfect solution to the majority function problem.

The comparison of the sequence of this perfect solution with the sequences of all its ancestors shows that it is a direct descendant of chromosome 0 of generation 18. Also worth pointing out is that not only the perfect solution but also all the individuals of generations 18 and 19 are descendants of one of the worst individuals of the initial population, chromosome 3; by changing this chromosome again and again during reproduction, a perfect solution to the problem at hand was created by generation 19.

The last event of inversion that led to the creation of this perfect solution is shown in Figure 3.16. As you can see, a small sequence of four elements "AaAb" was inverted in gene 2, creating a new sub-ET that is not only considerably different from its mother but is also slightly better (a fitness of 8 as opposed to 7). This new individual is in fact better than all its ancestors and encodes a perfect solution to the problem at hand.

Note that inversion has not only the power of making small or medium changes (such as the one that led to the creation of this perfect solution) in expression trees but has also the power of causing huge macromutations. For instance, chromosome 7 of generation 19 has a terminal at the start position of gene 1 as a result of a small inversion that occurred during reproduction (Figure 3.17). In this case, the daughter sub-ET turned out a total of 11 nodes smaller than its mother, remaining equally fit in the process. Again we can see that macromutations are not always bad: they are in fact essential to drive evolution into other, very distant peaks.

You have certainly noticed that the examples I have chosen in order to illustrate the mechanisms and effects of the modification operators have

```
Generation N: 0
01234567890123401234567890 1234
OaAOObOcababcabNObbAcAabccbbbc-[0]  = 5
NOcacbNbbcacbacANccOONbcbaabca-[1]  = 2
NNaAcbOaaaacbccAbNaNaNacccccbbb-[2] = 6
NOaOAAAbcbbcaaaAbAAOaccbbaaacc-[3]  = 2
AObAbbAccbbbbbabAaaNbObbccaaaab-[4] = 6
ObAObcAccbcaacbAbAacOAabbccaac-[5]  = 6
AcAccbNbccbaaccNabOcOcacacacac-[6]  = 4
AbANbbacbccaaabAOccaNNccccbaba-[7]  = 6
NAAccNAcbaaccaaNaccAOabaccccac-[8]  = 3
NAOaNaabbabcacaNAANNOAcbbcabcc-[9]  = 3
```

. . .

```
Generation N: 18
01234567890123401234567890 1234
AaOAANObcbbcaaaAOAaAbccbbaaacc-[0]  = 7
AAaAONObcbbcaaaAOAaAbccbbaaacc-[1]  = 7
OaAAONAbcbbcaaaOaAbAAccbbaaacc-[2]  = 7
AaOAANObcbbcaaaAOAcbAacbbaaacc-[3]  = 6
ONAAOaAbcbbcaaaAOAaAbccbbaaacc-[4]  = 4
AaOAANObcbbcaaabAaAOAccbbaaacc-[5]  = 7
AAaANOObcbbcaaaOcAbaAAcbbaaacc-[6]  = 7
AAaOONAbcbbcaaaObAcAaAcbbaaacc-[7]  = 7
OaAAONAbcbbcaaaOcbAAAacbbaaacc-[8]  = 5
AAaAONObcbbcaaaAOAaAbccbbaaacc-[9]  = 7
```

```
Generation N: 19
01234567890123401234567890 1234
AAaAONObcbbcaaaAOAaAbccbbaaacc-[0]  = 7
ANOAAaObcbbcaaaOaAbAAccbbaaacc-[1]  = 7
OaAAONAbcbbcaaaaOAbAAccbbaaacc-[2]  = 6
AaOAANObcbbcaaaAOAabAccbbaaacc-[3]  = 7
ONAaOAAbcbbcaaaAOAaAbccbbaaacc-[4]  = 3
OaAAONAbcbbcaaaOaAbcAAcbbaaacc-[5]  = 7
AaOAANObcbbcaaaAObAaAccbbaaacc-[6]  = 8
aAAAONObcbbcaaaAOAaAbccbbaaacc-[7]  = 7
AAaAONObcbbcaaaAOAaAcbcbbaaacc-[8]  = 7
OaANOAAbcbbcaaaOaAbAAccbbaaacc-[9]  = 6
```

Figure 3.15. An initial population and its later descendants created via inversion to solve the Majority(a, b, c) function problem. The chromosomes encode sub-ETs linked by OR. Note how different the later descendants are from their ancestors of the initial population. Note also the appearance of genes with a terminal at the start position in later generations, a feat that can only be achieved by inversion and mutation. The perfect solution found in generation 19 (chromosome 6) and its mother (chromosome 0 of generation 18) are shown in bold. The event of inversion that led to this perfect solution is shown in Figure 3.16.

a.
```
     01234567890123401234567890 1234
     AaOAANObcbbcaaaAOAaAbccbbaaacc-[m] = 7
     AaOAANObcbbcaaaAObAaAccbbaaacc-[d] = 8
```

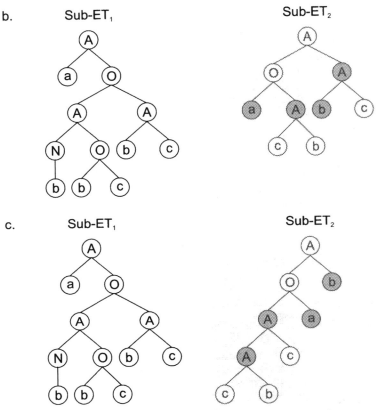

Figure 3.16. Illustration of inversion. **a)** The mother and daughter chromosomes with the inverted sequence shown in bold. **b)** The sub-ETs encoded by the mother chromosome (before inversion). **c)** The sub-ETs encoded by the daughter chromosome (after inversion) (the inverted nodes are shown in gray). Note that inversion changed significantly sub-ET$_2$, by arranging the nodes differently in the tree. Note also that with the inversion of this sequence, a perfect solution to the Majority(a, b, c) function was discovered (the sub-ETs are linked by OR).

always resulted in fitter descendants. Remember, however, that in the vast majority of cases, their actions result in less fit or even unviable individuals. But, as in nature, evolution happens because of these extremely rare, highly improbable events.

a.   0123456789012340123456789012340

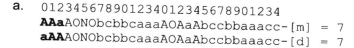

AAaAONObcbbcaaaAOAaAbccbbaaacc-[m] = 7
aAAAONObcbbcaaaAOAaAbccbbaaacc-[d] = 7

b.

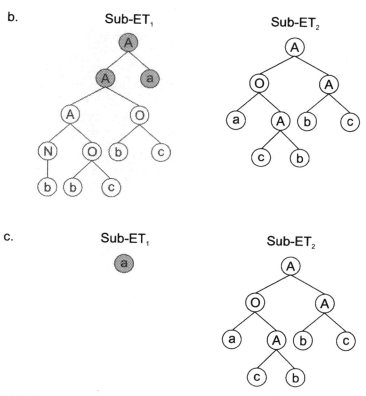

Figure 3.17. Illustration of a macromutation created by inversion. a) The mother and daughter chromosomes with the inverted sequence shown in bold. b) The sub-ETs encoded by the mother chromosome (before inversion). c) The sub-ETs encoded by the daughter chromosome (after inversion) (the inverted nodes are shown in gray). Note that this inversion event changed drastically the first sub-ET, shortening the daughter sub-ET by 11 nodes. Note also that the fitness remained the same despite this huge transformation (the sub-ETs are linked by OR).

3.3.4 Transposition and Insertion Sequence Elements

The transposable elements of gene expression programming are fragments of the genome that can be activated and jump to another place in the chromosome. And there are three kinds of transposable elements in GEP: (1) short fragments with either a function or terminal in the first position that transpose to the head of genes, except the root (insertion sequence elements or IS elements); (2) short fragments with a function in the first position that transpose to the root of genes (root IS elements or RIS elements); and (3) entire genes that transpose to the beginning of chromosomes.

The existence of IS and RIS elements is a remnant of the developmental process of GEP itself, as the first gene expression algorithm used only single-gene chromosomes and, in such systems, a gene with a terminal at the root was of little use. Consequently, transposition to the root was tightly controlled and only transposons with a function in the first position were allowed to transpose to the root. When multigenic chromosomes were introduced, this feature was maintained as these operators not only serve different purposes in evolution but are also important to the understanding of the functions and the requirements of the genetic operators. Indeed, the comparison of the transforming power of these operators (see a discussion of The Genetic Operators and Their Power in chapter 12) shows clearly that there is no need to be cautious while designing genetic operators, rejecting everything that seems to give rise to too drastic a modification. In fact, on its own, root transposition – undoubtedly the most disruptive operator in GEP – is capable of finding very good solutions very efficiently. Moreover, IS and RIS transposition are also capable of creating simple repetitive sequences in the genome and this, in itself, is not only curious but also extremely important to an efficient evolution.

Transposition of IS Elements

Any sequence in the genome can become an IS element and, therefore, these elements are randomly chosen throughout the chromosome. The transposon is then copied at the place of origin and the copy is afterwards inserted at a randomly chosen point in the head of a gene, except the start position. Thus, the transposition operator randomly chooses the chromosome, the start and termination points of the IS element, and the target site. Typically, a small IS transposition rate p_{is} of 0.1 is used as this operator is seldom used as the only source of genetic variation.

The workings of IS transposition can be analyzed in the evolutionary history presented in Figure 3.18. As an illustration, and because only IS transposition is being used to create genetic diversity, a much higher transposition rate of 1.0 was chosen. And as you can see, by itself, this operator can also make populations evolve and find good solutions to the problem at hand. Indeed, chromosome 8 of generation 13 is a perfect solution to the majority function problem.

```
Generation N: 0
01234567890123456789001234567890123456789 0
AbOabcOAOOaacaccacbccANaObOAbNaabcbcccbcbb-[0]  =  6
AaAOaOOaAAbbacabcabbcOAAcONaOAccabbbcabbbc-[1]  =  6
OaAbcNbaNaabccbbbcaaaNbcbNbAANbacabacacbcb-[2]  =  5
ObbbAONNNAccaccbabcbcAAcacObOAaabaaaacbaca-[3]  =  7
ONNAbObbONabcabbcbbcbAObaNbOAcbcccbcbaaabb-[4]  =  4
ANabONNNcabbcccabcbacAbObOcAaAbabacacaabcb-[5]  =  6
AONNObONOOaaacbaaababOAOAANaObacbbcccbbcaa-[6]  =  4
NNbabcNbNNcbaacaabcccAbNcANNbbObababbcabaa-[7]  =  6
NONNAaAbAOcacaaababccObaabbOabababacaacbab-[8]  =  6
NONbcANNONcabbacaccabAbaOcOAacNbbccbacacba-[9]  =  6


                    . . .


Generation N: 12
01234567890123456789001234567890123456789 0
ANabcaabbcbbcccabcbacAbOcAaAbaaabacacaabcb-[0]  =  7
AaANbAaANcbbcccabcbacAbOOcAaAbaaabacacaabcb-[1]  =  7
AAaANcaabbbbcccabcbacAbcccOOcAaabacacaabcb-[2]  =  6
ANababbcbObbcccabcbacAbOcAaAbaaabacacaabcb-[3]  =  7
ANacaabbcbbbcccabcbacAbOcAaAbaaabacacaabcb-[4]  =  6
AAbaaANcaabbcccabcbacAbOOcAaAbaabacacaabcb-[5]  =  7
ANababbcbObbcccabcbacAAbaabbObcabacacaabcb-[6]  =  6
ANababbcbObbcccabcbacAbOcaabcAaabacacaabcb-[7]  =  7
ANcababbcbbbcccabcbacAbccabObOcabacacaabcb-[8]  =  5
ANaabbcbObbbcccabcbacAbOcAaAbaaabacacaabcb-[9]  =  7

Generation N: 13
01234567890123456789001234567890123456789 0
ANaabbcbObbbcccabcbacAbOcAaAbaaabacacaabcb-[0]  =  7
ANabcaabbcbbcccabcbacAbOcbcAaAbabacacaabcb-[1]  =  6
ANacbabbcbbbcccabcbacAbOcaabcAaabacacaabcb-[2]  =  6
ANababbcbObbcccabcbacAbObOccaababacacaabcb-[3]  =  6
AaANbAaANcbbcccabcbacAbOAOcAaAbabacacaabcb-[4]  =  6
ANacababbcbbcccabcbacAbOcAaAbaaabacacaabcb-[5]  =  6
ANcababbcbbbcccabcbacAbcabObOcaabacacaabcb-[6]  =  5
ANcababbcbbbcccabcbacAbccabObOcabacacaabcb-[7]  =  5
AcabAbaaANbbcccabcbacAbOOcAaAbaabacacaabcb-[8]  =  8
AccaAbaaANbbcccabcbacAbOOcAaAbaabacacaabcb-[9]  =  7
```

Figure 3.18. An initial population and its later descendants created via IS transposition to solve the Majority(a, b, c) function problem. The chromosomes encode sub-ETs linked by OR. Note that none of the later descendants are identical to the ancestors of the initial population. Note also the appearance of repetitive sequences in the genome in later generations. The perfect solution found in generation 13 (chromosome 8) and its mother (chromosome 5 of generation 12) are shown in bold. The event of transposition that led to this perfect solution is shown in Figure 3.19.

The comparison of the sequence of this perfect solution with the sequences of all its ancestors shows that it is a direct descendant of chromosome 5 of generation 12. The event of IS transposition that led to the creation of this perfect solution is shown in Figure 3.19.

As you can see in Figure 3.19, during IS transposition, IS elements are copied into the head of the target gene. In this case, the sequence "cab" (positions 14-16 in gene 1) is activated and jumps to the insertion site, namely, bond 1 in gene 1 (between positions 0 and 1). As a result, a copy of the transposon appears at the site of insertion. Note also that a sequence with as many symbols as the IS element is deleted at the end of the head (in this case, the sequence "caa" is deleted). Thus, despite this insertion, the structural organization of chromosomes is maintained and, therefore, all the new individuals created by IS transposition are syntactically correct programs.

Note that IS transposition can also be a macromutator, causing very profound modifications in the expression trees. Obviously, the more upstream the insertion site the more drastic the change. For example, the sub-ET_1 shown in Figure 3.19 was shortened by two nodes due to transposition. But obviously IS transposition has also the power to make small adjustments in the sub-ETs, making IS transposition a very well balanced genetic operator.

Also interesting is that neutral sequences on the noncoding regions of genes might be activated by this operator and jump to another place in the genome where they will get expressed. This was indeed what happened in the example of Figure 3.19, where a sequence in a noncoding region jumped to the middle of a coding region, changing considerably the sub-ET.

Root Transposition

All RIS elements start with a function and, therefore, must be chosen among the sequences of the heads. Thus, a point is randomly chosen in the head and, from this point onwards, the gene is scanned until a function is found. This function becomes the first position of the RIS element. If no functions are found, the operator does nothing. So, the transposition operator randomly chooses the chromosome, the gene to be modified, and the start and termination points of the RIS element. Typically, a root transposition rate p_{ris} of 0.1 is used as this operator is seldom used as the only source of genetic variation.

In the example shown in Figure 3.20, a much higher RIS transposition rate of 1.0 was chosen as only this operator was used to create genetic modification. By the results shown in Figure 3.20, you can see that this operator on its own is also capable of making populations evolve efficiently. In this case, by

a.
```
012345678901234567890012345678901234567890
AAbaaANcaabbcccabcbacAbOOcAaAbaabacacaabcb-[m] = 7
AcabAbaaANbbcccabcbacAbOOcAaAbaabacacaabcb-[d] = 8
```

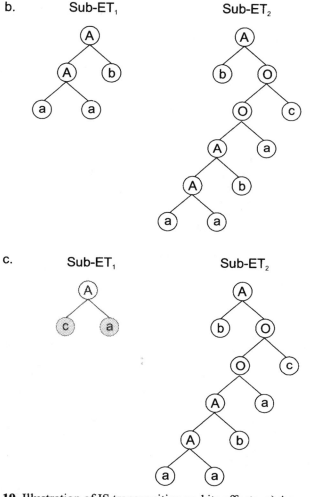

Figure 3.19. Illustration of IS transposition and its effects. **a)** An event of IS transposition with the transposon shown in bold. Note that a sequence with three elements is deleted at the end of the head of the target gene. Note also that, in this case, the transposon ends up duplicated in the daughter chromosome. **b)** The sub-ETs encoded in the mother chromosome (before IS transposition). **c)** The sub-ETs encoded in the daughter chromosome (after IS transposition) (the transposon nodes are shown in gray). Note that this insertion changed considerably sub-ET₁, creating a new sub-ET two nodes smaller than its mother. Note also that the transposon "cab" was in a noncoding region before becoming activated.

```
Generation N: 0
012345678901234567890012345678901234567890
NAccbAaOccaaaaabccccbOAAaANObAbababcacccab-[0] = 4
NOONAONbAAcbbaabbbbccAOONaOacAOaccbaacacac-[1] = 4
AbabNcOcacaaaccbbbcccNcAbOaObcOabbbccbbcaa-[2] = 3
NNAcaNNANbccbcbbbacbcOacAabaNccbbbacbcbbcb-[3] = 6
ONbcaNNcAAccbcbbcaabcAaOaabbbAbaabccabbbca-[4] = 5
NObcNNaaNbabccccaccacNbcAbcNaAbacaaabcabab-[5] = 2
NOAacOcbbAacbaacaaabcAOacNaaabOcacccabbcbb-[6] = 4
ONaAcaNcOacbbbacbccacOcOAOcNcANcbcaabbbaba-[7] = 4
AOAONbAObbbbcbbacbbcbOONAbANAcbbcbabcccaab-[8] = 4
ANbbcbaaccaaaaccbabcbAOaaacObaObabccaacaba-[9] = 6

                       . . .

Generation N: 16
012345678901234567890012345678901234567890
AOcAcAOccNaaaaabccccbAbaabAbaOAababcacccab-[0] = 7
AaAcAaOAaOaaaaabccccbAOAbaObAObababcacccab-[1] = 7
OcOccAaOccaaaaabccccbAaAAbaabAaababcacccab-[2] = 7
AaOccAcAaOaaaaabccccbAOObAAOAObababcacccab-[3] = 7
OAaaaAaAaAaaaaabccccbObAObAAbaOababcacccab-[4] = 6
OAaOaAaAcAaaaaabccccbAAbObAObAAababcacccab-[5] = 7
AaAaAcAaOAaaaaabccccbObAObAAbaOababcacccab-[6] = 7
OccAcAaOccaaaaabccccbObAAOAObAAababcacccab-[7] = 6
OccAaOccAcaaaaabccccbObAAOAObAAababcacccab-[8] = 6
OccOccAaOcaaaaabccccbAaAAbaabAaababcacccab-[9] = 7

Generation N: 17
012345678901234567890012345678901234567890
OccOccAaOcaaaaabccccbAaAAbaabAaababcacccab-[0] = 7
AaOAaAaAcAaaaaabccccbObAObAAbaOababcacccab-[1] = 6
OAAaAaAcAaaaaabccccbObAObAAbaOababcacccab-[2] = 6
AaOcOccAaOaaaaabccccbAaAAbaabAaababcacccab-[3] = 7
OaAaAOAaOaaaaabccccbAAbObAObAAababcacccab-[4] = 7
OccAaOccOcaaaaabccccbAaAAbaabAaababcacccab-[5] = 7
AaAaAcAaOAaaaaabccccbAbObAObAAbababcacccab-[6] = 7
OcOccAaOccaaaaabccccbAAAaAAbaabababcacccab-[7] = 7
AcAaaOAaOaaaaabccccbAAbObAObAAababcacccab-[8] = 8
AaOccAcAaOaaaaabccccbAAOAOObAAOababcacccab-[9] = 6
```

Figure 3.20. An initial population and its later descendants created via RIS transposition to solve the Majority(a, b, c) function problem. The chromosomes encode sub-ETs linked by OR. Note that none of the later descendants resemble their ancestors of generation 0; however, by the sequence of the tails, one can see that all of them are descendants of chromosome 0 of generation 0. Note also the appearance of repetitive sequences in later generations. The perfect solution found by generation 17 (chromosome 8) and its mother (chromosome 5 of generation 16) are shown in bold. The event of transposition that led to this perfect solution can be seen in Figure 3.21.

generation 17 a perfect solution to the Majority(a, b, c) function problem was discovered (chromosome 8).

The comparison of the sequence of this perfect solution with the sequences of all its ancestors shows that it is a direct descendant of chromosome 5 of generation 16. The event of RIS transposition that led to the creation of this perfect solution is shown in Figure 3.21. In this case, the RIS element "AcAaa" in gene 1 (positions 7-11) was transposed to the start position of that same gene. And as in IS transposition, in RIS transposition, a sequence with exactly the same length as the transposon is deleted at the end of the head of the gene being modified so that the structural organization of the chromosome is maintained. In this case, the sequence "AaAcA" was deleted.

It is worth emphasizing that this highly disruptive operator is, nevertheless, capable of forming simple, repetitive sequences like, for instance, the sequences $(Aa)_n$ or $(Oc)_n$ present in the later generations of Figure 3.20. Interestingly, DNA is also full of small repetitive sequences. In fact, in some eukaryotes more than 40% of DNA consists of small repetitive sequences. Most of these sequences are not even transcribed, but some genes are also interspersed with small islands of repetitive sequences that do get transcribed.

Gene Transposition

In gene transposition an entire gene works as a transposon and transposes itself to the beginning of the chromosome. In contrast to the other forms of transposition (IS and RIS transposition), in gene transposition, the transposon (the gene) is deleted at the place of origin. This way, the length of the chromosome is maintained.

Apparently, gene transposition is only capable of shuffling genes and, for sub-ETs linked by commutative functions, this contributes nothing to adaptation in the short run. Note, however, that when the sub-ETs are linked by a non-commutative function or are part of a cellular system, the order of the genes matters and, in those cases, gene transposition becomes a macromutator, generating most of the times less fitter or even unviable individuals. However, gene transposition becomes particularly interesting when used in conjunction with recombination, for it allows not only the duplication of genes but also a more generalized recombination of genes and smaller building blocks.

Thus, for illustration purposes and because we are evolving sub-ETs linked by a commutative function, we are going to use gene transposition together with gene recombination (see Gene Recombination in section 3.3.5 below). The last population shown in Figure 3.22 is the product of nine generations

a.
```
0123456789012345678900123456789012345 67890
OAaOaAaAcAaaaaabccccbAAbObAObAAababcacccab-[m] = 7
AcAaaOAaOaaaaaabccccbAAbObAObAAababcacccab-[d] = 8
```

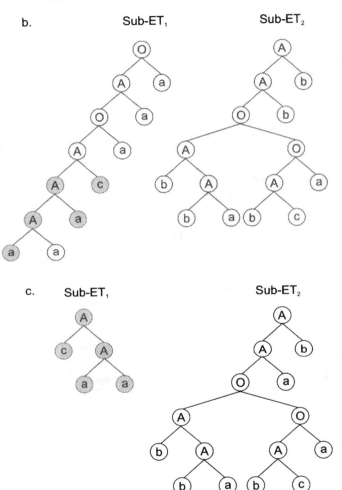

Figure 3.21. Illustration of RIS transposition and its effects. **a)** An event of RIS transposition with the transposon shown in bold. Note that a sequence with the same length as the transposon is deleted at the end of the head of the target gene. Note also that the transposon became, in this case, only partially duplicated in the daughter chromosome as its other elements disappeared at the end of the head. **b)** The sub-ETs encoded by the mother chromosome (before RIS transposition). **c)** The sub-ETs encoded by the daughter chromosome (after RIS transposition) (the transposon elements are shown in gray). Note that, in this case, root transposition changed drastically the sub-ET$_1$, shortening the daughter sub-ET in eight nodes and making it a better adapted individual in the process.

```
Generation N: 0
012345678900123456789001234567890
ObOaAbcabbcOcbccbbccbcAOANacccabc-[0]  = 5
AbObNbbbbaaAAbaNbccbbaOANNObabcab-[1]  = 4
OaaAOaaaaacOAbObcbccacNaAObaccbac-[2]  = 4
AOabccbbabcAcOAbacbcbbAcNAAcababb-[3]  = 7
AaAAbcbcbabOabbbbbaaccNbObNabbcbb-[4]  = 4
ANAANacbabaOOcNNcbbcacOAOabaabbbb-[5]  = 4
AAcaObaacaaAAbbAbaabcaNabObcbcccc-[6]  = 5
AbNNccacabbNaccNbcaabaOOAaNcbabaa-[7]  = 4
OcNaccabaacAacAAbbbccbANNAAbcbaac-[8]  = 3
AacbbbbccaaAObbacacabcOcaabcbbcab-[9]  = 5

                   . . .

Generation N: 7
012345678900123456789001234567890
AOabccbbabcAOabccbbabcANAANacbaba-[0]  = 7
AacbbbbccaaAObbacacabcOcaabcbbcab-[1]  = 5
AcNAAcababbAcOAbacbcbbAOabccbbabc-[2]  = 7
AOabccbbabcAOabccbbabcANAANacbaba-[3]  = 7
AcOAbacbcbbAOabccbbabcAcNAAcababb-[4]  = 7
AacbbbbccaaAObbacacabcOcaabcbbcab-[5]  = 5
AcNAAcababbOAOabaabbbbAcOAbacbcbb-[6]  = 6
AcOAbacbcbbAOabccbbabcAcNAAcababb-[7]  = 7
AOabccbbabcAcOAbacbcbbAcNAAcababb-[8]  = 7
AacbbbbccaaAObbacacabcOcaabcbbcab-[9]  = 5

Generation N: 8
012345678900123456789001234567890
AOabccbbabcAcOAbacbcbbAcNAAcababb-[0]  = 7
AOabccbbabcAOabccbbabcANAANacbaba-[1]  = 7
AcNAAcababbAcOAbacbcbbAOabccbbabc-[2]  = 7
AOabccbbabcAOabccbbabcAcOAbacbcbb-[3]  = 8
AcOAbacbcbbAOabccbbabcAcNAAcababb-[4]  = 7
ANAANacbabaAOabccbbabcAOabccbbabc-[5]  = 7
AcNAAcababbOAOabaabbbbANAANacbaba-[6]  = 6
AOabccbbabcAcOAbacbcbbAcNAAcababb-[7]  = 7
OcaabcbbcabAacbbbbccaaAObbacacabc-[8]  = 5
AcOAbacbcbbAOabccbbabcAcNAAcababb-[9]  = 7
```

Figure 3.22. An initial population and its later descendants created, via gene transposition and gene recombination, to solve the Majority(a, b, c) function problem. The chromosomes encode sub-ETs linked by OR. Note that the perfect solution found in generation 8 (chromosome 3) has a duplicated gene (genes 1 and 2). The ancestor genes of this individual were traced back to generation 0 and all their descendants are shown. Note how these genes got scattered throughout the genome, jumping to all conceivable places and often becoming duplicated.

of adaptation in which the only source of genetic variation was a gene transposition rate p_{gt} of 0.2 and a gene recombination rate p_{gr} of 0.2. This means
that the only source of novelty consisted, in this case, of moving around the
genes already present in the initial population (for a comparison, the initial
population is also shown in Figure 3.22). It is worth noticing how genes got
scattered throughout the chromosome, occupying all the positions available
to them. It is worth emphasizing that only gene transposition is capable of
moving genes around in the chromosome. Indeed, several events of gene
transposition had to occur in order to shuffle the genes so thoroughly. Such
an event is illustrated below:

```
01234567890012345678900123456789 0
AOabccbbabcAOabccbbabcANAANacbaba- [m]  =  7
ANAANacbabaAOabccbbabcAOabccbbabc- [d]  =  7
```

Here, gene 3 in the mother chromosome (chromosome 3 of generation 7)
moved to the beginning of the chromosome, forming chromosome 5 of generation 8 (see Figure 3.22). Structurally, these chromosomes are different
from one another but, mathematically, they encode equivalent expressions.
What is important is that chromosomes with the same kind of gene in different positions can recombine and give rise to new chromosomes with duplicated genes (note, however, that in this particular case of sub-ETs linked by
OR, one of the duplicated genes is neutral). For instance, the perfect solution
found in generation 8 (chromosome 3) has a duplicated gene. Indeed, as
shown in Figure 3.22, several chromosomes have duplicated genes (chromosomes 0 and 3 in generation 7 and chromosomes 1, 3, and 5 in generation 8).
Curiously enough, one of these chromosomes is a perfect solution to the
majority function problem. Its expression is shown in Figure 3.23.

Although I have chosen to illustrate the combined effect of gene transposition and gene recombination with an example that also resulted in the discovery of a perfect solution to the majority function problem, the transforming power of these operators is, however, very limited, especially when the
population sizes are very small (say, up to 500 individuals). It is worth emphasizing that, by themselves, these operators are unable to create new genes:
they only move existing genes around and recombine them in different ways.
A system creating diversity as such, could only evolve good solutions to
complex problems if it used gigantic populations, as all its genes would have
to be present in the initial population (see the Evolutionary Studies of chapter 12 for a discussion).

a. `012345678900123456789001234567890`
    `AOabccbbabcAOabccbbabc``AcOAbacbcbb-[3] = 8`

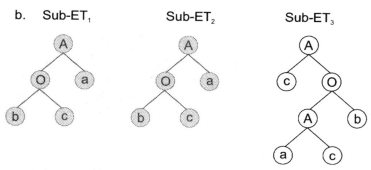

b.   Sub-ET₁   Sub-ET₂   Sub-ET₃

Figure 3.23. A perfect solution to the majority function with a duplicated gene (shown in bold). These, and all duplicated genes in gene expression programming, get duplicated only by the combined effect of gene transposition and recombination. **a)** The three-genic chromosome with the duplicated genes highlighted. **b)** The sub-ETs encoded in the chromosome.

3.3.5 Recombination

Gene expression programming uses three different kinds of recombination: one-point recombination, two-point recombination, and gene recombination. In all types of recombination, though, two chromosomes are randomly chosen and paired to exchange some material between them, resulting in the formation of two new individuals.

One-point Recombination

In one-point recombination the parent chromosomes are paired side by side and split up at exactly the same point. The material downstream of the re-combination point is afterwards exchanged between the two chromosomes. In GEP, an event of recombination always involves two parent chromosomes and always results in two new individuals. Usually the new daughter chromosomes are as different from each other as they are from their mothers.

Figure 3.24 shows three populations – an initial population and two later generations – obtained in a run in order to show the workings of one-point recombination. Thus, in this experiment, the only source of genetic variation was one-point recombination at a rate of $p_{1r} = 0.8$ so that the effects of this operator could be clearly analyzed. By the results shown in Figure 3.24, you

```
Generation N: 0
012345678012345678
OccbccbacAOAccaaaa-[0] = 6
AcbcacabcAcAAcacac-[1] = 7
AbNNabbccOaacaacbc-[2] = 6
OcbObcacaAAObbabbc-[3] = 6
ONbNaabbcOaOaaaccb-[4] = 6
AcaccbcaaNNcAcabcb-[5] = 6
AOcNaabacONOcccaac-[6] = 4
NOacccacaOOONbbbcb-[7] = 4
AAcNaabbcAONOccaca-[8] = 4
ObcbcbacbNbObbbaaa-[9] = 4
```

. . .

```
Generation N: 12
012345678012345678
AcbcacaccAcAAcacbc-[0] = 7
AcbcacacaAAObbabbc-[1] = 6
AcbcacacaAcAAcacbc-[2] = 7
AcbcacaccAcAAcacbc-[3] = 7
AcbcacacaAcAAbabbc-[4] = 6
AcbcbcacaAcAAcacbc-[5] = 7
NObcacacaAcAAcacbc-[6] = 4
AcacccacaAAOAcacbc-[7] = 7
AcbcacacaAAObcacbc-[8] = 6
AcbcacacaAcAAcacbc-[9] = 7
```

```
Generation N: 13
012345678012345678
AcbcacacaAcAAcacbc-[0] = 7
AObcacacaAcAAcacbc-[1] = 8
AcbcacaccAcAAcacbc-[2] = 7
AcbcacaccAcAAcacbc-[3] = 7
AcbcacaccAcAAcacbc-[4] = 7
AcbcacaccAcAAbabbc-[5] = 6
NcacccacaAAOAcacbc-[6] = 4
AcbcacacaAAObbabbc-[7] = 6
AcbcacacaAcAAcacbc-[8] = 7
AcbcacacaAAObbabbc-[9] = 6
```

Figure 3.24. An initial population and its later descendants created, via one-point recombination, to solve the Majority(a, b, c) function problem. The chromosomes encode sub-ETs linked by OR. The perfect solution found in generation 13 (chromosome 1) is a daughter of chromosomes 6 and 7 of the previous generation (also shown in bold). Their other daughter (chromosome 6) is also highlighted. Note that none of the later descendants resembles their ancestors of generation 0, but pretty much resemble one another in these relatively late stages of evolution. The event of one-point recombination that led to the creation of the perfect solution is shown in Figure 3.25.

can see that this operator on its own is also capable of making populations evolve efficiently. However, the conservative tendencies of recombination bear much more weight than its disruptive tendencies, a fact very well documented in Figure 3.24. As you can see, there is not much diversity left in the later generations of this experiment. But nevertheless, in this case, by generation 13 it was still possible to create a perfect solution to the Majority(a, b, c) function problem (chromosome 1) using just one-point recombination as the source of genetic modification.

The comparison of the sequence of this perfect solution with the sequences of all its ancestors shows that it is a direct descendant of chromosomes 6 and 7 of generation 12, and that chromosome 6 of generation 13 is its less fit twin. The event of one-point recombination that led to the creation of this perfect solution is shown in Figure 3.25. In this case, the parent chromo-

a.
```
012345678012345678
NObcacacaAcAAcacbc-[mA]  =  4
AcacccacaAAOAcacbc-[mB]  =  7

012345678012345678
AObcacacaAcAAcacbc  -[dA]  =  8
NcacccacaAAOAcacbc  -[dB]  =  4
```

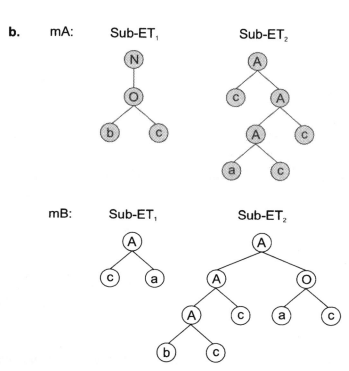

Figure 3.25. Illustration of one-point recombination and its effects. **a)** An event of one-point recombination. The daughter chromosomes are the result of crossing over point 1 (between positions 0 and 1 in gene 1). Note that both mothers and daughters differ among themselves. **b)** The sub-ETs encoded by the mother chromosomes (before recombination). **c)** The sub-ETs encoded by the daughter chromosomes (after recombination) (see next page). Note that the daughter dA is a perfect solution to the majority function problem.

c.   dA: Sub-ET₁   Sub-ET₂

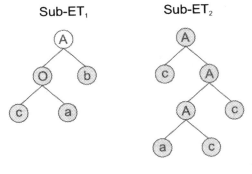

dB: Sub-ET₁   Sub-ET₂

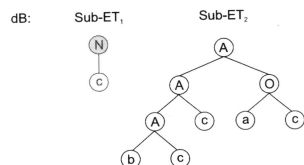

Figure 3.25. *Continued.*

somes crossed over point 1 (between positions 0 and 1) of gene 1, exchanging part of gene 1 and the entire gene 2.

The expression of these four chromosomes (parents and progeny) shows how profound the effects of one-point recombination can be. In this case, the first gene was split immediately after the start position, drastically reshaping the sub-ETs, whereas the second gene was left undisturbed. One of the newly created individuals (chromosome 6 of generation 13) is as mediocre as the worse of its progenitors (chromosome 6 of generation 12); the other (chromosome 1 of generation 13) is better than its parents and is, in fact, a perfect solution to the problem at hand.

It is worth emphasizing that GEP chromosomes can cross over any point in the genome, continually disrupting old building blocks and continually forming new ones. Furthermore, due to both the multigenic nature of GEP chromosomes and the existence of noncoding regions in most genes, entire genes and intact ORFs can be swapped between recombining chromosomes. Thus, the disruptive tendencies of one-point recombination (splitting of building blocks) coexist side by side with its more conservative tendencies (swapping of genes and ORFs), making one-point recombination (and of course two-point recombination too) a very well balanced genetic operator.

Also worth pointing out is the fact that, like all the other recombinational operators, when used together with gene transposition, one-point recombination is also capable of duplicating genes, which is also very important for an efficient and innovative evolution.

Two-point Recombination

In two-point recombination the parent chromosomes are paired side by side and two points are randomly chosen by which both chromosomes are split. The material between the recombination points is then exchanged between the two chromosomes, forming two new daughter chromosomes.

Figure 3.26 shows two-point recombination at work. An initial population and its immediate descendants obtained in a run are shown. In this experiment, the only source of genetic variation was a two-point recombination rate p_{2r} of 0.8. Note that a perfect solution was obtained early in generation 1 (chromosome 3). Indeed, when recombination is the only source of genetic variation, most of the times, perfect solutions are either found early on in the run or are not found at all, as populations become less and less diverse with time (see the Evolutionary Studies in chapter 12 for a discussion of the homogenizing tendencies of recombination). Note that chromosomes 1 and 3 of generation 1 are the daughters of chromosomes 4 and 6 of the initial population.

The event that led to the creation of this perfect solution in generation 1 is shown in Figure 3.27. In this case, the parent chromosomes exchanged the genetic material between point 7 (between positions 6 and 7 of gene 1) and point 14 (between positions 2 and 3 of gene 2). Note that the first gene is, in both parents, split downstream of the termination point. Note also that the second gene of chromosome 4 was also cut downstream of the termination point. Indeed, the noncoding regions of GEP chromosomes are ideal regions where chromosomes can be split to cross over without disrupting the ORFs. We have already seen that these regions are also ideal places where neutral mutations can accumulate and then such operators as two-point and one-point recombination can activate these neutral regions by integrating them in the coding regions of a new gene. Note, however, that gene 2 of chromosome 6 was split upstream of the termination point, changing profoundly the encoded sub-ET. Note also that when these chromosomes recombined, the noncoding region of chromosome 4 was activated and became part of the perfect solution found in generation 1 (chromosome 3).

```
Generation N: 0
012345678900123456789 0
ONOANbbcaaaNcONcaabccc-[0]  = 4
ANANbaacbbcNANOAabbcbc-[1]  = 5
NNNcOacbbabONcbacbbabb-[2]  = 4
ANAcbbababcNcbNbabcaab-[3]  = 2
AOaObbcccacObaNOacabab-[4]  = 6
OOAcOaabccaAacbNcbaccb-[5]  = 5
OcbbAcccbccAAOObacbaab-[6]  = 6
NbAANccbacaAbNNNbbcbbc-[7]  = 2
AaONbaababcAbNNAcacaaa-[8]  = 6
NOAcObaccbaAObbObcacca-[9]  = 4


Generation N: 1
012345678900123456789 0
AaONbaababcAbNNAcacaaa-[0]  = 6
OcbbAccccacObaObacbaab-[1]  = 5
AOaObaabccaAacbNacabab-[2]  = 6
AOaObbccbccAAONOacabab-[3]  = 8
ONOANbbcaaaNcONcaabccc-[4]  = 4
OOAcOaabccaAaNONcbaccb-[5]  = 5
OOAcObcccacObaNOcbaccb-[6]  = 5
ANANbaacbbcNAcbAabbcbc-[7]  = 3
ANANbaacbbcNANOAabbcbc-[8]  = 5
OaOANbbcaaaNcONcaabccc-[9]  = 5
```

Figure 3.26. An initial population and its immediate descendants created, via two-point recombination, to solve the Majority(a, b, c) function problem. The chromosomes encode sub-ETs linked by OR. The perfect solution found in generation 1 (chromosome 3) is a daughter of chromosomes 4 and 6 (shown in bold). Their other daughter (chromosome 1) is also highlighted. Note that chromosome 1 is less fit than both its mothers whereas chromosome 3 surpasses them greatly and is indeed a perfect solution to the majority function problem. The event of two-point recombination that led to the creation of this perfect solution is shown in Figure 3.27.

It is worth emphasizing that two-point recombination is more disruptive than one-point recombination in the sense that it recombines the genetic material more thoroughly, constantly destroying old building blocks and creating new ones (see the Evolutionary Studies of chapter 12 for a comparison with the other recombinational operators). But like one-point recombination, two-point recombination has also a conservative side and is good at swapping entire genes and ORFs. And finally, two-point recombination can also give rise to duplicated genes if used together with gene transposition.

a. 0123456789001234567890
AOaObbcccacObaNOacabab-[mA] = 6
OcbbAcccbccAAOObacbaab-[mB] = 6

0123456789001234567890
AOaObbccbccAAO**NOacabab**-[dA] = 8
OcbbAcc**ccacOba**Obacbaab-[dB] = 5

b.

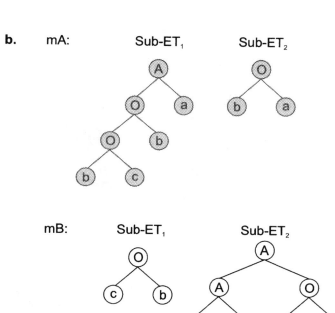

Figure 3.27. Illustration of two-point recombination and its effects. **a)** An event of two-point recombination. The daughter chromosomes were obtained after crossing over points 7 (between positions 6 and 7 in gene 1) and 14 (between positions 2 and 3 in gene 2). Note that both parents and offspring are different. **b)** The sub-ETs encoded by the mother chromosomes (before recombination). **c)** The sub-ETs encoded by the daughter chromosomes (after recombination) (see next page). Note that the daughter dA is much better than both its parents and is, in fact, a perfect solution to the majority function problem.

Notwithstanding, if the goal is to evolve good solutions quickly and efficiently, one-point or two-point recombination should never be used as the only source of genetic variation as they tend to homogenize populations, making them converge prematurely (see chapter 12 for a discussion of the

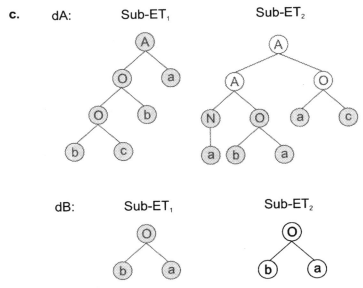

Figure 3.27. *Continued.*

homogenizing tendencies of recombination). However, together with mutation, inversion, and transposition, these operators work wonderfully and play
an important part in the creation of good solutions to virtually all problems.

Gene Recombination

In the third kind of GEP recombination, entire genes are exchanged between
two parent chromosomes, forming two new daughter chromosomes containing genes from both parents. Thus, the gene recombination operator randomly chooses the two parent chromosomes and the gene to be exchanged.
Typically, a small gene recombination rate p_{gr} of 0.3 is used as this operator
is seldom used as the only source of genetic variation.

The last two populations shown in Figure 3.28 are the outcome of an evolutionary process in which the only source of genetic variation was gene recombination at a much higher rate of 0.8. For this simple problem, again the majority function problem, it was possible to find a perfect solution only by shuffling the genes already present in the initial population. Obviously, we wouldn't
have been that lucky with a more complex problem, unless huge populations
had been used. Note also that, with this operator, genes never trade places in
the chromosome, always occupying the same position. Indeed, genes can only

```
Generation N: 0
012345678012345678012345678
OAOcaacacAOaAcbbcbAaAAbabbb-[0]  = 6
NNcaabbabONbcbacacOAOObccbb-[1]  = 4
NcaacaaacNcAaaccbcOcaaacbbb-[2]  = 4
OcbaacbccAOaccbccbNONNbcbbb-[3]  = 6
ONAObcbcaNANccacabNAbObacab-[4]  = 4
AOaaaabacAbbObcabaNcbAacacb-[5]  = 5
NONOabaaaNaObccbbbNccabcbba-[6]  = 2
NAaAaacbbNANbcbacaAAONacbaa-[7]  = 3
NaAAbcbbaNaOAcbacbNcAaababb-[8]  = 2
AcaabbcabNAcacaacbNNOAcbbbc-[9]  = 4


Generation N: 1
012345678012345678012345678
OcbaacbccAOaccbccbNONNbcbbb-[0]  = 6
OcbaacbccAOaccbccbNcbAacacb-[1]  = 4
NONOabaaaAOaccbccbNONNbcbbb-[2]  = 7
AcaabbcabNaObccbbbNccabcbba-[3]  = 4
AcaabbcabAOaAcbbcbNNOAcbbbc-[4]  = 6
NONOabaaaNaObccbbbNccabcbba-[5]  = 2
OAOcaacacNAcacaacbAaAAbabbb-[6]  = 4
NONOabaaaNAcacaacbNNOAcbbbc-[7]  = 4
AOaaaabacAbbObcabaNONNbcbbb-[8]  = 6
OAOcaacacAOaAcbbcbAaAAbabbb-[9]  = 6


Generation N: 2
012345678012345678012345678
NONOabaaaAOaccbccbNONNbcbbb-[0]  = 7
OAOcaacacAOaAcbbcbAaAAbabbb-[1]  = 6
AOaaaabacAbbObcabaNONNbcbbb-[2]  = 6
AcaabbcabAOaAcbbcbNONNbcbbb-[3]  = 8
NONOabaaaAOaccbccbAaAAbabbb-[4]  = 7
AOaaaabacAbbObcabaNNOAcbbbc-[5]  = 5
OAOcaacacAOaAcbbcbNccabcbba-[6]  = 4
NONOabaaaNaObccbbbAaAAbabbb-[7]  = 4
NONOabaaaNaObccbbbNONNbcbbb-[8]  = 3
NONOabaaaAOaccbccbNccabcbba-[9]  = 4
```

Figure 3.28. An initial population and its immediate descendants created via gene recombination to solve the Majority(*a*, *b*, *c*) function problem. The chromosomes encode sub-ETs linked by OR. The perfect solution found in generation 2 (chromosome 3) is a daughter of chromosomes 4 and 8 of the previous generation (also shown in bold). Their other daughter (chromosome 5) is also highlighted. Note that chromosome 5 is less fit than its parents, whereas chromosome 3 surpasses them greatly. In fact, it codes for a perfect solution to the majority function problem (see its expression in Figure 3.29).

jump from place to place when this operator works together with gene transposition (see Gene Transposition in section 3.3.4 above).

   Figure 3.29 shows the gene recombination event that led to creation of the
perfect solution found in generation 2 (chromosome 3). In this case, gene 3
was exchanged between chromosomes 4 and 8 of generation 1, obtaining

a.   012345678012345678012345678
      AcaabbcabAOaAcbbcbNNOAcbbbc-[mA] = 6
      AOaaaabacAbbObcabaNONNbcbbb-[mB] = 6

      012345678012345678012345678
      AcaabbcabAOaAcbbcbNONNbcbbb-[dA] = 8
      AOaaaabacAbbObcaba**NNOAcbbbc**-[dB] = 5

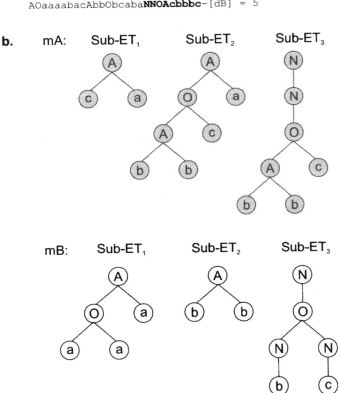

Figure 3.29. Illustration of gene recombination and its effects. a) An event of gene
recombination. In this case, the parent chromosomes exchanged gene 3, forming
two new daughter chromosomes. Note that both the parents and the offspring
differ among themselves. b) The sub-ETs encoded by the parents (before recombination). c) The sub-ETs encoded by the daughter chromosomes (after recombination) (shown in the next page). Note that the daughter dA is considerably better
than both its parents, encoding a perfect solution to the majority function problem.

two new daughter chromosomes (chromosomes 3 and 5 of generation 2). Note that chromosome 5 is less fit than its parents, whereas chromosome 3 surpasses them both considerably. Indeed, this new individual has maximum fitness and therefore codes for a perfect solution to the problem at hand.

It is worth emphasizing that gene recombination is unable to create new genes: the individuals created by this operator are different arrangements of existing genes. Obviously, if gene recombination were to be used as the unique source of genetic variation, more complex problems could only be solved using very large initial populations in order to provide for the necessary diversity of genes (see the Evolutionary Studies of chapter 12 for a discussion). However, GEP evolvability is based not only in the shuffling of genes (achieved by gene recombination and gene transposition), but also in the constant creation of new genetic material which is carried out essentially by mutation, inversion, and transposition (both IS and RIS transposition) and, to a lesser extent, by recombination (both one-point and two-point recombination).

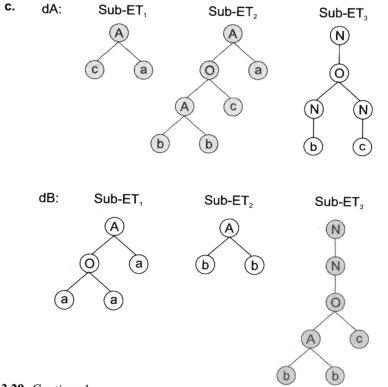

Figure 3.29. *Continued.*

3.4 Solving a Simple Problem with GEP

The aim of this section consists of studying a successful run in its entirety in order to understand how populations adapt, finding in the process perfect or good solutions to the problem at hand.

And the problem we are going to solve with gene expression programming, is a very simple problem of symbolic regression (also called function finding), where the goal is to find an expression that satisfactorily explains the dependent variable. The input into the system is a set of fitness cases in the form $(a_{(i,1)}, a_{(i,2)}, ..., a_{(i,n)}, y_i)$ where $a_{(i,1)} - a_{(i,n)}$ are the n independent variables and y_i is the dependent variable. We have already seen that the set of fitness cases consists of the environment in which populations of candidate solutions (computer programs that are, in fact, the output of the system) adapt, discovering, with time, good or perfect solutions to the problem at hand.

In the simple example of this section, the fitness cases were computer generated using a test function. This means that, for this particular problem, we know exactly what the perfect solution to our problem should look like. And this is indeed a good thing because then we will be able to undoubtedly recognize and rightly appreciate the amazing results produced by the blind action of the genetic operators. Keep in mind, however, that in real-world problems the target function is what we are trying to discover and is obviously unknown.

So, let's suppose that we are given a sampling of the numerical values from the curve:

$$y = \frac{a^2}{2} + 3a \tag{3.19}$$

over 10 randomly chosen points in the real interval [-10, +10] and we wanted to find a function fitting those values within a certain error. In this case, we are given a sample of data in the form of 10 pairs (a_i, y_i), where a_i is the value of the independent variable in the given interval and y_i is the respective value of the dependent variable (Table 3.2). These 10 pairs are the fitness cases (the input to the system) that will be used as the selection environment, against which the fitness of each individual will be evaluated generation after generation. And, hopefully, after a certain number of generations, the individual better adapted to this environment will be a perfect solution to our problem.

There are five major steps in preparing to use gene expression programming, and the first is to choose the fitness function. For this problem we will measure the fitness by equation (3.3a), using an absolute error of 100 as the

Table 3.2
Set of 10 random computer generated fitness cases
used in the simple problem of symbolic regression.

a	f(a)
6.9408	44.909752
-7.8664	7.3409245
-2.7861	-4.4771234
-5.0944	-2.3067443
9.4895	73.493805
-9.6197	17.410214
-9.4145	16.072905
-0.1432	-0.41934688
0.9107	3.1467872
2.1762	8.8965232

selection range and a precision for the error equal to 0.01. Thus, for the 10 fitness cases of Table 3.2 f_{max} = 1000.

The second major step is to choose the set of terminals **T** and the set of functions **F**. For this problem, the terminal set consists obviously of the independent variable, giving T = {a}. The choice of the appropriate function set is not so obvious, but a good guess can be done so that all the necessary mathematical operators are included. For this problem, we will make things simple and use the four arithmetic operators, thus giving F = {+, -, *, /}.

The third major step is to choose the chromosomal architecture: the length of the head and the number of genes. In this problem we will use an h = 7 and three genes per chromosome.

The fourth major step is to choose the kind of linking function. In this case we will link the sub-ETs by addition.

And finally, the fifth major step in preparing to use gene expression programming is to choose the set of genetic operators and their rates. In this case we will use a combination of all the modification operators introduced in the previous section (mutation, inversion, the three kinds of transposition, and the three kinds of recombination).

The parameters used per run are summarized in Table 3.3. For this problem, a small population of 20 individuals was chosen so that all the individuals created in the evolutionary process could be completely analyzed, without however filling this book with pages and pages of encoded individuals.

Table 3.3
Parameters for the simple symbolic regression problem.

Number of generations	50
Population size	20
Number of fitness cases	10 (Table 3.2)
Function set	+ - * /
Terminal set	a
Head length	7
Number of genes	3
Linking function	+
Chromosome length	45
Mutation rate	0.044
Inversion rate	0.1
IS transposition rate	0.1
RIS transposition rate	0.1
Gene transposition rate	0.1
One-point recombination rate	0.4
Two-point recombination rate	0.2
Gene recombination rate	0.1
Fitness function	Eq. (3.3a)
Selection range	100
Precision	0.01

However, as we will see, one of the advantages of gene expression programming is that it is capable of solving relatively complex problems using small population sizes and, thanks to the compact Karva notation, it is easy to analyze every single individual from a run.

Figure 3.30 shows the progression of average fitness and the fitness of the best individual of the successful run we are going to analyze in this section. As you can see in Figure 3.30, in this run, a perfect solution was found in generation 10.

The initial population of this run and the fitness of each individual in the particular environment of Table 3.2, is shown below (the best of generation is shown in bold):

```
Generation N: 0
012345678901234012345678901234012345678901234
-/a+*+*aaaaaaaa*++*/a+aaaaaaaa+++a*--aaaaaaaa-[ 0] = 285.2363
+a-*/+aaaaaaaaa+/*+/+*aaaaaaaa+/++++/aaaaaaaa-[ 1] = 324.4358
```

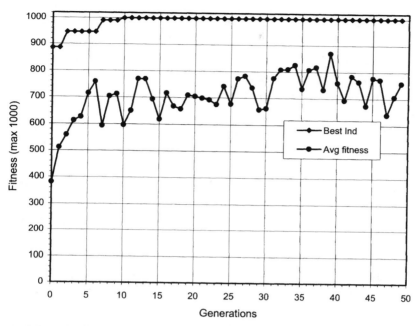

Figure 3.30. Progression of average fitness of the population and the fitness of the best individual for a successful run of the experiment summarized in Table 3.3.

```
--a+a/*aaaaaaaa+a-/aa/aaaaaaaa***-+a+aaaaaaaa-[ 2]  = 697.6004
++//a*aaaaaaaaa/aa/+*+aaaaaaaa-++*/+-aaaaaaaa-[ 3]  = 711.1687
/*+-*a/aaaaaaaa*+a/-*aaaaaaaa*////-aaaaaaaa-[ 4]  = 730.5334
-a/a/+*aaaaaaaa*+-*a*-aaaaaaaa/*/**aaaaaaaa-[ 5]  = 256.4514
/*+*/++aaaaaaaa-/*-a-*aaaaaaaa*++--/*aaaaaaaa-[ 6]  = 821.6851
**+//++aaaaaaaa*aa+++/aaaaaaaa+++a/aaaaaaaa-[ 7]  = 816.0122
/*-/+++aaaaaaaa//-+aa/aaaaaaaa+a+--a-aaaaaaaa-[ 8]  = 0
-+/a-/*aaaaaaaa-a-a+a/aaaaaaaa-/+*a//aaaaaaaa-[ 9]  = 729.2897
--//+/*aaaaaaaa-***/a+aaaaaaaa+-*/*+/aaaaaaaa-[10]  = 376.2071
**+/+*aaaaaaaa/+*-/-*aaaaaaaa*a**a+aaaaaaaa-[11]  = 0
-aa-+/+aaaaaaaa+a/*-a*aaaaaaaa-a*//++aaaaaaaa-[12]  = 0
-a+a+//aaaaaaaa*-aa/+/aaaaaaaa--/*a+-aaaaaaaa-[13]  = 0
+/*a+/-aaaaaaaa-/*--/-aaaaaaaa**a/**aaaaaaaa-[14]  = 0
-a+*a**aaaaaaaa+-***a*aaaaaaaa-/+-+a/aaaaaaaa-[15]  = 294.7556
/+-***-aaaaaaaa+a/a/+*aaaaaaaa*-*-a*-aaaaaaaa-[16]  = 886.7593
/a++/--aaaaaaaa*-a/a-/aaaaaaaa*/+*+*/aaaaaaaa-[17]  = 392.185
+a-+/+/aaaaaaaa*--a**/aaaaaaaa/---+/-aaaaaaaa-[18]  = 311.389
/*/aa*-aaaaaaaa/-+--+*aaaaaaaa*/-+*a/aaaaaaaa-[19]  = 0
```

Note that six out of 20 individuals are unviable and thus have fitness 0, meaning that either they could not solve a sole fitness case within the chosen selection range or they returned calculation errors such as division by zero. And as you can easily check by drawing their expression trees, in this particular case, it was a division by zero that caused all these individuals to die without leaving offspring: indeed, all the six unviable chromosomes code for programs where a division by zero took place and therefore were labeled unfit to reproduce.

It is worth pointing out that, in gene expression programming, all calculation errors are handled similarly: every time a program gives a calculation error (divisions by zero, square roots or logarithms of negative numbers, and so forth) it is made unviable. This means that it will not be able to pass on its genes to the next generation. This is a good alternative to the creation of useless programs with dubious protected mathematical operators as is usually done in genetic programming.

As shown in the initial population above, the best individual of this generation, chromosome 16, has fitness 886.7593. Let's now analyze its performance more closely. Table 3.4 compares the results predicted by this program with the target value. Note that none of the fitness cases was solved within the chosen precision of 0.01. However, in all cases, the absolute error was within the chosen selection range of 100.

The expression of the best individual of this generation and the corresponding mathematical expression is shown in Figure 3.31. Note that gene 3 contributes nothing to the overall solution as the whole sub-ET encodes zero, and therefore may be considered a neutral gene. Note also the creation of

Table 3.4
Fitness evaluation of a model (best program of generation 0).

Target	Model	Error	Fitness
44.91	33.0282	11.8818	88.1182
7.341	25.0737	17.7327	82.2673
-4.477	3.09508	7.57208	92.42792
-2.307	9.88206	12.18906	87.81094
73.494	56.5148	16.9792	83.0208
17.41	38.6496	21.2396	78.7604
16.073	36.9019	20.8289	79.1711
-0.419	1.86705	2.28605	97.71395
3.147	3.32539	0.17839	99.82161
8.897	6.54412	2.35288	97.64712

a.
```
012345678901234012345678901234012345678901234
/+-***-aaaaaaaa+a/a/+*aaaaaaaa*-*-a*-aaaaaaaa
```

b.

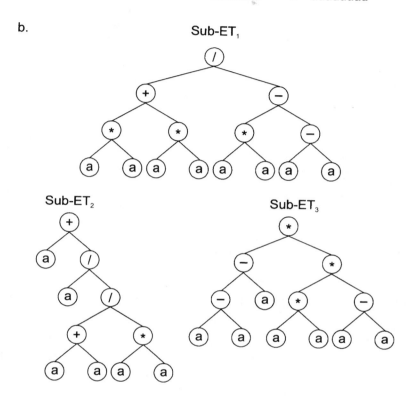

Sub-ET₁

Sub-ET₂   Sub-ET₃

c.

$$y = (2) + \left(a + \frac{a^2}{2} \right) + (0)$$

Figure 3.31. Best individual of generation 0 (chromosome 16). It has a fitness of 886.7593 evaluated against the set of fitness cases presented in Table 3.2. **a)** The chromosome of the individual. **b)** The sub-ETs codified by each gene. **c)** The corresponding mathematical expression after linking with addition (the contribution of each sub-ET is shown in brackets). Note that gene 3 is a neutral gene as it contributes nothing to the overall solution.

useful numerical constants in sub-ET₁ and sub-ET₂ as a result of simple arithmetic manipulations.

The descendants of the individuals of the initial population are shown below (the best of generation are shown in bold):

```
Generation N: 1
012345678901234012345678901234012345678901234
/+-***-aaaaaaaa+a/a/+*aaaaaaaa*-*-a*-aaaaaaaa-[ 0] = 886.7593
++//a*aaaaaaaaa-a-++a/aaaaaaaa*-/+*a/aaaaaaaa-[ 1] = 366.9803
+a/+/+/aaaaaaaa*--a**/aaaaaaaa/----/*aaaaaaaa-[ 2] = 313.1345
-+/a-/*aaaaaaaa/aa/+*+aaaaaaaa-++*/+/aaaaaaaa-[ 3] = 650.7792
-a+*a**aaaaaaaa+-***a*aaaaaaaa-/+-*aaaaaaaaaa-[ 4] = 297.1939
++//a*aaaaaaaaa**+//*/aaaaaaaa/--*+/-aaaaaaaa-[ 5] = 462.7996
*aa+++/aaaaaaaa/aa/+++aaaaaaaa-/+*a//aaaaaaaa-[ 6] = 756.707
--a+a/*aaaaaaaaa-/a*/aaaaaaaa***--a+aaaaaaaa-[ 7] = 697.6004
**+*+*/aaaaaaaa/a++/--aaaaaaaa*-a/a-/aaaaaaaa-[ 8] = 199.4485
/*+*-*+aaaaaaaa-/*-a-*aaaaaaaa*-+-+/-aaaaaaaa-[ 9] = 794.29
-/a+*+*aaaaaaaa*++*/a+aaaaaaaa+++a*--aaaaaaaa-[10] = 285.2363
/+-***-aaaaaaaa+a/a/+*aaaaaaaa*-*-a*-aaaaaaaa-[11] = 886.7593
++//a*-aaaaaaaa/aa/+*+aaaaaaaa-++*/+-aaaaaaaa-[12] = 0
/a++/--aaaaaaaa*/-a/a-aaaaaaaa*/+*+*/aaaaaaaa-[13] = 446.7045
-+a-/*aaaaaaaaa-a-a+a/aaaaaaaa-/+*a//aaaaaaaa-[14] = 730.5334
/*+**a/aaaaaaaa*+a--*aaaaaaaaa*////-aaaaaaaa-[15] = 334.3612
-a/a/+*aaaaaaaa*+-*a*-aaaaaaaa/*/+*+a/aaaaaaaa-[16] = 314.1658
-+/a-/*aaaaaaaa-a-a+a/aaaaaaaa+++a/aaaaaaaaaa-[17] = 716.5283
+a-+/+/aaaaaaaa*--+a**+aaaaaaaa-++*/+aaaaaaaa-[18] = 361.1277
/+-***-aaaaaaaa+a/a/a*aaaaaaaa-**-*-aaaaaaaa-[19] = 738.8981
```

Note that despite the global improvement (compare the average fitness of both populations in Figure 3.30), none of the descendants surpassed the best individual of the previous generation. In fact, the best individuals of generation 1 have exactly the same genetic makeup as the best of generation 0. The first one, chromosome 0, was generated by elitism. The other one, chromosome 11, managed to get reproduced without change.

In the next generation a new individual was created, chromosome 8, considerably better than the best individual of the previous generation. Its expression is shown in Figure 3.32. The complete population is shown below (the best of generation is shown in bold):

```
Generation N: 2
012345678901234012345678901234012345678901234
/+-***-aaaaaaaa+a/a/+*aaaaaaaa*-*-a*-aaaaaaaa-[ 0] = 886.7593
/+-***-aaaaaaaa+a/a/+*aaaaaaaa*-*-a*-aaaaaaaa-[ 1] = 886.7593
/+-**+-aaaaaaaa+a/a/+*aaaaaaaa*---a*-aaaaaaaa-[ 2] = 361.2925
/+a/**-aaaaaaaa+a/a/+*aaaaaaaa*-**a//aaaaaaaa-[ 3] = 0
*-*-a*-aaaaaaaa/+-*a+aaaaaaaaa***--a+aaaaaaaa-[ 4] = 822.5905
+a/+/+/aaaaaaaa*--a**/aaaaaaaa/----/*aaaaaaaa-[ 5] = 313.1345
/a++*--aaaaaaaa*/-a/a-aaaaaaaa*/+*+*/aaaaaaaa-[ 6] = 449.5403
*+-***-aaaaaaaa+a/a/a*aaaaaaaa-**-*-aaaaaaaa-[ 7] = 256.8001
/*+*-*+aaaaaaaa+a/a/+*aaaaaaaa-/+*a//aaaaaaaa-[ 8] = 945.8432
-+a-/*aaaaaaaaa---a+a/aaaaaaaa-/+*a//aaaaaaaa-[ 9] = 611.6426
*/*/**+aaaaaaaa/a++/--aaaaaaaa*-a/a-/aaaaaaaa-[10] = 475.9649
```

a. 012345678901234012345678901234012345678901234
/*+*-*+aaaaaaaa+a/a/+*aaaaaaaa-/+*a//aaaaaaaa

b.

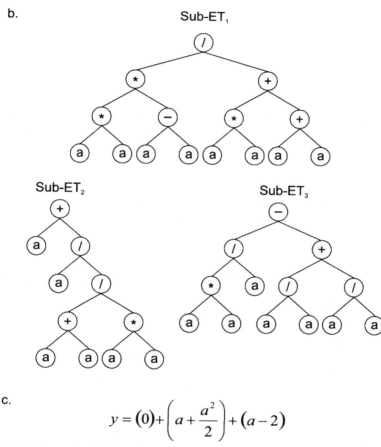

Sub-ET₁

Sub-ET₂

Sub-ET₃

c.

$$y = (0) + \left(a + \frac{a^2}{2} \right) + (a - 2)$$

Figure 3.32. Best individual of generation 2 (chromosome 8). It has a fitness of 945.8432 evaluated against the set of fitness cases presented in Table 3.2. **a)** The chromosome of the individual. **b)** The sub-ETs codified by each gene. Note that sub-ET₂ was already present in the best individual of generation 0 (see Figure 3.31). **c)** The corresponding mathematical expression after linking with addition (the contribution of each sub-ET is shown in brackets). Note again the existence of a neutral gene (gene 1) encoding zero.

```
/+-***-aaaaaaaa+a/a/+*aaaaaaaa*-a-a*-aaaaaaaa-[11]  = 372.5987
-+a-/*aaaaaaaa-a-a+a/aaaaaaaa*-+-+/-aaaaaaaa-[12]  = 611.6426
/a++/--aaaaaaaa*--a/a/aaaaaaaa*/+*+*/aaaaaaaa-[13]  = 458.072
*aa+/+/aaaaaaaa*/-a/a-aaaaaaaa/+*+//aaaaaaaa-[14]  = 471.6405
aa-/aa/aaaaaaaa--a+**-aaaaaaaa+**a/a/aaaaaaaa-[15]  = 804.8516
```

```
++//aaaaaaaaaaa**+//*/aaaaaaaa//-a+/-aaaaaaaa-[16] = 723.8981
/+-***-aaaaaaaa+a/a/+*aaaaaaaa*-*-a*-aaaaaaaa-[17] = 886.7593
/+-***-aaaaaaaa//*-a-*aaaaaaaa*-*-a*-aaaaaaaa-[18] = 0
/+-***-aaaaaaaa/a+a/a/aaaaaaaa*-*-a*-aaaaaaaa-[19] = 831.525
```

In the next four generations the best individuals are either clones of the best individual of generation 2 or minor variations of this individual (these chromosomes are highlighted in all populations):

```
Generation N: 3
012345678901234012345678901234012345678901234
/*+*-*+aaaaaaaa+a/a/+*aaaaaaaa-/+*a//aaaaaaaa-[ 0] = 945.8432
/+-***-aaaaaaaa+a/a/+*aaaaaaaa*-*aa*-aaaaaaaa-[ 1] = 886.7593
*/*/**+aaaaaaaa/a/a/aaaaaaaa*/*-a-/aaaaaaaa-[ 2] = 789.7097
/+-***-aaaaaaaa/a+-++/aaaaaaaa-a-a*+aaaaaaaa-[ 3] = 830.1644
/+-***-aaaaaaaa-a/a/+*aaaaaaaa*aa-a*-aaaaaaaa-[ 4] = 886.7593
/a++/--aaaaaaaa*--a/a/aaaaaaaa/++*a//aaaaaaaa-[ 5] = 558.8105
/+-***-aaaaaaaa+a/a/+*aaaaaaaa*-*/a*/aaaaaaaa-[ 6] = 352.5538
-+a-/*aaaaaaaa-a-a+a/aaaaaaaa//-a+/-aaaaaaaa-[ 7] = 691.6004
+a/+/+/aaaaaaaa*--a-*/aaaaaaaa--/----aaaaaaaa-[ 8] = 0
/*+*-*+aaaaaaaa+a/a/+*aaaaaaaa-/+*a//aaaaaaaa-[ 9] = 945.8432
/a-*/*-aaaaaaaa+a/a/+*aaaaaaaa*-*--*-aaaaaaaa-[10] = 836.6745
/+-***-aaaaaaaa+a/a/+*aaaaaaaa*-a-a*-aaaaaaaa-[11] = 372.5987
/a++*--aaaaaaaa*/-a/a-aaaaaaaa+**a/a/aaaaaaaa-[12] = 766.7137
--a+**-aaaaaaaaa-/aa/aaaaaaaa*/+*+*/aaaaaaaa-[13] = 420.6074
+a/+/+/aaaaaaaa*--a-*/aaaaaaaa*/+*+*-aaaaaaaa-[14] = 353.6146
-+a-/*aaaaaaaa---a+a/aaaaaaaa-/+*a//aaaaaaaa-[15] = 611.6426
*-*-a*-aaaaaaaa/+-*a+aaaaaaaa+**--a+aaaaaaaa-[16] = 466.7956
/*+*-*+aaaaaaaa+a/a/+*aaaaaaaa-----/*aaaaaaaa-[17] = 569.2436
/+-/*+-aaaaaaaa+a/a/+*aaaaaaaa*---a*-aaaaaaaa-[18] = 356.1858
-a-a+a/aaaaaaaa-+a-/*aaaaaaaaa+-+-+/-aaaaaaaa-[19] = 611.6426
```

```
Generation N: 4
012345678901234012345678901234012345678901234
/*+*-*+aaaaaaaa+a/a/+*aaaaaaaa-/+*a//aaaaaaaa-[ 0] = 945.8432
-+-***-aaaaaaaa-a/a/+*aaaaaaaa-----/*aaaaaaaa-[ 1] = 569.2436
/+-*-*+aaaaaaaa/a/+a/aaaaaaaaa/----/*aaaaaaaa-[ 2] = 817.2341
*/*/**+aaaaaaaa/a/a/a/aaaaaaaa-/+-a-/aaaaaaaa-[ 3] = 779.7097
/*+***-aaaaaaaa/a+--+/aaaaaaaa-a-a*+aaaaaaaa-[ 4] = 750.3948
/*+*-*+aaaaaaaa+a/a/+*aaaaaaaa-/+*a*/aaaaaaaa-[ 5] = 585.3681
/*+--*+aaaaaaaa+a/a/+*aaaaaaaa*/**a/-aaaaaaaa-[ 6] = 891.116
/a+-*--aaaaaaaa*/-a/a-aaaaaaaa+**a/a/aaaaaaaa-[ 7] = 766.7137
**--/a-aaaaaaaa/a/+*--aaaaaaaa*/+/+**aaaaaaaa-[ 8] = 0
/+-***-aaaaaaaa-a/a/+*aaaaaaaa*aa-a*-aaaaaaaa-[ 9] = 886.7593
/*+*-a+aaaaaaaa+a/a/+*aaaaaaaa-/+*a//aaaaaaaa-[10] = 945.8432
/*+/-*+aaaaaaaa+a/a/+*aaaaaaaa-//*a//aaaaaaaa-[11] = 945.8432
*/-a-a/aaaaaaaaa-/aa/aaaaaaaa+**a/a/aaaaaaaa-[12] = 0
-+a-/*aaaaaaaa-a-a+a/aaaaaaaa//-a+//aaaaaaaa-[13] = 0
```

```
+-a+-+/aaaaaaaa-a/a+a/aaaaaaaa-+a-/*aaaaaaaa-[14] = 763.2632
/a++/--aaaaaaaa*--a-a/aaaaaaaa/+/*a/-aaaaaaaa-[15] = 0
/a++*--aaaaaaaa*/-a/a-aaaaaaaa+**a/a/aaaaaaaa-[16] = 766.7137
/*+*-*+aaaaaaaa+a/a/+*aaaaaaaa*aa-a*-aaaaaaaa-[17] = 565.8646
/*+*-*+aaaaaaaa---a+a/aaaaaaaa-//*a+/aaaaaaaa-[18] = 738.811
a+a-/*aaaaaaaa---a+a/aaaaaaaa-/+*a//aaaaaaaa-[19] = 816.363
```

Generation N: 5
```
01234567890123401234567890123401234567890123401234
/*+/-*+aaaaaaaa+a/a/+*aaaaaaaa-//*a//aaaaaaaa-[ 0] = 945.8432
a+a-/--aaaaaaaa-a/a/+*aaaaaaaa*aa-a*-aaaaaaaa-[ 1] = 945.5575
/*+*-*+aaaaaaaa+a/a/+*aaaaaaaa*aa-a*-aaaaaaaa-[ 2] = 565.8646
/*+*-*+aaaaaaaa---a+a-aaaaaaaa*-a-a*-aaaaaaaa-[ 3] = 367.2566
-+-***-aaaaaaaa-a/a/+*aaaaaaaa---*a++aaaaaaaa-[ 4] = 564.0838
aa/a/+*aaaaaaaa/+-***/aaaaaaaa///-*/*aaaaaaaa-[ 5] = 819.3729
/*+*-*+aaaaaaaa/a/+a/aaaaaaaa*aa-a*-aaaaaaaa-[ 6] = 758.7488
+-a+-+/aaaaaaaa-a/a+a/aaaaaaaa-+a-/*aaaaaaaa-[ 7] = 763.2632
/*+*-*+aaaaaaaa++/a-+*aaaaaaaa-/+aa//aaaaaaaa-[ 8] = 798.3864
/a+-*--aaaaaaaa*/-a/aaaaaaaaaa+**a/a/aaaaaaaa-[ 9] = 799.0481
/a+-*--aaaaaaaa*/-aaa-aaaaaaaa*aa-a*-aaaaaaaa-[10] = 747.9244
-/+*a//aaaaaaaa*+*-*+aaaaaaaa+aa/+*aaaaaaaa-[11] = 754.3026
/+-***aaaaaaaa---a+a/aaaaaaaa-/+*a//aaaaaaaa-[12] = 803.7867
*/*/**/aaaaaaaa/a/a/aaaaaaaa-/+-a-/aaaaaaaa-[13] = 816.5905
a+a-/*aaaaaaaa---a+a+aaaaaaaa-/+*a//aaaaaaaa-[14] = 785.1168
/*+*-*+aaaaaaaa+a/a/+*aaaaaaaa---+*a//aaaaaaaa-[15] = 568.9833
/*+*-*+aaaaaaaa/a/a/+*aaaaaaaa*aa-a*-aaaaaaaa-[16] = 721.8472
a+a-/*aaaaaaaa---a+//aaaaaaaa+*-/+*aaaaaaaa-[17] = 752.1143
/*+*-**aaaaaaaa*a/a/+*aaaaaaaa-/+*a//aaaaaaaa-[18] = 408.8309
/+-***-aaaaaaaa-a/a/+*aaaaaaaa+**a/a/aaaaaaaa-[19] = 593.1216
```

Generation N: 6
```
01234567890123401234567890123401234567890123401234
/*+/-*+aaaaaaaa+a/a/+*aaaaaaaa-//*a//aaaaaaaa-[ 0] = 945.8432
/*+*--+aaaaaaaa/a/a/+*aaaaaaaa*aa-a*-aaaaaaaa-[ 1] = 721.8472
---*a++aaaaaaaa-+aa-+*aaaaaaaa*aa-**-aaaaaaaa-[ 2] = 752.1143
a+a-/*aaaaaaaa---a+a+aaaaaaaa-/a*a//aaaaaaaa-[ 3] = 817.4047
/+-***aaaaaaaa---a/a/aaaaaaaa+aa/+*aaaaaaaaa-[ 4] = 800.2618
*/+*a//aaaaaaaa*a*-*+aaaaaaaa-aa/+*aaaaaaaaa-[ 5] = 789.7097
/*+*-*+aaaaaaaa++/a-+*aaaaaaaa-/+aa//aaaaaaaa-[ 6] = 798.3864
a//a/+*aaaaaaaa/a*a/+aaaaaaaaa/a/-//*aaaaaaaa-[ 7] = 817.1696
a+a-/*-aaaaaaaa-a/a/**aaaaaaaa-a/a/+*aaaaaaaa-[ 8] = 594.5831
-/**a//aaaaaaaa*+*-*+aaaaaaaa+/+*a//aaaaaaaa-[ 9] = 796.5607
aa/a/+*aaaaaaaa/+-**a/aaaaaaaa///-*/*aaaaaaaa-[10] = 783.2302
*a*a/+*aaaaaaaa-a/aa+aaaaaaaaa---*a++aaaaaaaa-[11] = 778.0015
a+a-/-aaaaaaaa---a+a+aaaaaaaa-/+*a//aaaaaaaa-[12] = 785.1168
-+-***-aaaaaaaa-a/a/+*aaaaaaaa-/-*a++aaaaaaaa-[13] = 945.5575
/*+*-*+aaaaaaaa/+-***/aaaaaaaa*aa-a*-aaaaaaaa-[14] = 708.8708
a+a-/--aaaaaaaa-a/a/+*aaaaaaaa*a-/--aaaaaaaaa-[15] = 594.5831
```

```
/*+*-*+aaaaaaaa/a/a/+*aaaaaaaa//a-a*-aaaaaaaa-[16]  = 818.3482
-+-***-aaaaaaaa/*+*-**aaaaaaaa-/+*a//aaaaaaaa-[17]  = 760.707
/+-***-aaaaaaaa-a/a/+*aaaaaaaa*+**a/aaaaaaaa-[18]  = 374.7753
a+a-/*aaaaaaaaa---a+a+aaaaaaaa-/+*a//aaaaaaaa-[19]  = 785.1168
```

In generation 7 a new individual was created that exhibited an improvement in fitness (chromosome 12). Its expression is shown in Figure 3.33, but the whole population is shown below:

```
Generation N: 7
01234567890123401234567890123401234567890123401234567890123401234
/*+/-*+aaaaaaaa+a/a/+*aaaaaaaa-//*a//aaaaaaaa-[ 0]  = 945.8432
/*+/-*+aaaaaaaa/+-**a/aaaaaaaa///-*/*aaaaaaaa-[ 1]  = 800.2618
/+-*-*-aaaaaaaa-a/a*+*aaaaaaaa*+**a/aaaaaaaa-[ 2]  = 356.2026
---*a++aaaaaaaa-a*a/+aaaaaaaa-/-*a++aaaaaaaa-[ 3]  = 701.6004
/a/-//-aaaaaaaa//a/+*aaaaaaaa/**a/+-aaaaaaaa-[ 4]  = 0
a+a-/--aaaaaaaa+a*a/+aaaaaaaa/+/-a/*aaaaaaaa-[ 5]  = 360.1841
a+a-/-aaaaaaaaa---a+a+aaaaaaaa-/+*a//aaaaaaaa-[ 6]  = 785.1168
/+-***-aaaaaaaa-a/a/+*aaaaaaaa-/-*a++aaaaaaaa-[ 7]  = 610.0637
///a*+*aaaaaaaa-a/a/+*aaaaaaaa/a/-//*aaaaaaaa-[ 8]  = 621.5726
/*+*---aaaaaaaa/a/a/+*aaaaaaaa**a-**-aaaaaaaa-[ 9]  = 0
/+a/a/+aaaaaaaa-a/a/+*aaaaaaaa-**/-*aaaaaaaa-[10]  = 333.7551
/*+/-*+aaaaaaaa+a/a/+*aaaaaaaa-//*a//aaaaaaaa-[11]  = 945.8432
aa/a/+*aaaaaaaa+a/a/+*aaaaaaaa-//*a//aaaaaaaa-[12]  = 989.9987
a+-***-aaaaaaaa-+aa-+*aaaaaaaa*aa-a--aaaaaaaa-[13]  = 789.7097
/*+*+-+aaaaaaaa/+/+/+*aaaaaaaa*aa-a*-aaaaaaaa-[14]  = 388.9994
/+-***-aaaaaaaa-a/a/+*aaaaaaaa+***+**aaaaaaaa-[15]  = 343.4301
/*+*-*+aaaaaaaa++/a-+*aaaaaaaa-/+aa//aaaaaaaa-[16]  = 798.3864
aa/a/+*aaaaaaaa/+-a*a/aaaaaaaa////*/*aaaaaaaa-[17]  = 805.9502
---*a++aaaaaaaa-aa+aa-aaaaaaaa*aa-**-aaaaaaaa-[18]  = 471.685
a/-a/+*aaaaaaaa-a/-/+aaaaaaaaa*a-/--aaaaaaaa-[19]  = 816.5905
```

In the next two generations no improvement in fitness occurred:

```
Generation N: 8
01234567890123401234567890123401234567890123401234567890123401234
aa/a/+*aaaaaaaa+a/a/+*aaaaaaaa-//*a//aaaaaaaa-[ 0]  = 989.9987
a+a-/-aaaaaaaaa---a+a+aaaaaaaa-/+*a//aaaaaaaa-[ 1]  = 785.1168
/*+/-/+aaaaaaaa+a/a/+*aaaaaaaa-//+a//aaaaaaaa-[ 2]  = 889.116
aa/a/+*aaaaaaaa/a/a/+*aaaaaaaa-//*a//aaaaaaaa-[ 3]  = 782.9793
a+a-/-aaaaaaaaa---a+a+aaaaaaaa-/+*a//aaaaaaaa-[ 4]  = 785.1168
/*+/-*+aaaaaaaa+aaa/**aaaaaaaa-//*a++aaaaaaaa-[ 5]  = 779.7097
/+-*-**aaaaaaaa-a/a*+*aaaaaaaa////*/*aaaaaaaa-[ 6]  = 0
aa/a/+*aaaaaaaa+a/a/+*aaaaaaaa-//*a//aaaaaaaa-[ 7]  = 989.9987
++/a-+*aaaaaaaa/*/a/+aaaaaaaa/a/-//*aaaaaaaa-[ 8]  = 794.4986
/*+/-*+aaaaaaaa+a/a/+*aaaaaaaa-//*a//aaaaaaaa-[ 9]  = 945.8432
///a*+-aaaaaaaa-a+*-*+aaaaaaaa-/+aa//aaaaaaaa-[10]  = 0
/a+/-*+aaaaaaaa/*+/-*+aaaaaaaa+a/a/+*aaaaaaaa-[11]  = 886.7593
/*+/-*+aaaaaaaa/+-**a/aaaaaaaa///-*/*aaaaaaaa-[12]  = 800.2618
```

a. 012345678901234012345678901234012345678901234
aa/a/+*aaaaaaaa+a/a/+*aaaaaaaa-//*a//aaaaaaaa

b.

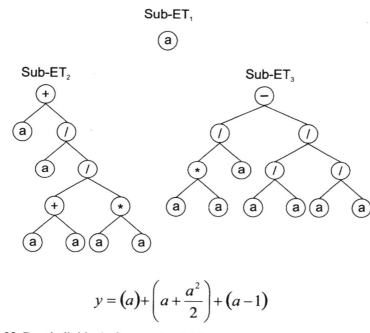

Sub-ET₁

Sub-ET₂   Sub-ET₃

c.

$$y = (a) + \left(a + \frac{a^2}{2} \right) + (a - 1)$$

Figure 3.33. Best individual of generation 7 (chromosome 12). It has a fitness of 989.9987 evaluated against the set of fitness cases presented in Table 3.2. **a)** The chromosome of the individual. **b)** The sub-ETs codified by each gene. Note that sub-ET₂ was already present in the previous best of generation individuals (see Figures 3.31 and 3.32). **c)** The corresponding mathematical expression after linking with addition (the contribution of each sub-ET is shown in brackets).

```
/*+*-*+aaaaaaaa++/a-+*aaaaaaaa-/+aa//aaaaaaaa-[13]  =  798.3864
/+-***-aaaaaaaa-////+*aaaaaaaa-a/*a//aaaaaaaa-[14]  =  807.1407
/+-***-aaaaaaaa-a/a/+*aaaaaaaa+***+**aaaaaaaa-[15]  =  343.4301
a/-a/+*aaaaaaaa-a/-/+*aaaaaaaa*a-/--aaaaaaaa-[16]   =  730.5334
/+-***-aaaaaaaa-a*a/++aaaaaaaa*//*a+**aaaaaaaa-[17] =  355.8565
aa/a/+*aaaaaaaa+a/a/+*aaaaaaaa-//*a//aaaaaaaa-[18]  =  989.9987
///a*+*aaaaaaaa-a/a/+*aaaaaaaa/a/-//*aaaaaaaa-[19]  =  621.5726
```

```
Generation N: 9
012345678901234012345678901234012345678901234
aa/a/+*aaaaaaaa+a/a/+*aaaaaaaa-//*a//aaaaaaaa-[ 0]  =  989.9987
/*+/-*+aaaaaaaa+aaa/**aaaaaaaa-//*a++aaaaaaaa-[ 1]  =  779.7097
aa/a++*aaaaaaaa/a/a/+*aaaaaaaa-/+/*a/aaaaaaaa-[ 2]  =  788.0954
```

```
a+a-/-aaaaaaaaa---a+a+aaaaaaaa/*+/-*+aaaaaaaa-[ 3] = 816.5905
++/a-+*aaaaaaaa-////+*aaaaaaaa/a/-//*aaaaaaaa-[ 4] = 819.1696
+aaa/+*aaaaaaaa-/-/a+/aaaaaaaa-/+-a-/aaaaaaaa-[ 5] = 817.4047
/+-***-aaaaaaaa-//-/+*aaaaaaaa-//*a//aaaaaaaa-[ 6] = 805.3593
aa/a/+*aaaaaaaa/a/a/+*aaaaaaaa-a/*a/-aaaaaaaa-[ 7] = 791.2383
/*+/-*+aaaaaaaa+aaa/**aaaaaaaa-//a*++aaaaaaaa-[ 8] = 789.0481
/aaa+-*aaaaaaaa-a/-/+*aaaaaaaa*a-/--aaaaaaaa-[ 9] = 697.6004
/+-***-aaaaaaaa-a/a/+*aaaaaaaa+***+**aaaaaaaa-[10] = 343.4301
aa/a/+*aaaaaaaa/a/a/+*aaaaaaaa-//*a/aaaaaaaa-[11] = 799.0481
a+a-/-aaaaaaaaa---a+a+aaaaaaaa+/+*a//aaaaaaaa-[12] = 817.6932
/*++-*+aaaaaaaa/+***a/aaaaaaaa///*a/*aaaaaaaa-[13] = 370.8166
/*+*-*+aaaaaaaa++/a-+*aaaaaaaa-/+aa//aaaaaaaa-[14] = 798.3864
/*+/-*+aaaaaaaa/+-*-a*aaaaaaaa///-*/*aaaaaaaa-[15] = 812.4845
a+a-/-aaaaaaaaa---a+a+aaaaaaaa//-*a//aaaaaaaa-[16] = 0
a*/a/+*aaaaaaaa+a/a/+*aaaaaaaa-//*a//aaaaaaaa-[17] = 989.9987
-//*a//aaaaaaaa/a//a/+aaaaaaaa/a/a/+*aaaaaaaa-[18] = 327.7421
/*+/-/+aaaaaaaa+a/a/+*aaaaaaaa-//+a//aaaaaaaa-[19] = 889.116
```

Finally, by generation 10 an individual with maximum fitness, matching exactly the target function (3.19), was created:

```
Generation N: 10
012345678901234012345678901234012345678901234
a*/a/+*aaaaaaaa+a/a/+*aaaaaaaa-//*a//aaaaaaaa-[ 0] = 989.9987
+/+a-//aaaaaaaa/*+*-*+aaaaaaaa++/a-+*aaaaaaaa-[ 1] = 0
/*+/-*+aaaaaaaa+a///+*aaaaaaaa/a/-//*aaaaaaaa-[ 2] = 800.3348
/*+/-*aaaaaaaa+-aa/**aaaaaaaa-//-a++aaaaaaaa-[ 3] = 785.1168
/+aa/a/aaaaaaaa+a/a/+*aaaaaaaa///-*/*aaaaaaaa-[ 4] = 886.7593
/+-*-**aaaaaaaa-/+-/+*aaaaaaaa-//*a//aaaaaaaa-[ 5] = 0
-//*a++aaaaaaaa/*+/-*+aaaaaaaa*+aaa/*aaaaaaaa-[ 6] = 474.2741
a+a-/-aaaaaaaaa---a+a+aaaaaaaa-//*a/aaaaaaaa-[ 7] = 796.9538
aa/a/+*aaaaaaaa/a/a/**aaaaaaaa--/*a/-aaaaaaaa-[ 8] = 0
a+a/+/+aaaaaaaa+a-a/+*aaaaaaaa//+/-*+aaaaaaaa-[ 9] = 0
a*/a/+*aaaaaaaa+a/a/+*aaaaaaaa-//*a//aaaaaaaa-[10] = 989.9987
-*/a/+*aaaaaaaa+a/a/+*aaaaaaaa-///-*+aaaaaaaa-[11] = 0
a*/a/+*aaaaaaaa+a/a/+*aaaaaaaa///*a//aaaaaaaa-[12] = 1000
a+a-/*aaaaaaaa-+-a+a+aaaaaaaa/*+*a//aaaaaaaa-[13] = 381.2412
aa/a/+*aaaaaaaa+a/a/+*aaaaaaaa-//*a+/aaaaaaaa-[14] = 891.116
a+a-/-aaaaaaaaa---*+a+aaaaaaaa+/+*a//aaaaaaaa-[15] = 738.8981
aa/a/+*aaaaaaaa/aaa/**aaaaaaaa-//*a++aaaaaaaa-[16] = 804.8516
++/a-+*aaaaaaaa-//a/+*aaaaaaaa-/+*a//aaaaaaaa-[17] = 769.7097
/*+/-*+aaaaaaaa/+-*-a*aaaaaaaa-//*a//aaaaaaaa-[18] = 798.2679
/*+-/-aaaaaaaaa---a+a+aaaaaaaa-//+a-/aaaaaaaa-[19] = 829.525
```

Note that the best individual of this generation (chromosome 12) is a descendant, via mutation, of the best individual of the previous generation: their chromosomes differ in just one position (the "-" at position 0 in gene 3 was replaced by "/"). And as you can see by its expression in Figure 3.34,

a. 012345678901234012345678901234012345678901234
a*/a/+*aaaaaaaa+a/a/+*aaaaaaaa///*a//aaaaaaaa

b.

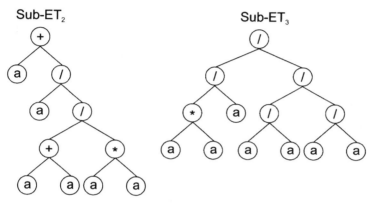

c.

$$y = (a) + \left(a + \frac{a^2}{2} \right) + (a) = \frac{a^2}{2} + 3a$$

Figure 3.34. Perfect solution to the simple problem of symbolic regression. This program was discovered in generation 10 and has maximum fitness. **a)** The chromosome of the individual. **b)** The sub-ETs codified by each gene. Note that sub-ET$_1$ and sub-ET$_2$ were already present in the previous best of generation solution (see Figure 3.33), and sub-ET$_2$ has an even longer pedigree, going back to the initial population (see Figure 3.31). **c)** The corresponding mathematical expression after linking with addition (the contribution of each sub-ET is shown in brackets). Note that it matches exactly the target function (3.19).

this individual with maximum fitness codes indeed for a perfect solution to the problem at hand, matching perfectly the target function (3.19).

So, we have seen how a random collection of random chromosomes grew into mature programs, giving their first tentative steps in a hostile environment into which they were thrown. These highly unfit individuals were then selected to reproduce with modification, giving birth to new individuals that adapted quickly and incredibly well to this unchanging environment. And as you could follow, this wondrous adaptation was made possible by the blind modification on the chromosomes carried out by the genetic operators and the inexorable hand of selection. In the next chapters we will see how these principles can be efficiently applied to solve a great variety of complex problems, much better and much quicker than any human mind.

4 The Basic GEA
in Problem Solving

In this chapter we will see how the basic gene expression algorithm can be used to solve complex problems from very different fields. We will start by exploring the workings of the fundamental parameters of the algorithm by solving a simple cubic polynomial. Then we will continue our exploration by analyzing the performance of the algorithm on a complex test function of five arguments. Furthermore, we will also discuss the power of the algorithm to extract knowledge from noisy data, not only by mining a noisy computer-generated dataset but also by mining complex real-world data, including mining a dataset with 51 input attributes in order to decide whether to approve or not a credit card. We will also see how the basic GEA can be used to diagnose diseases and classify different types of plants. Particularly interesting is the three-class prediction of the iris data that will be tackled using two different approaches: the first consisting of the conventional way of partitioning the data into three separate datasets so that three different models are created and afterwards combined to make the final prediction; and the second consisting of a three-genic system evolving three different models at the same time, in which each model is responsible for classifying a certain type of plant.

Also in this chapter we will learn how gene expression programming can be used to find parsimonious solutions. Indeed, finding parsimonious solutions is not only a matter of concern in mathematics but also, and perhaps most importantly, in logic. And, in section 4.3, Logic Synthesis and Parsimonious Solutions, we will see how gene expression programming can be used to find parsimonious solutions with aplomb. For that purpose, special fitness functions with parsimony pressure will be described and a simple suite of Boolean functions will be solved in order to find their most parsimonious representations using five well-known universal systems: the classical Boolean system of ANDs, ORs, and NOTs; the NAND and NOR systems; the

Cândida Ferreira: *Gene Expression Programming*, Studies in Computational Intelligence (SCI) **21**, 121–180 (2006)
www.springerlink.com

Reed-Muller system of ANDs, XORs, and NOTs; and the MUX system. Furthermore, we will see how to enrich the evolutionary toolkit of GEP through the use of user defined functions in order to design parsimonious solutions to complex modular functions using smaller building blocks.

And finally, we will see how gene expression programming can be used to evolve cellular automata rules for the density-classification task. The rules evolved by gene expression programming have accuracy levels of 82.513% and 82.55%, exceeding all human-written rules and the rule evolved by genetic programming. And most impressive of all, we will see that this was achieved using computational resources more than four orders of magnitude smaller than those used by the GP technique.

4.1 Symbolic Regression

We have already seen how gene expression programming can be used to do symbolic regression in the simple example of section 3.4. Here, we will analyze more complex problems of symbolic regression in order to evaluate the performance of the algorithm. The first is a simple test function that can be exactly solved by the basic GEA and, therefore, is ideal for showing the importance of the fundamental parameters of the algorithm, such as the population size, the number of generations, fitness function, chromosome size, number of genes, head size, and the linking function. The second consists of a complex test function with five arguments that shows how gene expression programming can be efficiently applied to model complex realities with great accuracy. And the third problem was chosen to illustrate how gene expression programming can be efficiently used for mining relevant information from extremely noisy data.

4.1.1 Function Finding on a One-dimensional Parameter Space

The target function of this section is the simple cubic polynomial:

$$y = a^3 + a^2 + a + 1 \tag{4.1}$$

This function was chosen not only because it allows the execution of hundreds of runs in a few seconds but also because it can be solved exactly by the algorithm and, therefore, is ideal for rigorously evaluating the performance in terms of success rate. Consequently, it can be used to illustrate how to choose the appropriate settings for a problem, including the fitness function, the number of genes, the head length, the function set and the linking function. This kind of analysis is useful for developing an intuitive understanding of the fundamental parameters of the algorithm. The parameters chosen for this problem are summarized in Table 4.1; how and why they were chosen is discussed below.

Consider we are given a sampling of numerical values from the test function (4.1) over 10 random points chosen from the interval [-10, 10] (Table 4.2), and we wanted to find a function fitting those values within 0.01 of the correct value. We could, therefore, evaluate the fitness by equation (3.3a) to make sure that all the solutions with maximum fitness match indeed the target function.

Table 4.1
Settings for the polynomial function problem.

Number of runs	100
Number of generations	50
Population size	30
Number of fitness cases	10 (Table 4.2)
Function set	+ - * /
Terminal set	a
Head length	6
Gene length	13
Number of genes	4
Linking function	+
Chromosome length	52
Mutation rate	0.0385
Inversion rate	0.1
One-point recombination rate	0.3
Two-point recombination rate	0.3
Gene recombination rate	0.3
IS transposition rate	0.1
RIS transposition rate	0.1
Gene transposition rate	0.1
Fitness function	Equation (3.3a)
Selection range	100
Precision	0.01
Success rate	100%

Table 4.2
Set of fitness cases used in the polynomial function problem.

a	f(a)
6.9408	44.909752
-7.8664	7.3409245
-2.7861	-4.4771234
-5.0944	-2.3067443
9.4895	73.493805
-9.6197	17.410214
-9.4145	16.072905
-0.1432	-0.41934688
0.9107	3.1467872
2.1762	8.8965232

First of all, the set of functions and the set of terminals must be chosen. For the sake of simplicity, in this case we will choose just the basic arithmetic operators to design this model, thus giving $F = \{+, -, *, /\}$; and the terminal set will include just the independent variable, giving $T = \{a\}$. Next, the structural organization of the chromosomes, namely, the length of the head h and the number of genes, must also be chosen. It is wise to start with short, single-gene chromosomes and then gradually increase h. Figure 4.1 shows such an analysis for this problem. Note how the success rate increases abruptly in the beginning, from a small success rate of 29% for the most compact organization (a gene length g of 13) to a high success rate of 86% obtained with a moderately redundant organization ($g = 29$). Note also that, from this point onward, the success rate starts to decrease progressively. As you can see, for each problem, it is possible to guess more or less accurately the ideal head length in order to create a search landscape not only easy to navigate through but also full of riches.

It is worth pointing out that GEP can be used to search for the most parsimonious solution to a problem by choosing smaller and smaller head sizes (we will see in section 4.3, Logic Synthesis and Parsimonious Solutions, how to do this more efficiently by using special fitness functions with parsimony pressure). As shown in Figure 4.1, it was not possible to find a correct solution to this problem using a head length of five. Only with $h = 6$ was it possible to evolve a perfect solution. In this case, these perfect solutions are

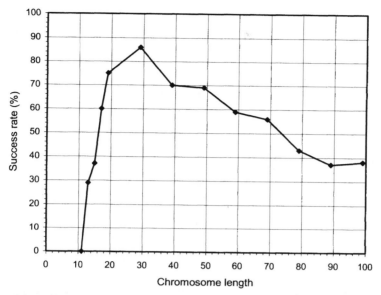

Figure 4.1. Variation of success rate with chromosome length. For this analysis, chromosomes composed of one gene were used, $G = 50$, and $P = 30$. The success rate was evaluated over 100 identical runs.

also the most parsimonious solutions to the problem at hand. For instance, the following chromosomes with 13 nodes each:

```
0123456789012
*++/*/aaaaaaa-[1]
*++*//aaaaaaa-[2]
```

both code for perfect parsimonious solutions to the target function (4.1).

Note that gene expression programming can evolve solutions efficiently using large values of h, i.e., is capable of dealing with highly redundant information. As shown in Figure 4.1, for each problem, there is a chromosome length that allows the most efficient evolution. And, at least for simple functions, this ideal chromosome length can be easily found. Note also that the most compact genomes are not the most efficient. This suggests that a certain redundancy is fundamental to the efficient evolution of good programs (see a discussion of The Role of Neutrality in Evolution in chapter 12).

As shown in Table 4.1, for this analysis, a population size P of 30 individuals and an evolutionary time G of 50 generations were used. Unless otherwise stated, in all the experiments of this section, only mutation and two-point recombination were used in order to simplify the analysis. The mutation rate was equivalent to two one-point mutations per chromosome and the re-combination rate was equal to 0.7. Furthermore, in all the experiments of this section, the same set of 10 random fitness cases was used (see Table 4.2). The fitness was evaluated by equation (3.3a) and, in this case, a selection range of 100 and a precision of 0.01 were chosen. Thus, for a set of 10 fitness cases, f_{max} = 1000.

In the next experiment we are going to analyze the relationship between success rate and population size P (Figure 4.2). This parameter is extremely important not only in terms of evolution but also in terms of processing time. And given the tradition (and necessity) of using huge population sizes in genetic programming, it is important to show that the evolutionary system of gene expression programming can evolve efficiently using small populations sizes of just 20-100 individuals.

It is worth pointing out that, in this experiment, a medium, far from ideal, value for the head (h = 19) was used in order to have some resolution in the

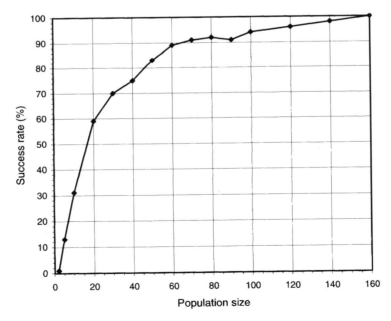

Figure 4.2. Variation of success rate with population size in unigenic systems. For this analysis, G = 50, and a medium, far from ideal, value of 39 for chromosome length (h = 19) was used. The success rate was evaluated over 100 identical runs.

plots. Indeed, not only the h chosen was not the most advantageous (see Figure 4.1), but also single-gene chromosomes were used. Remember, though, that, for most problems, the multigenic system of gene expression programming is much more efficient than the unigenic one, as most problems are better modeled by using multiple, smaller building blocks or terms.

Suppose that, after the analyses shown in Figures 4.1 and 4.2, we were unable to find a satisfactory solution to the problem at hand or the system was not evolving efficiently. Then we could try a multigenic system and also try different ways of linking the sub-ETs. For instance, we could start by choosing multiple genes with a head length of six, encoding sub-expression trees linked by addition. Figure 4.3 shows such an analysis for this problem. In this experiment, the mutation rate was equivalent to two one-point mutations per chromosome and, therefore, varies according to chromosome length, $p_{1r} = p_{2r} = p_{gr} = 0.3$, and $p_i = p_{is} = p_{ris} = p_{gt} = 0.1$.

It is worth pointing out that gene expression programming can cope very well with an excess of genes: the success rate for the 10-genic system is still very high (59%). Again we can see that a certain amount of redundancy

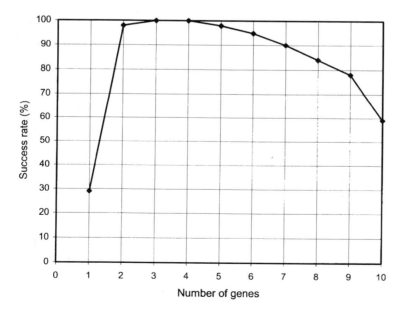

Figure 4.3. Variation of success rate with the number of genes. For this analysis $G = 50$, $P = 30$, $h = 6$ (a gene length of 13), and the sub-ETs were linked by addition. The success rate was evaluated over 100 identical runs.

allows a greater efficiency. In this case, chromosomes encoding between two and six sub-ETs linked by addition are ideal to create fruitful solution landscapes for this problem.

In Figure 4.4 another important relationship is shown: how the success rate depends on the number of generations G. For this analysis, single-gene chromosomes with a very large and, therefore, far from ideal, chromosome length $g = 79$ ($h = 39$) were used so that we could have resolution in the plots for high values of G. The general parameters used in this analysis are exactly the same as in the analysis shown in Figure 4.1. As Figure 4.4 emphasizes, in GEP, populations can adapt and evolve indefinitely because new material is constantly being introduced in the genetic pool by the genetic operators of mutation, inversion, transposition, and recombination. As a comparison, recall that, in genetic programming, trying to prolong evolution beyond the 50 generations mark is futile, as this system relies solely on tree-crossover and after 50 generations not much genetic diversity remains.

Finally, suppose that the multigenic system with sub-ETs linked by addition could not evolve a satisfactory solution to the problem at hand. Then we could choose another linking function or let the system evolve the linking

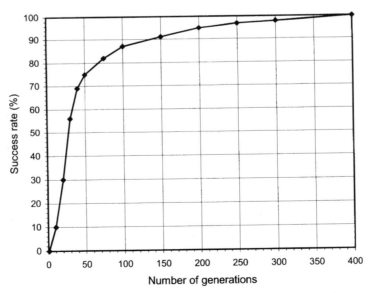

Figure 4.4. Variation of success rate with the number of generations. For this analysis $P = 30$ and a chromosome length of 79 (a single-gene chromosome with $h = 39$) was used. The success rate was evaluated over 100 identical runs.

functions on its own by choosing a cellular system. In the analysis shown in Figure 4.5, multiplication was used to link the sub-ETs. The remaining parameters are exactly the same as in the analysis shown in Figure 4.3.

As expected, for this polynomial function, the algorithm excelled with multigenic chromosomes encoding sub-ETs posttranslationally linked by addition. In real-world problems this testing of waters is done until a feel for the best chromosomal structure and composition is developed, and this usually entails making 3 or 4 preparatory runs, not the exhaustive analysis done here. Typically, one usually experiments with a couple of chromosomal organizations and tests different function sets. By observing such indicators as best and average fitnesses, it is easy to see whether the system is evolving efficiently or not. Then, after choosing the appropriate settings, one just lets the system evolve the best possible solution on its own.

Consider, for instance, the multigenic system composed of four genes linked by addition. As shown in Figure 4.3, the success rate has, in this case, the maximum value of 100%, meaning that, in all the runs, a perfect solution was found. Let's analyze with more detail a successful run of this experiment. The parameters used per run are summarized in Table 4.1 above.

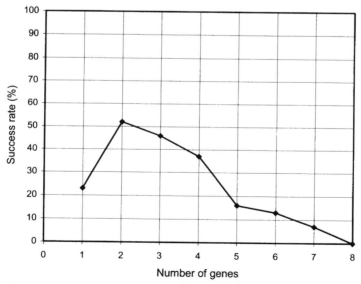

Figure 4.5. Variation of success rate with the number of genes. For this analysis, $G = 50$, $P = 30$, and $h = 6$ (corresponding to a gene size of 13). The sub-ETs were linked by multiplication. The success rate was evaluated over 100 identical runs.

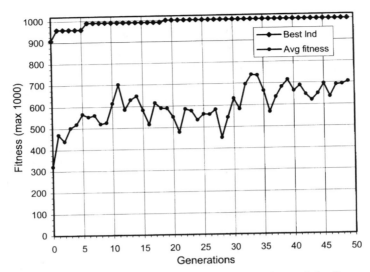

Figure 4.6. Progression of average fitness of the population and the fitness of the best individual for run 0 of the experiment summarized in Table 4.1.

The evolutionary dynamics of this successful run (run 0) is shown in Figure 4.6. And as you can see, a perfect solution was found in generation 19.

The chromosomes and fitnesses of all the individuals of the initial population of this run are shown below (the best individual of this generation is shown in bold):

```
Generation N: 0
012345678901201234567890120123456789012012345678901201234567890120123456789012
**//++aaaaaaa/a//**aaaaaaa++/-**aaaaaaa+/+-aaaaaaaaa-[ 0] = 479.9631
*/-//+aaaaaaa/*+a/aaaaaaaa--aa//aaaaaaa-+/++*aaaaaaa-[ 1] = 354.6601
-+a/a/aaaaaaa/-/+-*aaaaaaa*-*/+/aaaaaaa*/a-/aaaaaaa-[ 2] = 373.0389
---//+aaaaaaa*+/a--aaaaaaa++aa+aaaaaaaa**+++-aaaaaaa-[ 3] = 193.9366
/--aa/aaaaaaa+-a+--aaaaaaa//+a/-aaaaaaa-/+a*+aaaaaaa-[ 4] = 287.6966
-*/a/+aaaaaaa*+/+-/aaaaaaa/-*++*aaaaaaa+++-a/aaaaaaa-[ 5] = 355.6841
*aaaa*aaaaaaa*a-/-*aaaaaaa/--*+/aaaaaaa++*-*aaaaaaa-[ 6] = 408.3963
--/++aaaaaaa*a//++aaaaaaa///a-aaaaaaaa*+a++*aaaaaaa-[ 7] = 0
---**aaaaaaaa*-++++aaaaaaa*/aa+aaaaaaaa--aa/aaaaaaa-[ 8] = 319.8496
/-/+--aaaaaaa+---/*aaaaaaa*--*-+aaaaaaa/*aa+*aaaaaaa-[ 9] = 0
-**/-+aaaaaaa*+aa-/aaaaaaa-/-a*+aaaaaaa*+-+++aaaaaaa-[10] = 324.3444
+a/-/-aaaaaaa-+*/+-aaaaaaa/-/a/-aaaaaaa+-++-aaaaaaa-[11] = 0
+a+*--aaaaaaa-+-*+-aaaaaaa+a///-aaaaaaa+//a+-aaaaaaa-[12] = 412.0092
/a/-/-aaaaaaa*+/*-*aaaaaaa**a/*+aaaaaaa*--+aaaaaaaa-[13] = 215.5287
***/*/aaaaaaa-*//-aaaaaaaa//-a-aaaaaaaa/*/-+-aaaaaaa-[14] = 0
```

```
-a+/a+aaaaaaa--*a/+aaaaaaa-**-+aaaaaaaa-/+++/aaaaaaa-[15]  = 420.3384
//+a-*aaaaaaa/a-a*+aaaaaaa**+/-aaaaaaaa/+-a/-aaaaaaa-[16]  = 0
+-aa/+aaaaaaa-*-**aaaaaaaa*//*-/aaaaaaa*+*a/aaaaaaaa-[17]  = 0
*/-+a*aaaaaaa+*-+-/aaaaaaa/a-//aaaaaaaa-+-/+aaaaaaaa-[18]  = 0
+-++a-aaaaaaa+-a*-aaaaaaaa/*a---aaaaaaa+*a+a*aaaaaaa-[19]  = 789.56
+*aa++aaaaaaa*++a+/aaaaaaa+*/**/aaaaaaa*/*/*-aaaaaaa-[20]  = 171.441
/a**a/aaaaaaa*+-++/aaaaaaa*//-++aaaaaaa----/-aaaaaaa-[21]  = 403.2058
+a*--aaaaaaaa+/-//-aaaaaaa--++a*aaaaaaa+***aaaaaaaaa-[22]  = 779.5309
*////*aaaaaaa-//-a/aaaaaaa/*/a-+aaaaaaa+-/*+*aaaaaaa-[23]  = 296.2591
/a-++-aaaaaaa+-**a*aaaaaaa*+-a*aaaaaaaa-/+aa-aaaaaaa-[24]  = 873.554
*a++/aaaaaaaa/-a--*aaaaaaa***a+/aaaaaaa*/*+a-aaaaaaa-[25]  = 313.3738
-++/+/aaaaaaa*+a+*/aaaaaaa/+-+-/aaaaaaa-+//a+aaaaaaa-[26]  = 908.0693
*/***/aaaaaaa+-*---aaaaaaa/+a/a*aaaaaaa*+*+/*aaaaaaa-[27]  = 195.4805
+*a+/aaaaaaaa**a+-*aaaaaaa/*+a+/aaaaaaa+***a-aaaaaaa-[28]  = 748.5801
-/---+aaaaaaa//+*+-aaaaaaa-/---aaaaaaaa+-++*-aaaaaaa-[29]  = 0
```

Note that eight of the 30 randomly generated individuals are unviable, that is, have zero fitness. This means that, either they were unable to solve a single fitness case within the chosen selection range, or they returned mathematical errors such as a division by zero. As you can easily check, in this case, the reason why they all received zero fitness is that they all perform divisions by zero.

Consider, for instance, the best individual of this generation, chromosome 26, with fitness 908.0693:

```
0123456789012012345678901201234567890120123456789012
-++/+/aaaaaaa*+a+*/aaaaaaa/+-+-/aaaaaaa-+//a+aaaaaaa        (4.2a)
```

As its expression shows, two of the four terms of the target function (the first and second terms) are already present in this solution (the contribution of each gene is shown in brackets):

$$y = (3a) + (a^3 + a^2 + a) + \left(\frac{2a}{1-a}\right) + (a-1) \qquad (4.2b)$$

In the next generation (generation 1) a new individual with better traits than the best individual of generation 0 was created:

```
0123456789012012345678901201234567890120123456789012
+a+*-aaaaaaaa+/-//-aaaaaaa--++a*aaaaaaa+***aaaaaaaaa        (4.3a)
```

It has a fitness of 958.3982 evaluated against the set of fitness cases presented in Table 4.2. Its expression shows that three of the terms of this intermediate solution match exactly the target function (4.1); in fact, only the third term is missing (the contribution of each gene is shown in brackets):

$$y = (a^2 + a) + (1-a) + (-a^2) + (a^3 + a^2) \qquad (4.3b)$$

As you can see in Figure 4.6, for the next four generations no improvement in best fitness was observed. But in generation 6 a better solution with fitness 990 was created:

```
0123456789012012345678901201234567890120123456789012
+-/-a-aaaaaaa+-a**aaaaaaaa*+-a*aaaaaaaa+*a+a*aaaaaaa
```
(4.4a)

As its expression shows, three of the terms of this intermediate solution match exactly the target function (4.1); in this case, only the last term is missing (the contribution of each gene is shown in brackets):

$$y = (-a) + (a) + (0) + (a^3 + a^2 + a)$$
(4.4b)

For the next 12 generations no improvement in best fitness occurred (see Figure 4.6). But in generation 19 a perfect solution with maximum fitness was found:

```
0123456789012012345678901201234567890120123456789012
+*a/-aaaaaaaa*-/++aaaaaaaa-/***-aaaaaaa+***/*aaaaaaa
```
(4.5a)

As its expression shows, it matches exactly the target function (4.1) (the contribution of each gene is shown in brackets):

$$y = (a) + (0) + (1) + (a^3 + a^2)$$
(4.5b)

The detailed analysis of these best-of-generation programs shows that some of their components are redundant or neutral, like the addition/subtraction of zero or the multiplication/division by one. However, the existence of these redundant clusters, be they small neutral clusters in the sub-ETs or entire neutral genes like gene 3 in chromosome (4.4) above or gene 2 in chromosome (4.5), is important to the evolution of fitter individuals (compare, in Figures 4.1 and 4.3, the success rate of a compact unigenic system with a head length of six with other less compact systems, either with more genes or head lengths greater than six).

In summary, gene expression programming is a very flexible system, with easily navigable solution spaces. Consequently, default settings can be used to solve most problems, needing just small adjustments here and there to make the most of any problem. Thus, small populations of just 30 individuals (evolving for as long as necessary as GEP populations never become stagnant) undergoing the typical degree of genetic modification (see, for instance, Table 4.1); a chromosome architecture of three genes with $h = 10$

linked by addition and a function set composed of the basic arithmetic operators plus the sqrt(x), exp(x), ln(x) functions to spice things a bit; and a standard fitness function such as the one based on the mean squared error (see
section 3.2.2, Fitness Functions for Symbolic Regression), is a good starting
point for most problems of symbolic regression.

4.1.2 Function Finding on a Five-dimensional Parameter Space

The goal of this section is to show how gene expression programming can be
used to model complex realities with high accuracy. The test function chosen is the following five parameter function:

$$y = \frac{\sin(a) \cdot \cos(b)}{\sqrt{\exp^c}} + \tan(d - e) \tag{4.6}$$

where a, b, c, d, and e are the independent variables, and exp is the irrational
number 2.71828183.

Consider we were given a sampling of the numerical values from this
function over 100 random points in the interval [0, 1] and we wanted to find
a function fitting those values as accurately as possible. We can use, for
instance, the mean squared error to design the fitness function and evaluate
each candidate solution by equation (3.4a), giving $f_{max} = 1000$.

The set of 100 fitness cases used in this complex task could very well be
unrepresentative of the problem domain, and the program designed by the
algorithm would be modeling a reality other than the reality of function (4.6).
To solve this dilemma, it is common to use a testing set with a reasonable
amount of sample cases. This dataset is not used during the learning process
and therefore can be used to check the usefulness of the model, or in other
words, its generalizing capabilities. For this problem, and because it does
not delay evolution, we will be generous and use a testing set of 200 computer generated samples. In real-world problems where samples are costly, it is
common practice to use between 30-35% of samples for testing.

The domain of this problem suggests, besides the basic arithmetical functions, the use of sqrt(x), exp(x), sin(x), cos(x) and tan(x) in the function set,
which, for simplicity, will be represented, respectively, by the symbols "Q",
"E", "S", "C", and "T". Thus, for this problem, the function set consisted of
$F = \{+, -, *, /, Q, E, S, C, T\}$, and the set of terminals consisted obviously of
the independent variables, giving $T = \{a, b, c, d, e\}$.

For this problem, two genes per chromosome were used, each with a head length of 12 and, therefore, encoding sub-ETs with a maximum of 25 nodes each. The sub-ETs were posttranslationally linked by addition. Both the performance and the parameters used per run are summarized in Table 4.3.

As you can see in Table 4.3, small populations of just 50 individuals were used, but they were left to evolve for 5000 generations, enough time to discover good solutions, as the average best-of-run R-square of 0.99043015 indicates.

The best-of-experiment solution has a fitness of 999.992 and an R-square of 0.99998525, and was discovered in generation 4946 of run 31:

```
012345678901234567890123401234567890123456789012345678901234
*S*aCCQbSce/bccbabbadceeeT-deb/-EceTebeccbbebedcad                    (4.7a)
```

which, mathematically, corresponds to the following expression:

$$y = \sin(a) \cdot \cos(b) \cdot \cos \sqrt{\sin(c)} + \tan(d - e) \qquad (4.7b)$$

Table 4.3
Settings for the five-parameter function problem.

Number of runs	100
Number of generations	5000
Population size	50
Number of fitness cases	100
Function set	+ - * / Q E S C T
Terminal set	a b c d e
Head length	12
Gene length	25
Number of genes	2
Linking function	+
Chromosome length	50
Mutation rate	0.044
Inversion rate	0.1
IS transposition rate	0.1
Gene transposition rate	0.1
One-point recombination rate	0.3
Two-point recombination rate	0.3
Gene recombination rate	0.3
RIS transposition rate	0.1
Fitness function	Equation (3.4a)
Average best-of-run fitness	997.661
Average best-of-run R-square	0.99043015

Note that three of the terms of this model match exactly the target function (4.6). Note, however, that the term $\cos\sqrt{\sin(c)}$ was discovered by GEP as an excellent approximation to $1/\sqrt{\exp^c}$. Indeed, in the testing set, this model has a fitness of 999.9907 and an R-square of 0.99998534, which tells us that this model has excellent generalizing capabilities and is indeed an almost perfect match to the target function (4.6). Indeed, as we will see again and again in this book, gene expression programming can be used to find very good solutions to problems of great complexity.

4.1.3 Mining Meaningful Information from Noisy Data

Tools for mining knowledge from huge databases are crucial in a world where data is constantly increasing. The quantity of data is so big that to find the meaningful factors in the sea of data becomes a Herculean task and new technologies have been developed to extract relevant knowledge from these huge databases. Gene expression programming is one of these emerging technologies and is ideal for separating the wheat from chaff. In this section we are going to illustrate very clearly how this can be success-fully achieved with a function finding problem where nine out of ten vari-ables are meaningless.

The test function we are going to use in this experiment is the already familiar function of section 4.1.1, with the difference that the meaningful parameter is to be discovered among a total of 10 variables. In Table 4.4 are shown both the performance and the parameters used per run in this experi-ment. And as you can see by the high success rate obtained (77%), gene expression programming was not overwhelmed by the quantity of irrelevant data and found its way around this huge amount of irrelevant information very efficiently.

The first perfect solution was created in generation 61 of run 0. Its chro-mosome is shown below (the sub-ETs are linked by addition):

```
0123456789012
*a*aa-hgadadc
-ah*d-gcfjcbd
/--gcgciijeeg
h+eeehbeddbfd
*aadaabcecfgb
```
$$(4.8a)$$

Table 4.4
Settings used in the 10-dimensional data mining problem.

Number of runs	100
Number of generations	1000
Population size	50
Number of fitness cases	100
Function set	+ - * /
Terminal set	a b c d e f g h i j
Head length	6
Gene length	13
Number of genes	5
Linking function	+
Chromosome length	65
Mutation rate	0.044
Inversion rate	0.1
IS transposition rate	0.1
RIS transposition rate	0.1
One-point recombination rate	0.3
Two-point recombination rate	0.3
Gene recombination rate	0.3
Gene transposition rate	0.1
Fitness function	Equation (3.3b)
Selection range	100%
Precision	0.01%
Success rate	77%

where a represents the meaningful variable and b-j represent the remaining meaningless variables. As its expression shows, this chromosome encodes a function equivalent to the target function (4.1) (the contribution of each gene is shown in brackets):

$$y = (a^3) + (a - h) + (1) + (h) + (a^2) \qquad (4.8b)$$

4.2 Classification Problems

In this section we will use the basic gene expression algorithm to model three complex real-world problems, namely, the diagnosis of breast cancer, the assignment of a credit card, and the classification of Fisher's irises. The first two datasets were obtained from PROBEN1 – a set of neural network bench-mark problems and benchmarking rules (Prechelt 1994). Both the technical report and the datasets are available through anonymous FTP from the Neural Bench archive at Carnegie Mellon University (machine **ftp.cs.cmu.edu**, directory **/afs/cs/project/connect/bench/contrib/prechelt**) and from the machine **ftp.ira.uka.de** in directory **/pub/neuron**. The file name in both cases is **proben1.tar.gz**. The last dataset, the iris dataset (Fisher 1936), is perhaps the most famous dataset used in data mining and is also freely available online.

4.2.1 Diagnosis of Breast Cancer

In this diagnosis task the goal is to classify a tumor as either benign (0) or malignant (1) based on nine different cell analysis (input attributes or termi-nals) – clump thickness, uniformity of cell size, uniformity of cell shape, marginal adhesion, single epithelial cell size, bare nuclei, bland chromatin, normal nucleoli, and mitoses.

The model presented here was obtained using the **cancer1** dataset of PROBEN1 where the binary 1-of-m encoding in which each bit represents one of the m-possible output classes was replaced by a 1-bit encoding ("0" for benign and "1" for malignant). The first 350 samples were used for train-ing and the last 174 were used for testing the performance of the model in real use. This means that absolutely no information from the testing set sam-ples or the testing set performance are available during the adaptive process. Thus, the classification error on the testing set will be used to evaluate the generalization performance of the evolved models.

For this problem, F = {+, +, -, -, *, *, /, LT, GT, LOE, GOE, ET, NET} (the last six functions are comparison functions of two arguments which return 1 if the condition is true or 0 if false, representing, respectively, less than, greater than, less or equal to, greater or equal to, equal to, and not equal to); the set of terminals consisted of the nine attributes used in this problem and were represented by T = {d_0, ..., d_8} which correspond, respectively, to clump thickness, uniformity of cell size, uniformity of cell shape, marginal adhe-sion, single epithelial cell size, bare nuclei, bland chromatin, normal nu-cleoli, and mitoses.

In classification problems where the output (the dependent variable) is often binary, it is important to set criteria to convert predicted values (usu-ally real-valued numbers) into zero or one. This is the 0/1 rounding threshold R that converts the output of an individual program into 1 if the output is equal to or greater than R, or into 0 otherwise. For this problem we are going to use a rounding threshold of 0.5.

For evaluating the fitness we will use equation (3.14), which is based both on the sensitivity and specificity. Thus, for this problem, $f_{max} = 1000$.

For this problem, chromosomes composed of three genes with an $h = 7$ and sub-ETs linked by addition were used. The parameters used per run are summarized in Table 4.5. Note that the number of generations is not shown, as in real-world situations one usually performs several dozens of runs until a good solution has been found. Also note that small populations of 30 individuals were used, as this allows a quick and efficient evolution.

Given the complexity of the problem, I used the software Automatic Problem Solver (APS) by Gepsoft to model this function because it reflects more accurately the effort and strategies that go into solving a complex real-world problem. For one, it allows the easy optimization of intermediate solutions and, for another, it also allows the easy checking of the evolved models against

Table 4.5
Settings used in the breast cancer problem.

Population size	30
Number of training samples	350
Number of testing samples	174
Function set	+ + - - * * / LT GT LOE GOE ET NET
Terminal set	d0 - d8
Head length	7
Gene size	15
Number of genes	3
Linking function	+
Chromosome length	45
Mutation rate	0.044
Inversion rate	0.1
IS transposition rate	0.1
RIS transposition rate	0.1
One-point recombination rate	0.3
Two-point recombination rate	0.3
Gene recombination rate	0.3
Gene transposition rate	0.1
Fitness function	Equation (3.14)
Rounding threshold	0.5

a testing set and the immediate evaluation of standard statistical parameters such as the confusion matrix, the sensitivity, the specificity, the classification error (the percent of incorrectly classified samples) and the classification accuracy (the percent of samples correctly classified).

In one run a very good solution with a fitness of 957.0898 and a classification error of 2.571% and a classification accuracy of 97.429% was found (the genes are shown separately and a dot is used to separate the different elements; the heads are shown in bold):

```
*.+.+.d8.d5.d0.NET.d6.d6.d2.d4.d0.d4.d5.d1
+.*.*.d6.d1.*.LOE.d5.d1.d8.d7.d2.d4.d8.d1
*.d1.d3.d0.GT./.*.d3.d3.d8.d4.d7.d4.d8.d6          (4.9a)
```

In terms of number of hits, this model classifies correctly 341 out of 350 fitness cases in the training set and 173 out of 174 in the testing set. This corresponds to a testing set classification error of 0.575% and a classification accuracy of 99.425%, even better than the classification accuracy on the training set, what tells us that the model (4.9) above is indeed a very good model for diagnosing breast cancer.

The confusion matrices obtained both for the training and the testing sets are shown in Figure 4.7. With their help one can easily evaluate such important parameters as the sensitivity, the specificity, the positive predictive value, and the negative predictive value, all of them important in the medical field. Thus, in the testing set, the sensitivity, evaluated by equation (3.11), is equal to 98.462%; the specificity, evaluated by equation (3.12), is equal to 100%; the PPV, evaluated by equation (3.15), is also 100%; and the NPV, evaluated by equation (3.16), is equal to 99.091%.

Note that for the expression of this chromosome to be complete the 0/1 rounding threshold $R = 0.5$ must be taken into account. With Gepsoft APS we can automatically convert the model (4.9) above into a fully expressed computer program, such as the C++ function below:

```cpp
int apsModel(double d[])
{
    const double ROUNDING_THRESHOLD = 0.5;
    double dblTemp = 0.0;
    dblTemp  = ((d[8]+d[5])*(d[0]+(d[6]!=d[6]?1:0)));
    dblTemp += ((d[6]*d[1])+((d[5]*d[1])*(d[8]<=d[7]?1:0)));
    dblTemp += (d[1]*d[3]);
    return (dblTemp >= ROUNDING_THRESHOLD ? 1:0);
}
```
                                                          (4.9b)

Training				Testing		
	Yes	No			Yes	No
Yes	120	1		Yes	64	1
No	8	221		No	0	109

Figure 4.7. The confusion matrices obtained with the model (4.9) on the training and testing sets. Note that the true positives and the true negatives occupy the main diagonal, and their sum corresponds to the number of hits ($120 + 221 = 341$ on the training set and $64 + 109 = 173$ on the testing set).

Indeed, all models evolved by gene expression programming can be immediately converted into virtually any programming language through the use of grammars, including the universal representation of parse trees (Figure 4.8). These trees can then be used to grasp immediately the mathematical intricacies of the evolved models and therefore are ideal for extracting knowledge from data. For instance, you can easily see in Figure 4.8 that the expression encoded in the first gene can be simplified by removing the neutral block addition-of-zero, obtaining:

```
int apsModel(double d[])
{
    const double ROUNDING_THRESHOLD = 0.5;
    double dblTemp = 0.0;
    dblTemp  = ((d[8]+d[5])*d[0]);
    dblTemp += ((d[6]*d[1])+((d[5]*d[1])*(d[8]<=d[7]?1:0)));
    dblTemp += (d[1]*d[3]);
    return (dblTemp >= ROUNDING_THRESHOLD ? 1:0);
}
```
(4.9c)

As you can clearly see in Figure 4.8, not all the cell analyses seem to be relevant to an accurate diagnosis of breast cancer: indeed, attribute d_2 uniformity of cell shape and attribute d_4 single epithelial cell size are not used in the design of this extremely accurate model. This is, indeed, one of the great advantages of gene expression programming: the possibility of extracting knowledge almost instantaneously as the models evolved by GEP can be represented in any conceivable language, including the universal diagram representation of expression trees.

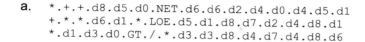

a. `*.+.+.d8.d5.d0.NET.d6.d6.d2.d4.d0.d4.d5.d1`
    `+.*.*.d6.d1.*.LOE.d5.d1.d8.d7.d2.d4.d8.d1`
    `*.d1.d3.d0.GT./.*.d3.d3.d8.d4.d7.d4.d8.d6`

b.

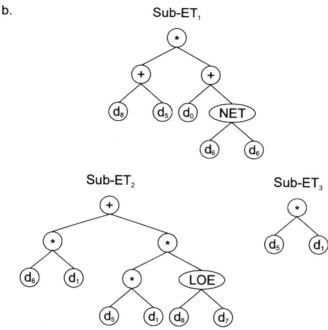

Figure 4.8. Model evolved by GEP to diagnose breast cancer. **a)** The three-genic chromosome encoding sub-ETs linked by addition. **b)** The sub-ETs codified by each gene. Note that the expression of this chromosome is only complete after linking with addition and applying the rounding threshold, which in this case is equal to 0.5 (see model 4.9b).

4.2.2 Credit Screening

In credit screening the goal is to determine whether to approve or not a customer's request for a credit card. Each sample in the dataset represents a real credit card application and the output describes whether the bank granted the credit card or not. This problem has 51 input attributes, all of them unexplained in the original dataset for confidentiality reasons.

The model presented here was obtained using the **card1** dataset of PROBEN1 where the binary 1-of-*m* encoding was again replaced by a 1-bit encoding ("1" for approval and "0" for non-approval). The first 345 examples were used for training and the last 172 were used for testing.

For this problem, the function set $F = \{+, -, *, LT, GT, LOE, GOE, ET, NET\}$ (the last six functions are comparison functions of two arguments which return 1 if the condition is true or 0 if false, representing, respectively, less than, greater than, less or equal to, greater or equal to, equal to, and not equal to), in which each function was weighted 5 times; and the set of terminals included all the 51 attributes which were represented by d_0- d_{50}. The 0/1 rounding threshold R was equal to 0.5 and the fitness was based on the number of hits and was evaluated by equation (3.8). Thus, in this case, $f_{max} = 345$.

For this problem, chromosomes composed of 10 genes with an $h = 7$ and sub-ETs linked by addition were used. The parameters used per run are summarized in Table 4.6. Note again that, despite the high dimensionality of the problem, small populations of just 30 individuals were used, as this allows a quick and efficient evolution.

In one run, by generation 11446 a very good solution with a classification error of 10.145% and a classification accuracy of 89.855% was created (the

Table 4.6
Settings used in the credit screening problem.

Population size	30
Number of training samples	345
Number of testing samples	172
Function set	$(+ - * LT\ GT\ LOE\ GOE\ ET\ NET)_5$
Terminal set	d0 - d50
Head length	7
Gene size	15
Number of genes	10
Linking function	+
Chromosome length	150
Mutation rate	0.044
Inversion rate	0.1
IS transposition rate	0.1
RIS transposition rate	0.1
Gene transposition rate	0.1
One-point recombination rate	0.3
Two-point recombination rate	0.3
Gene recombination rate	0.3
Fitness function	Equation (3.8)
Rounding threshold	0.5

genes are shown separately and a dot is used to separate the different elements; the heads are shown in bold):

```
NET.d41.d8.+.d12.d8.GOE.d37.d44.d9.d38.d28.d47.d0.d9
d25.d25.d12.d44.d18.NET.ET.d16.d42.d41.d50.d14.d39.d50.d12
-.NET.+.LT.GT.NET.LOE.d20.d46.d16.d40.d48.d34.d41.d4
ET.GOE.d46.d27.d13.d7.LOE.d25.d9.d30.d2.d40.d12.d49.d41
NET.*.d42.d18.d7.+.d20.d2.d41.d35.d42.d28.d13.d5.d38
*.d24.GOE.d38.d43.d24.d7.d24.d37.d2.d45.d49.d24.d47.d34
d38.d28.d0.GOE.d10.GOE.d27.d49.d40.d0.d9.d44.d45.d27.d19
GOE.d21.+.GOE.NET.LOE.d31.d47.d42.d5.d27.d22.d49.d33.d25
+.d2.d23.d32.LT.NET.d44.d40.d31.d22.d21.d3.d15.d38.d1
d17.d20.LT.ET.d37.-.d34.d42.d11.d48.d19.d18.d16.d23.d3
```
                                                                    (4.10a)

It has a fitness of 310 evaluated against the training set of 345 fitness cases and a fitness of 151 evaluated against the testing set of 172 samples. This corresponds to a testing set classification error of 12.209% and a classification accuracy of 87.791%, almost identical to the classification error and classification accuracy obtained in the training set, which is indicative of the excellent generalizing capabilities of the evolved model. Below is shown the fully expressed individual translated into a C++ function automatically generated with Gepsoft APS:

```
int apsModel(double d[])
{
    const double ROUNDING_THRESHOLD = 0.5;
    double dblTemp = 0.0;
    dblTemp  = (d[41]!=d[8]?1:0);
    dblTemp += d[25];
    dblTemp += (((d[20]<d[46]?1:0)!=(d[16]>d[40]?1:0)?1:0) -
                ((d[48]!=d[34]?1:0)+(d[41]<=d[4]?1:0)));
    dblTemp += ((d[27]>=d[13]?1:0)==d[46]?1:0);
    dblTemp += ((d[18]*d[7])!=d[42]?1:0);
    dblTemp += (d[24]*(d[38]>=d[43]?1:0));
    dblTemp += d[38];
    dblTemp += (d[21]>=(((d[5]<=d[27]?1:0)>=d[31]?1:0)+
                (d[47]!=d[42]?1:0))?1:0);
    dblTemp += (d[2]+d[23]);
    dblTemp += d[17];
    return (dblTemp >= ROUNDING_THRESHOLD ? 1:0);
}
```
                                                                    (4.10b)

Note that not all the attributes integrate the model evolved by GEP, which indicates that they are apparently irrelevant to the decision at hand. In fact, of the 51 attributes only 25 are used in this extremely accurate model. This is indeed a good example of how gene expression programming can be successfully used for extracting knowledge from huge databases and designing good predictive models.

4.2.3 Fisher's Irises

In this classification problem the goal is to classify three different types of irises based on four measurements: sepal length, sepal width, petal length, and petal width. The iris dataset contains fifty examples each of three types of iris: *Iris setosa, Iris versicolor*, and *Iris virginica* (Fisher 1936).

Classification problems with more than two classes, say n classes, can be solved by GEP using two different approaches. The first one requires decomposing the data into n separate 0/1 classification problems. And the second explores the multigenic nature of gene expression programming to solve problems of multiple outputs in one go, that is, chromosomes composed of n different genes are used to design a classification model composed of n different sub-models, where each sub-model is responsible for the identification of a particular class.

The second approach is very appealing and the GEP system with multiple outputs (GEP-MO) is indeed very efficient at solving relatively complex problems of this kind. But, for really complex problems (say, problems with more than 10 different classes and/or more than 50 attributes), the first approach, although more time-consuming, is a better choice as it is much more flexible in terms of both the structure and the composition of each sub-model. Both approaches will be compared below on the iris dataset.

Decomposing a Three-class Problem

The classification of data into n distinct classes C requires processing the data into n separate 0/1 classification problems as follows:

1. C_1 versus NOT C_1
2. C_2 versus NOT C_2
...
n. C_n versus NOT C_n

Then n different sub-models are evolved separately and afterwards combined in order to create the final model.

For the iris data we are going to decompose our problem into three separate 0/1 classification problems. The first one is *Iris setosa* versus NOT *Iris setosa*; the second is *Iris versicolor* versus NOT *Iris versicolor*; and the last is *Iris virginica* versus NOT *Iris virginica*.

For all these sub-problems, F = {+, -, *, /} and the set of terminals includes obviously all the four attributes which were represented by d_0- d_3, corresponding, respectively, to sepal length, sepal width, petal length, and petal width. The 0/1 rounding threshold was set to 0.5 and the fitness was based on the number of hits and was evaluated by equation (3.8). Thus, for this problem, f_{max} = 150.

For all the sub-problems, I started with three-genic chromosomes with an h = 8 and sub-ETs linked by addition. The first dataset (*Iris setosa* versus NOT *Iris setosa*) was almost instantaneously classified without errors and I soon found out that a very simple structure is required to classify correctly this dataset. The model below perfectly classifies all the irises into *Iris setosa* and NOT *Iris setosa*:

```
int apsModel(double d[])
{
      const double ROUNDING_THRESHOLD = 0.5;
      double dblTemp = 0;
      dblTemp = (d[1]-d[2]);
      return (dblTemp >= ROUNDING_THRESHOLD ? 1 : 0);
}
```
(4.11)

As you can see, only the difference between the sepal width and the petal length is relevant to distinguish *Iris setosa* from *Iris versicolor* and *Iris virginica*.

The classification of the remaining datasets was also extremely accurate, but on both cases only 149 out of 150 samples were correctly classified. And in both cases, the sub-models failed only to classify sample 84. The model below distinguishes *Iris versicolor* from the other two irises:

```
int  apsModel(double  d[])
{
      const  double  ROUNDING_THRESHOLD  =  0.5;
      double  dblTemp  =  0;
      dblTemp    =  (d[3]*(((d[0]*d[3])-d[1])+((d[1]*d[2])-2))));
      dblTemp  +=  (((d[2]-(d[2]+d[2]))-(d[0]/d[0]))/d[3]);
      dblTemp  +=  (((d[0]-(d[2]*d[3]))*(d[2]-d[1]))*d[0]);
      return  (dblTemp  >=  ROUNDING_THRESHOLD  ?  1  :  0);
}
```
(4.12)

And the next model distinguishes *Iris virginica* from *setosa* and *versicolor*:

```
int apsModel(double d[])
{
    const double ROUNDING_THRESHOLD = 0.5;
    double dblTemp = 0;
    dblTemp  = (d[1]/(d[0]*(d[0]/d[3])));
    dblTemp += (d[2]-d[1]);
    dblTemp += ((d[2]-(((d[0]+d[3])/d[1])/d[3]))-d[2]);
    return (dblTemp >= ROUNDING_THRESHOLD ? 1 : 0);
}
```
(4.13)

So, by combining the three models above and representing them by y_1, y_2, and y_3, the following classification rules are obtained:

IF ($y_1 = 1$ **AND** $y_2 = 0$ **AND** $y_3 = 0$) **THEN** *setosa*;
IF ($y_1 = 0$ **AND** $y_2 = 1$ **AND** $y_3 = 0$) **THEN** *versicolor*;
IF ($y_1 = 0$ **AND** $y_2 = 0$ **AND** $y_3 = 1$) **THEN** *virginica*;   (4.14)

which classify correctly 149 out of 150 irises and, therefore, this model, with a classification accuracy of 99.33% and a classification error of 0.667%, is extremely robust and one of the best that can be learned from this dataset.

Multiple Genes for Multiple Outputs

The multigenic nature of gene expression programming can also be explored to find solutions to problems with multiple outputs in one go. For that, chromosomes composed of *n* different genes are used to evolve *n* different sub-models that classify *n* different classes.

It is worth pointing out that, in this kind of multiple output system, the evolution of one sub-model encoded in a particular gene is intricately connected with the evolution of the other sub-models encoded in the other genes. So, a requirement for an efficient evolution consists of designing a good fitness function that will allow not only the creation of good models but also an efficient evolution. For instance, for a three-class problem, there are already $2^3 = 8$ different outputs for the three sub-models encoded in the chromosome, and just three of them are clear hits (100, 010, and 001 for classes A, B, and C, respectively).

So, how to go about designing the fitness function? There are at least three different fitness functions based on the number of hits that produce good results. The first one, let's call it *simple hits*, will look at the output of each sub-model independently of the output of the other sub-models, and reward all partial hits; for instance, outputs such as 101 or 000 for class A (output 100) will score four simple hits. The second one, let's call it *clear hits*, will only reward clear hits and, for instance, will not differentiate outputs such as 111 or 001 for class B (output 010). And the third one, let's call it *ambiguity/sharpness*, will reward clear hits and partial hits differently and will also distinguish between different degrees of ambiguity; for instance, outputs such as 000 or 111 for class C (output 001) will score differently, with the former being less ambiguous with just one mishit.

The ambiguity/sharpness fitness function performs slightly better than the other two (and the first is slightly worse than the second) and, therefore, we are going to use this fitness function in our three-class problem. So, let's start by explaining what is understood by sharpness and ambiguity.

The sharpness S is very easy to compute and corresponds to the number of partial hits; for instance, for a sample belonging to class A (output 100), the output 101 will have $S = 2$ and the output 001 will have $S = 1$. The ambiguity A is a measure of sharpness and the higher the sharpness the smaller the ambiguity. And by definition, maximum sharpness (which obviously corresponds to the number of classes n) corresponds to a degree of ambiguity one, that is, when $S = n$, then $A = 1$; $S = n-1$ corresponds to $A = 2$; $S = n-2$ corresponds to $A = 3$; and cases where $S = 0$ contribute nothing to fitness. More formally, the fitness $f_{(ij)}$ of an individual program i for sample case j is evaluated by the equation:

$$f_{(ij)} = \frac{1}{A^3} \tag{4.15}$$

Thus, when $A = 1$ (that is, when we have a clear hit and sharpness has the maximum value) $f_{(ij)} = 1$. Consequently, for the ambiguity/sharpness fitness function, maximum fitness is equal to the number of sample cases.

For this problem, we will use the same function set used in the previous section, that is, F = {+, -, *, /} and also the same set of terminals, which obviously includes all the four attributes of the iris data – sepal length, sepal width, petal length, and petal width. For this three-class problem, we will obviously use chromosomes composed of three genes, each encoding a different sub-model. The same 0/1 rounding threshold of 0.5 was chosen for all the sub-models to convert their outputs into 0 or 1. The head size for all the three genes is equal to 10, corresponding to a maximum sub-program length of 21 nodes. The fitness will be evaluated by equation (4.15) and, therefore, for this problem with 150 different plants, $f_{max} = 150$.

Since this is the first time that the GEP-MO system is being put to the test, we are going to evaluate its performance with an experiment with 100 runs, and since the algorithm is unable to classify correctly all the sample cases of the iris dataset, the performance will be evaluated in terms of average best-of-run number of hits (clear hits). Both the performance and the parameters used per run are summarized in Table 4.7.

One of the best solutions discovered in this experiment, has a fitness of 148.162 and was created in generation 12452 of run 53 (the genes are shown separately and a dot is used to separate the different elements; the head is shown in bold):

```
-.d1.d2.+.d1.d2.d2./.d1.d0.d0.d3.d0.d3.d0.d3.d0.d2.d0.d3.d1
/./.*.*.+.-.d2.-.d2.d1.d3.d2.d1.d0.d3.d3.d3.d1.d0.d0.d1
+.-.d3.*.+.*.-.*.d0.+.d3.d2.d1.d0.d3.d3.d1.d3.d1.d2.d2            (4.16)
```

In terms of number of hits, this model classifies correctly 148 out of 150 fitness cases. This corresponds to a classification error of 1.333% and a classification accuracy of 98.667%, thus is slightly worse than the model

Table 4.7
Settings used in the iris problem with multiple outputs.

Number of runs	100
Number of generations	20,000
Population size	30
Number of fitness cases	150
Function set	+ - * /
Terminal set	d0-d3
Head length	10
Gene length	21
Number of genes	3
Chromosome length	63
Mutation rate	0.044
Inversion rate	0.1
IS transposition rate	0.1
RIS transposition rate	0.1
One-point recombination rate	0.3
Two-point recombination rate	0.3
Gene recombination rate	0.3
Gene transposition rate	0.1
Fitness function	Equation (4.15)
Rounding threshold	0.5
Average best-of-run number of hits	146.48

designed using the first approach (let's call it the three-step approach), which classifies correctly 149 out of 150 sample cases.

Although this model was not created by Gepsoft APS, we can still use this software to automatically translate the Karva code of each sub-model into a conventional programming language. So, the first sub-model encoded in chromosome (4.16), which is responsible for the classification of the *setosa* variety, is translated into the following C++ function:

```
int apsModel(double d[])
{
    const double ROUNDING_THRESHOLD = 0.5;

    double dblTemp = 0.0;
    dblTemp = (d[1]-d[2]);

    return (dblTemp >= ROUNDING_THRESHOLD ? 1:0);
}
```
$$(4.17)$$

And as was discovered with the three-step approach, only the difference between the sepal width and the petal length is relevant to distinguish *Iris setosa* from the other two irises.

The identification of the *versicolor* variety was carried out by the second sub-model of program (4.16). In fact, this sub-model is the worst of the three sub-models, being able to classify correctly just 148 plants. More specifically, it failed to classify samples 73 and 84 (recall that the model created with the three-step approach was also unable to classify sample 84). Its expression is shown below:

```
int apsModel(double d[])
{
    const double ROUNDING_THRESHOLD = 0.5;

    double dblTemp = 0.0;
    dblTemp = ((((d[0]-d[3])*d[2])/(d[1]+d[3]))/
              ((d[2]-d[1])*d[2]));

    return (dblTemp >= ROUNDING_THRESHOLD ? 1:0);
}
```
$$(4.18)$$

And finally, the third sub-model is responsible for the identification of the *virginica* variety. It distinguishes perfectly a total of 149 plants and fails only to classify sample 84. Its code is shown below:

```
int apsModel(double d[])
{
    const double ROUNDING_THRESHOLD = 0.5;

    double dblTemp = 0.0;
    dblTemp = (((((d[3]+d[1])*d[3])*(d[2]-d[1]))-
             ((d[0]*d[3])+d[0]))+d[3]);

    return (dblTemp >= ROUNDING_THRESHOLD ? 1:0);
}
```
(4.19)

So, in terms of the classification rules evolved by the GEP-MO system, it is also necessary to combine the three sub-models above – programs (4.17), (4.18), and (4.19) – and by representing them, respectively, by y_1, y_2, and y_3, the following classification rules are obtained:

IF ($y_1 = 1$ **AND** $y_2 = 0$ **AND** $y_3 = 0$) **THEN** *setosa*;
IF ($y_1 = 0$ **AND** $y_2 = 1$ **AND** $y_3 = 0$) **THEN** *versicolor*;
IF ($y_1 = 0$ **AND** $y_2 = 0$ **AND** $y_3 = 1$) **THEN** *virginica*;   (4.20)

which classify correctly 148 out of 150 plants and, therefore, has a classification error of 1.333% and a classification accuracy of 98.667%. Thus, this model is slightly worse than the model (4.14) designed with the three-step approach.

4.3 Logic Synthesis and Parsimonious Solutions

In this section we are going to analyze a different kind of symbolic regression, one that is used in logic synthesis or, in other words, in the design of logical functions. And the design of logical functions is intricately connected with finding the most parsimonious solutions to a problem, as this is fundamental for the creation of even faster machines. And we will see in this section that gene expression programming is the ideal tool for finding these parsimonious solutions.

On the one hand, gene expression programming handles all logical systems the same way, that is, it works as efficiently with the classical Boolean system of NOTs, ANDs, and ORs gates, as with less intuitive systems such as the NAND system, the NOR system, the MUX system, the Reed-Muller system, or whatever system one wants to experiment with. The reason for this plasticity is that gene expression programming does not rely on logical rules or theorems such as the De Morgan's theorems or Karnaugh maps to design parsimonious solutions; rather, it relies on the same evolutionary principles we've been discussing throughout this book of constantly creating and experimenting with functional building blocks and, therefore, is impervious to which kinds of gates one chooses to build logical circuits.

On the other hand, logic synthesis problems are special in the sense that they all can be solved exactly, that is, it is always possible to design a perfect solution to the problem at hand. So the question is: How to make sure that we are looking at the most parsimonious solution? And what can be done to find the elusive parsimonious solution?

Despite giving no certain answers to the first question, gene expression programming can help us find, if not the most parsimonious, at least one of the most parsimonious solutions to the problem at hand, no matter what kind of universal system one chooses to work with. How this is achieved is explained in the next section.

4.3.1 Fitness Functions with Parsimony Pressure

The question of parsimony is intricately connected with neutrality and we now know that less compact systems (that is, systems with more room for the existence of neutral blocks) evolve much more efficiently than more compact ones (see a discussion of The Role of Neutrality in Evolution in chapter 12). This means, obviously, that if we try to optimize a solution not only based on its efficiency at the task at hand but also on its size, one must be careful in not hindering evolution completely by applying too high a pressure on parsimony.

But in logic synthesis problems one can have the best of both worlds and let the system evolve freely at first by using a good, far from compact organization until a perfect solution is found and only then start selecting solutions for their size as well. This way, the discovery of perfect solutions is completely unconstrained and as soon as a perfect solution is found the system will start applying pressure on the size of the evolving programs and, if there is a way of representing the perfect solution more compactly, it will most probably be found, for the system is relentless and will only stop until no more simplifications can be done.

So, in order to design such a fitness function, one must distinguish raw fitness from overall fitness (referred just as fitness for simplicity). As a measure of raw fitness one can use any of the fitness functions described in section 3.2.3 (number of hits, hits with penalty, sensitivity/specificity, and positive predictive value/negative predictive value fitness functions) and then complement them with a parsimony term.

Consider, for instance, the number of hits fitness function evaluated by equation (3.8). In this case, the raw fitness rf_i of an individual program i corresponds to the number of hits and, therefore, raw maximum fitness rf_{max} consists of the total number of fitness cases. And the overall fitness f_i (or fitness with parsimony pressure) is evaluated by the formula:

$$f_i = rf_i \cdot \left(1 + \frac{1}{5000} \cdot \frac{S_{max} - S_i}{S_{max} - S_{min}}\right) \tag{4.21}$$

where S_i is the size of the program, S_{max} and S_{min} represent, respectively, maximum and minimum program sizes and are evaluated by the formulas:

$$S_{max} = G(h + t) \tag{4.22}$$

$$S_{min} = G \tag{4.23}$$

where G is the number of genes, and h and t are the head and tail sizes (note that, for simplicity, the linking function was not taken into account). Thus, when $rf_i = rf_{max}$ and $S_i = S_{min}$ (highly improbable, though, as this can only happen for very simple functions as this means that all the sub-ETs are composed of just one node), $f_i = f_{max}$, with f_{max} evaluated by the formula:

$$f_{max} = 1.0002 \cdot rf_{max} \tag{4.24}$$

4.3.2 Universal Logical Systems

In logic there are quite a few widely studied and commonly used universal systems that have been used to design a wide range of logical functions or circuits. Why one is chosen and not another is most of the times a matter of efficiency, which mostly depends on the compactness of the solutions one can design with it.

By definition, a universal logical system is any set of logical functions that can be used to describe any other logical function. The most famous is the classical Boolean system of {NOT, AND, OR}, an extremely convenient system because we also think in terms of ANDs, ORs, and NOTs. But, in truth, this system is quite redundant, for it is possible to describe any logical function using just NOTs and ANDs or NOTs and ORs. So, if one creates a system, for instance, with ANDs and NOTs plus another function, say XOR, it will also be a universal logical system. In fact, such system does exist – the Reed-Muller system – as it seems to be advantageous in the design of certain logical circuits.

But perhaps the most important is understanding and exploring the most compact universal logical systems, that is, universal logical systems composed of just one basic function (say, any function of two or three arguments) plus any combination of the simple units {NOT, 0, 1}. This is, in fact, what constitutes what is called a universal logical module (ULM). Thus, for the sixteen functions of two arguments there are four well studied ULMs: {AND, NOT}, {OR, NOT}, {NAND}, and {NOR}. But there are also other less studied ULMs among the functions of two arguments: the systems {LT, NOT}, {GT, NOT}, {LOE, NOT}, and {GOE, NOT}. These systems are very interesting because they tell us something very important about the {NAND} and {NOR} systems: they are what they are (ULMs composed of just one unit), not because they are the inversion of AND or OR, both of them ULMs themselves, as this is no recipe for creating another ULM. In fact, by inverting LT one gets GOE and vice versa; and GT is the negation of LOE and vice versa. And neither of these functions are universal logical modules like NAND or NOR. Furthermore, besides the ones described above, among the functions of two arguments no more ULMs can be found or, more specifically, neither XOR nor NXOR are ULMs, and not even the combination of both these functions with the simple units {NOT, 0, 1} forms a universal logical system.

There is another concept I would like to introduce before starting to use all these interesting universal systems: the concept of a NAND/NOR-like module (NLM). The {NAND} and {NOR} systems are special in the sense that they do not require any other element (be it an inverter, 0, or 1) to describe any other logical function. And although they are certainly unique among the 16 functions of two arguments, there are lots of other functions with these same properties and it is important to have a name for them. For instance, among the 256 functions of three arguments there are a total of 56 NLMs, all of them listed in Table 4.8. Also shown are the parsimonious solutions to the NAND function designed with each one of these NLMs. For simplicity, the character "F" is used to represent all the NLMs in the K-expressions. The numbering scheme used for identifying the rules is the same proposed by Wolfram (1983) for naming the logical functions with three arguments in connection with one-dimensional cellular automata.

Table 4.8
NAND-like modules of three inputs and their respective NAND solutions.

Rule	Rule Table	NAND Solution (K-expression)	Size
1	00000001	FFFFaaFbbFFFFaaabbbaabaaabbb	28
3	00000011	FFFabFaFFbbbbbbbaab	19
5	00000101	FFaFFbFbbFabababbab	19
7	00000111	FFbaFbbbbb	10
9	00001001	FFaaFbabbb	10
11	00001011	FFaaFabaab	10
13	00001101	FFaaFabbab	10
17	00010001	FbFFbbFbFFabbbaaabb	19
19	00010011	FaFFaFFbFabaaabbaaa	19
21	00010101	FabFaaFaba	10
25	00011001	Fabb	4
27	00011011	Fabb	4
29	00011101	Fbaa	4
31	00011111	Fabb	4
33	00100001	FaFabFabaa	10
35	00100011	FbFbaFbaaa	10
37	00100101	Faba	4
39	00100111	Faba	4
41	00101001	FbFbaFbbbb	10
45	00101101	Faba	4
47	00101111	Fbab	4
49	00110001	FaFaaFbbbb	10
53	00110101	Faba	4
55	00110111	Fbab	4
57	00111001	Fbaa	4
59	00111011	Fbaa	4
61	00111101	Fabb	4
63	00111111	Fbaa	4
65	01000001	FbbFabFaaa	10
67	01000011	Fbba	4
69	01000101	FbbFbaFbbb	10
71	01000111	Faab	4
73	01001001	FFFababbbb	10
75	01001011	Fbba	4
79	01001111	Faab	4
81	01010001	FaaFbaFbaa	10
83	01010011	Fbba	4
87	01010111	Fbba	4
89	01011001	Fbaa	4
91	01011011	Fabb	4
93	01011101	Fabb	4
95	01011111	Fbba	4

Table 4.8 *Continued.*

Rule	Rule Table	NAND Solution (K-expression)	Size
97	01100001	FaFabFaaab	10
99	01100011	Faab	4
101	01100101	Faba	4
103	01100111	Faab	4
107	01101011	Fbba	4
109	01101101	Fbab	4
111	01101111	Fbba	4
115	01110011	Fbba	4
117	01110101	Faba	4
119	01110111	Fbba	4
121	01111001	Fbaa	4
123	01111011	Fbba	4
125	01111101	Fabb	4
127	01111111	Fbba	4

While building this list, I found it surprising not finding functions such as the widely used 3-multiplexer (MUX) (rule 172) and IF functions (if $a = 1$, then b, else c) (rule 202) among them. I later found out that, despite being ULMs, these functions are not NLMs: indeed, the MUX function can form a universal system with {MUX, NOT} and {MUX, 0, 1} and, therefore, is by definition a ULM; and the IF function can form a universal system with {IF, NOT} and {IF, 0, 1} and, therefore, is also a ULM.

These functions, the MUX and the IF functions, are two of the functions of three arguments that we are going to use as examples to design the most parsimonious representations using five different universal logical systems. But we will also use other functions of three arguments as examples: the already familiar majority function (rule 232) and the closely related minority function (rule 23); the more challenging even-parity (rule 105) and odd-parity (rule 150) functions; and four interesting NLMs: rule 39 ($a'b' + b'c$), rule 27 ($a' + b') \cdot (b' + c)$, rule 115 (If $a < b$, then ($a \cdot c$), else ($b \cdot c$)'), and rule 103 (If $a > b$, then ($a \cdot c$), else ($b \cdot c$)'). The majority and minority functions were chosen not only because of their simplicity but also because they are ULMs forming the following universal systems {MAJ, NOT, 0}, {MAJ, NOT, 1}, {MIN, 0}, and {MIN, 1}. The even-parity and odd-parity functions were chosen because they are hard to describe using most conventional universal systems such as the {NOT, AND, OR}, the {NAND}, or {NOR} systems (obviously it is elementary to describe these functions using the Reed-Muller system of NOTs, ANDs, and XORs).

Also as an illustration, we are also going to find the most parsimonious representations to all the interesting functions of two arguments (that is, AND, OR, NAND, NOR, LT, GT, LOE, GOE, XOR, and NXOR) for all the five universal systems chosen for this exercise: the classical Boolean system {AND, OR, NOT}, the {NAND} system, the {NOR} system, the Reed-Muller system {AND, XOR, NOT}, and the {MUX, 0, 1} system.

Boolean Logic

The Boolean universal system of {AND, OR, NOT} is the most popular of all universal systems and most of logic is based on it. There is nothing magical about this system, though, and the reason for its popularity and widespread use is the fact that it reflects our way of thinking: we can easily follow an argument described in terms of ANDs, ORs, and NOTs, but will have some difficulty following an argument described in terms of LTs and NOTs or even NANDs or NORs. Notwithstanding, this system is fairly compact as you can see by the sizes of the evolved solutions presented in Table 4.9.

The parsimonious solutions presented in Table 4.9 were discovered using small populations of 30 individuals. And in all cases, chromosomes composed of just one gene were used so that no constraints whatsoever are imposed on the architecture of the solutions.

For all the two-input functions, a head length of 10 was used, corresponding to maximum program length of 21 nodes, more than sufficient to realize all the functions of two arguments using ANDs, ORs, and NOTs. The fitness was evaluated by equation (3.8) prior to the discovery of a perfect solution, and after that by equation (4.21) in order to design more parsimonious configurations. And by letting the system run for a total of 20,000 generations it was ensured that the most parsimonious solution was indeed found.

For all the three-input functions, a larger head length of 20 was chosen (corresponding to maximum program length of 41) in order to cover for all kinds of functions, and a little more time (100,000 generations) was given for finding and pruning the perfect solutions.

By comparing the five universal systems in terms of compactness (see Tables 4.9 – 4.12), it is worth pointing out that, for the two-input functions, the {AND, OR, NOT} system is slightly less compact than the Reed-Muller system {AND, XOR, NOT} (46 against 43 total nodes) and also less compact in the three-input functions, even after removing the even- and odd-parity functions, giving a total of 71 nodes for the classical Boolean system and 69 for the Reed-Muller system.

Table 4.9

Parsimonious solutions designed with the universal system {AND, OR, NOT}.

Rule Name	Rule #	Rule Table	Parsimonious Solution	Size
AND	8	1000	Aab	3
OR	14	1110	Oab	3
NAND	7	0111	NAab	4
NOR	1	0001	NOab	4
LT	2	0010	ANba	4
GT	4	0100	AaNb	4
LOE	11	1011	ONba	4
GOE	13	1101	OaNb	4
XOR	6	0110	AONabAab	8
NXOR	9	1001	OANabOab	8
MUX	172	10101100	OAAacNba	8
IF	202	11001010	OAAabNca	8
MAJ	232	11101000	AOOaAbcbc	9
MIN	23	00010111	NAOOaAbcbc	10
EVEN	105	01101001	AOOAOANabNcONcOabAabab	22
ODD	150	10010110	OAAONAcaOAabbcOOAbacac	22
NLM39	39	00100111	ONAONcacb	9
NLM27	27	00011011	NAOOaNbcc	9
NLM115	115	01110011	NAObNca	7
NLM103	103	01100111	AONObANcbca	11

The comparison of the classical Boolean system with the MUX system is also interesting (see Table 4.13). The MUX system is considerably more compact with a total of just 82 nodes for the 10 functions of three arguments as opposed to 115 nodes required for the classical Boolean system; indeed, even after discounting the MUX and IF functions, the MUX system requires a total of 74 nodes to realize all the remaining functions, whereas the classical Boolean system requires a total of 99 nodes. For the functions of two arguments, though, the Boolean system is more compact with a total of 46 nodes, whereas the MUX system requires a total of 52 nodes.

It is worth emphasizing, however, that the goal of this study does not consist in discussing the compactness of the different logical systems (although for the functions of two arguments the discussion is as good as any as the entire set of relevant functions was used), but rather in showing that gene expression programming can be successfully used to discover the most parsimonious expressions irrespective of the universal system one chooses to work with, as the problems chosen for this study clearly illustrate.

Nand Logic

The NAND system is arguably the most important in industry as most digital circuits are built using NAND gates. It could easily have been the NOR system to get this privilege, but it just happened this way or perhaps the NAND system is overall slightly more compact than the NOR system (compare Tables 4.10 and 4.11). Indeed, on the two-input functions, these systems exhibit the same compactness, both requiring a total of 84 nodes. But on the three-input functions, the NAND system performs slightly better than the NOR system, with a total of 150 nodes as opposed to 158.

The parsimonious solutions presented in Table 4.10 were discovered using the same evolutionary strategies and the same chromosomal organizations used in the previous section, with the difference that now only NAND

Table 4.10
Parsimonious solutions designed with the NAND ("D") system.

Rule Name	Rule #	Rule Table	Parsimonious Solution	Size
AND	8	1000	DDDabab	7
OR	14	1110	DDDaabb	7
NAND	7	0111	Dab	3
NOR	1	0001	DDDDDDaaabbaa = DD1DDaabb	13
LT	2	0010	DDDDbDbabbb = DD1Dbab	11
GT	4	0100	DDDaDbDabbb = DD1aDab	11
LOE	11	1011	DaDab	5
GOE	13	1101	DDbaa	5
XOR	6	0110	DDDaDDbabab	11
NXOR	9	1001	DDDDDabaabb	11
MUX	172	10101100	DDDDbacab	9
IF	202	11001010	DDDabDcac	9
MAJ	232	11101000	DDDDbacDDabcc	13
MIN	23	00010111	DDDDDDDaaccacbb	15
EVEN	105	01101001	DDDDDDDDDacDDbbabbcDDacaacc	27
ODD	150	10010110	DDDDDDcabDDDDaDbDabDDacbcacbb	29
NLM39	39	00100111	DDDDDDcaabcbb	13
NLM27	27	00011011	DDDDDDcacbbac	13
NLM115	115	01110011	DDbaDcc	7
NLM103	103	01100111	DDDDDDcaDccbbab	15

was used in the function set, which, for simplicity, was represented by "D". It is worth pointing out that no other element such as zero or one were used in order to find the most parsimonious expressions to the functions presented in Table 4.10 and, as you can easily check by drawing the expression trees, the use of 1 would have made some of these solutions considerably more compact as a total of five nodes are required to express 1 using just NAND gates ((*a* nand *a*) nand *a*).

The attractiveness of the NAND system (and the NOR system, for that matter) resides not on its compactness (the classical Boolean system or the Reed-Muller system are much more compact, for example) but rather on the fact that just one kind of gate can be used to realize any kind of logical function, which, on practical terms, is a great advantage. But designing parsimonious solutions using just NAND (or NOR) gates is not as intuitive as using the classical Boolean functions of ANDs, ORs, and NOTs and, in fact, most NAND circuits are designed initially with ANDs, ORs, and NOTs and only later converted into NAND circuits by applying the De Morgan's theorems. But this is not the most efficient approach because it is possible to design much more compact expressions by using the NAND gate directly as your basic building block. And this is the reason why new techniques such as gene expression programming might be useful for helping in the design of even more compact and, consequently, faster digital circuits.

Nor Logic

The NOR system is, after the NAND system, the second most important system in industry. There are, however, specialists that claim that this system is better than the NAND system for a wide spectrum of applications, as there are functions for which the NOR system produces more compact solutions than the NAND system. But of course the reverse is also true and, overall, both systems are most probably very similar. However, there are cases for which the use of one system or the other might have drastic consequences in terms of processing time. This is one of the reasons why having the right tools to design and choose the most parsimonious configurations for a particular set of logical functions might bring large benefits.

The parsimonious solutions presented in Table 4.11 were created using the same evolutionary strategies and the same chromosomal organizations used in the previous sections, with the difference that now just the one function NOR was used in the function set, which, for simplicity, was represented by "R". Again note that no other element such as zero or one were

Table 4.11
Parsimonious solutions designed with the NOR ("R") system.

Rule Name	Rule #	Rule Table	Parsimonious Solution	Size
AND	8	1000	RRRaabb	7
OR	14	1110	RRRabab	7
NAND	7	0111	RRRRRRaaabbaa = RRORRaabb	13
NOR	1	0001	Rab	3
LT	2	0010	RaRbb	5
GT	4	0100	RRbaa	5
LOE	11	1011	RRRRbaRabaa = RRORbab	11
GOE	13	1101	RRRaRaRabaa = RR0aRab	11
XOR	6	0110	RRRabRRaabb	11
NXOR	9	1001	RRRaRRbbbaa	11
MUX	172	10101100	RRRabRcac	9
IF	202	11001010	RRRRbacab	9
MAJ	232	11101000	RRRaRbcRRabac	13
MIN	23	00010111	RRRRRRRabccacbb	15
EVEN	105	01101001	RRRRRRRaaRRRRbcRRbcabacbbcc	27
ODD	150	10010110	RRRRRRbRRacRRaRRcRRacabbbabbc	29
NLM39	39	00100111	RRRRRRcacbbac	13
NLM27	27	00011011	RRRRRRcaaccbb	13
NLM115	115	01110011	RRRaRRRabbbcc	13
NLM103	103	01100111	RRRRcRRRcbbccRbab	17

used in order to design these parsimonious solutions and, as happened with the NAND system where the use of 1 could have improved the compactness of three expressions, here the use of 0 would have also reduced the length of three expressions, as the zero element was discovered and used in the design of these parsimonious expressions. Indeed, as you can easily verify by drawing the expression trees of the solutions presented in Table 4.11, the zero module that was discovered and integrated in some of the parsimonious solutions makes use of a total of five nodes ((a nor a) nor a) that could have been easily replaced by just the one "0".

It is interesting to note that, for the two-input functions, the NOR system and the NAND system are exactly the same in terms of compactness of solutions (both add up to a total of 84 nodes) (compare Tables 4.10 and 4.11). However, for the three-input functions, the NAND system performs slightly better than the NOR system with a total of 150 nodes, whereas the NOR system requires a total of 158 nodes. This again shows that for each set of logical functions it is possible to find out which of the universal systems is more appropriate to design the digital circuits.

Reed-Muller Logic

The Reed-Muller system {AND, XOR, NOT} has strong adepts in the industry as there is a wide variety of applications where the use of this system is responsible for dramatic improvements in terms of processing time. Curiously enough, we will see that, for the functions used in this study, this system surpasses all others in terms of compactness.

The Reed-Muller system with ANDs, XORs, and NOTs is very similar to the classical Boolean system with ANDs, ORs, and NOTs in the sense that they both use the same kind and number of basic building blocks: two functions of two arguments and one function of one argument. And for gene expression programming, the algorithm we are going to use to design the most parsimonious representations, this is almost all that matters as this means that similar patterns will be followed to evolve the shape and size of all the candidate solutions. Hence, in this study, we are going to use exactly the same kind of chromosomal architectures and evolutionary strategies used in the classical Boolean system study.

The parsimonious solutions designed by gene expression programming using the Reed-Muller functions {AND, XOR, NOT} are shown in Table 4.12. And as you can see, the Reed-Muller system surpasses all the other systems in terms of compactness (compare especially with the classical Boolean system in Table 4.9 and the MUX system in Table 4.13). Indeed, for the functions of two arguments, of all the systems analyzed here, the Reed-Muller system is the system with better performance, requiring a total of 43 nodes, whereas the second best, the classical Boolean system (see Table 4.9), requires a total of 46 nodes. The MUX system is also fairly compact on the two-input functions, requiring a total of 52 nodes (see Table 4.13), thus more compact than the NAND and NOR systems, both requiring a total of 84 nodes (see Tables 4.10 and 4.11).

On the functions of three arguments, the results obtained with the Reed-Muller system are also interesting and, even after discounting the even- and odd-parity functions, this system still remains more compact than the classical Boolean system, requiring a total of 69 nodes to realize the remaining eight functions of three arguments, whereas the classical Boolean system needs an extra two nodes.

Table 4.12
Parsimonious solutions designed with the universal system {AND, XOR, NOT}.

Rule Name	Rule #	Rule Table	Parsimonious Solution	Size
AND	8	1000	Aab	3
OR	14	1110	XAbaNb	6
NAND	7	0111	NAab	4
NOR	1	0001	ANNab	5
LT	2	0010	ANba	4
GT	4	0100	AaNb	4
LOE	11	1011	NAaNb	5
GOE	13	1101	NANba	5
XOR	6	0110	Xab	3
NXOR	9	1001	NXab	4
MUX	172	10101100	XAbaXbc	7
IF	202	11001010	XAcaXbc	7
MAJ	232	11101000	XAcXXacbc	9
MIN	23	00010111	XNAAabXcab	10
EVEN	105	01101001	XNbXac	6
ODD	150	10010110	XXcab	5
NLM39	39	00100111	XaNAXcab	8
NLM27	27	00011011	NXAbXcab	8
NLM115	115	01110011	NANbAaNc	8
NLM103	103	01100111	XNAAbcANaNbc	12

Mux Logic

The MUX system is becoming extremely popular due to the new emerging technology of field programmable gate arrays (FPGAs). The advantage of this technology is that one is no longer restricted to the usual NAND or NOR gates, as virtually any kind of universal logical system can be used to build logical circuits. This is indeed the era of programmable gate arrays and evolvable hardware, and engineers all over the world are trying to find new systems that might prove more efficient in the design of digital circuits. So, being able to find the most parsimonious representation for any kind of logical function using any kind of universal system is essential for the advancement of this field, and now we know that gene expression programming is particularly well suited for doing just that as it handles all kinds of universal systems with the same efficiency.

We have seen already that the universal logical systems {MUX, NOT} and {MUX, 0, 1} are both universal, and for this study we are going to use the last one as 0's and 1's come very cheaply in digital circuits as they correspond to the ground and power. The parsimonious solutions presented in Table 4.13 were discovered using the same evolutionary strategies and the same head sizes used in the previous sections, but here, since we are using a three-argument function in the function set, the gene sizes are correspondingly larger. Thus, for the two-input functions, a gene size of 31 was used, whereas for the three-input functions, a gene size of 61 was used. For the sake of simplicity, in the K-expressions shown in Table 4.13, the MUX function is represented by the character "M", and zero and one are represented respectively by "0" and "1".

As you can see in Table 4.13, overall, the MUX system is extremely compact, requiring a total of 134 nodes to describe all the 20 functions. Indeed,

Table 4.13
Parsimonious solutions designed with the universal system {MUX, 0, 1}.

Rule Name	Rule #	Rule Table	Parsimonious Solution	Size
AND	8	1000	Mbba	4
OR	14	1110	Maba	4
NAND	7	0111	MM10bba	7
NOR	1	0001	MM10bab	7
LT	2	0010	Mab0	4
GT	4	0100	Mba0	4
LOE	11	1011	Ma1b	4
GOE	13	1101	Mb1a	4
XOR	6	0110	MabMba0	7
NXOR	9	1001	MaMbb1a	7
MUX	172	10101100	Mabc	4
IF	202	11001010	Macb	4
MAJ	232	11101000	MMcbacb	7
MIN	23	00010111	MM10Mbcabc	10
EVEN	105	01101001	MMMbacMb10ca0	13
ODD	150	10010110	MMcMabMc10ba0	13
NLM39	39	00100111	MM10cab	7
NLM27	27	00011011	MM10cba	7
NLM115	115	01110011	Mb1Mca0	7
NLM103	103	01100111	MM10bMccab	10

the Reed-Muller system is the only one that performs better than the MUX system in terms of compactness, requiring just 123 nodes to realize all the 20 functions (see Table 4.12). All the remaining systems are considerably more verbose than the MUX system, with the classical Boolean system requiring a total of 161 nodes (see Table 4.9), the NAND system a total of 234 (see Table 4.10), and the NOR system a total of 242 nodes (see Table 4.11).

So, perhaps among the logical functions of three arguments, not just the NLMs presented in Table 4.8 but also ULMs such as the MUX, the IF, the MAJ, and MIN functions analyzed here, there are other compact universal systems that can be exploited by the emergent FPGA technology in order to build more efficient logical circuits. And helping us making these decisions with amazing speed is the emerging technique of gene expression programming as the parsimonious solutions designed by GEP can be automatically translated into VHDL or Verilog and, therefore, can be immediately integrated into an FPGA design.

4.3.3 Using User Defined Functions as Building Blocks

Sometimes finding a good parsimonious representation to a complex function is so hard that the only way around it is trying to explore some pattern in its definition. Indeed, most such functions are modular in nature and, by finding simple modules and then use them as building blocks, it is possible most of the times to design good parsimonious solutions (by good parsimonious solutions it is meant solutions at least more parsimonious than the sum-of-products or product-of-sums solutions).

Gene expression programming can be used not only to discover patterns in logical functions by finding small building blocks, but also to integrate these building blocks in the design of more complex functions. And, in GEP, there are two different ways of designing more complex functions with smaller building blocks: the first one is totally automatic and the system not only discovers the building blocks but also uses them immediately to build more complex solutions (we will see how this is done in chapter 6, Automatically Defined Functions in Problem Solving); and the second, is much more prosaic and involves the well-known technique of using user defined functions (UDFs). How gene expression programming can be used to create complex structures composed of smaller UDFs is explained next with two modular problems: the odd-n-parity functions and the n-exactly-one-on functions. But let's first analyze how UDFs work and how they are implemented in GEP.

In gene expression programming, UDFs may represent functions of several variables although their arity in terms of expression rules is equal to zero. That is, nodes with UDFs behave as leaf nodes or terminals. So, the implementation of UDFs in gene expression programming can be done using at least two different methods: either they are treated as terminals and are used both in the heads and tails of genes or they are treated as functions and are used exclusively in the heads. Either way, the algorithm works very well. Thus, it is a matter of taste which one to choose. I chose the latter as it seems more consistent and less confusing. Consequently, UDFs are only used in the heads of genes, they can occupy the first positions in the genes of the individuals of the initial population, and they can also occupy the first position of an RIS element.

Let's now see how genes containing UDFs are expressed. Consider, for instance, the chromosome below:

```
012345678901234567890123456789012345678901234567890
AOOOdANUcAdOUcAAUOcddcbdaddaddcbdadbcdcbd
```
(4.25)

where "U" represents a user defined function. Its expression results in the following expression tree:

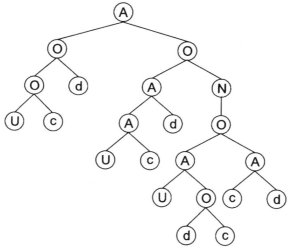

This program encodes a perfect solution to the odd-4-parity function using XOR as a UDF. The difference between this user defined XOR and a normal XOR is that the arguments to the user defined XOR are fixed during the definition of the function, whereas the arguments to the normal XOR are flexible and depend on the particular configuration on the expression tree. For instance, in the chromosome (4.25) above, the arguments to the UDF are by definition *a* and *b*, and by replacing "U" in the expression tree above by this user defined XOR, we obtain:

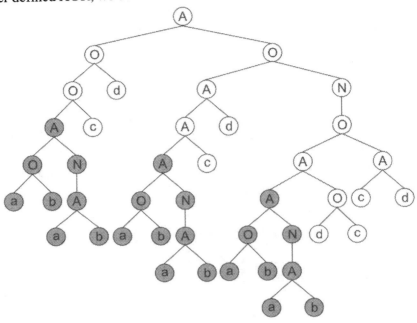

Compare this solution with the following solution to the odd-4-parity function discovered using a normal XOR instead of a rigid, user defined XOR:

```
0123456
XXXbcad
```
$$(4.26)$$

Nevertheless, UDFs are extremely useful, especially if they are carefully crafted to the problem at hand, for they can help us in the discovery of patterns in modular functions. The two problems we will be analyzing next clearly illustrate this point.

Odd-parity Functions

When expressed in binary form, a number is said to have odd parity if it has an odd number of 1's. And there is a class of so called odd-parity functions which return "1" or "true" if an *n*-bit number is odd, "0" or "false" otherwise. So, the simplest odd-parity function is the identity function of one argument or the odd-1-parity function. Another simple and familiar odd-parity function is the XOR function, which can also be called odd-2-parity function. We will start our exploration of odd-parity functions from here and then go on finding solutions to more complex odd-parity functions (up to odd-6-parity) by trying to discover and explore patterns in their design.

So, suppose we were using the classical Boolean system of ANDs, ORs, and NOTs, and we wanted to discover parsimonious solutions to a wide range of odd-parity functions (say, up to the odd-11-parity function). Of course we could try the direct approach and try to use gene expression programming to discover a solution to each of these functions with just ANDs, ORs, and NOTs in the function set, but we would have soon found out that these functions are really hard to design and it would be almost impossible to find a perfect solution to the higher order parity functions this way (Table 4.14). As you can see, tremendous resources were already necessary to solve the odd-4-parity function and higher order parity functions are even harder to design

Table 4.14
Parameters for the odd-n-parity problem without UDFs.

	Odd-2	Odd-3	Odd-4
Number of runs	100	100	100
Number of generations	50	10000	100000
Population size	30	30	30
Number of fitness cases	4	8	16
Function set	A O N	A O N	A O N
Terminal set	a b	a b c	a b c d
Head length	7	10	20
Gene length	15	21	41
Number of genes	2	2	2
Linking function	A	A	A
Chromosome length	30	42	82
Mutation rate	0.044	0.044	0.044
Inversion rate	0.1	0.1	0.1
IS transposition rate	0.1	0.1	0.1
RIS transposition rate	0.1	0.1	0.1
One-point recombination rate	0.3	0.3	0.3
Two-point recombination rate	0.3	0.3	0.3
Gene recombination rate	0.3	0.3	0.3
Gene transposition rate	0.1	0.1	0.1
Fitness function	Eq. (3.8)	Eq. (3.8)	Eq. (3.8)
Success rate	100%	68%	3%

and, consequently, very difficult to solve using this approach. So, a good alternative consists of trying to discover some kind of pattern in these functions and then use this information to design all the functions we need. Let's see how this works for the odd-parity functions.

We have already designed the most parsimonious solution to the XOR function using the function set F = {AND, OR, NOT} in the section Boolean Logic (see Table 4.9). As you can see by its expression below, it requires a total of eight nodes:

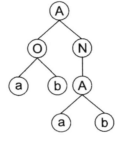

Now we could try and use this solution as a UDF to solve the next odd-parity function (the odd-3-parity function) in order to find out if there is some kind of pattern in their structure that we could exploit. As you can see in Table 4.15, the inclusion of this function as a UDF in the function set is responsible for a considerable increase in performance (notice that, in the absence of the UDF, not only the number of generations is 10 times higher but also the success rate is considerably lower). And this happens every time a lower order odd-parity function is used as a UDF to design a higher order function (compare Tables 4.14 and 4.15). And this most probably means that there is a pattern we can explore to design good parsimonious solutions to any parity function. Let's then see if we can find this pattern.

In order to do so, it helps a lot to be able to find the most parsimonious expressions so that all the unnecessary bits would disappear and the underlying order becomes clearly visible. For that an evolutionary strategy similar to the one used in section 4.3.2, Universal Logical Systems, is used. Thus, a flexible head length is chosen to allow an efficient evolution without restrictions regarding program size, and then, after a perfect solution has been found, a fitness function with parsimony pressure is applied in order to trim the perfect solution. The results of such an analysis are shown in Figure 4.9.

And as you can see in Figure 4.9, the pattern is perfectly clear and very simple, providing a simple and efficient way of designing any parity function one wishes for using just NOTs, ANDs, and ORs gates.

Table 4.15
Parameters for the odd-n-parity problem with UDFs.

	Odd-3	Odd-4	Odd-5	Odd-6
Number of runs	100	100	100	100
Number of generations	1000	1000	1000	1000
Population size	30	30	30	30
Number of fitness cases	8	16	32	64
Function set	A O N	A O N	A O N	A O N
User defined functions	Odd-2	Odd-3	Odd-4	Odd-5
Terminal set	a b c	a b c d	a b c d e	a b c d e f
Head length	7	7	7	7
Gene length	15	15	15	15
Number of genes	2	2	2	2
Linking function	A	A	A	A
Chromosome length	30	30	30	30
Mutation rate	0.044	0.044	0.044	0.044
Inversion rate	0.1	0.1	0.1	0.1
IS transposition rate	0.1	0.1	0.1	0.1
RIS transposition rate	0.1	0.1	0.1	0.1
One-point recombination rate	0.3	0.3	0.3	0.3
Two-point recombination rate	0.3	0.3	0.3	0.3
Gene recombination rate	0.3	0.3	0.3	0.3
Gene transposition rate	0.1	0.1	0.1	0.1
Fitness function	Eq. (3.8)	Eq. (3.8)	Eq. (3.8)	Eq. (3.8)
Success rate	98%	94%	79%	69%

Exactly-one-on Functions

This series of n-exactly-one-on functions also begins with the identity function (or 1-exactly-one-on function) and the XOR (or 2-exactly-one-on function). But this is where the similarity ends: the 3-exactly-one-on function is very different from the odd-3-parity function and if there is some pattern in the design of this series of functions one must again start by studying the structure of the 2-exactly-one-on function and all the other functions up in the hierarchy while using functions of lower order as UDFs.

The exactly-one-on functions are very interesting and challenging for two main reasons. The first one is that the sum-of-products solutions to these functions are fairly compact already as the number of 1's in the output bits is very small and grows very slowly according to the linear equation $y = n$, where y is the number of 1's in the output bits and n corresponds to the order

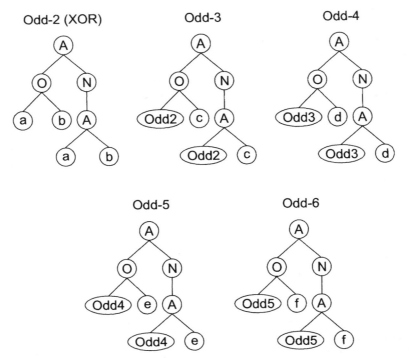

Figure 4.9. Finding patterns to design parsimonious solutions to odd-*n*-parity functions. Note that the simple structure that emerges here can be used to design parsimonious solutions to any odd-parity function.

of the function. The second reason is also related to the proportion of 0's and 1's in the output bits, more specifically, with the fact that the proportion of 0's increases exponentially according to the rule $y = 2^n - n$. And this poses a real problem not only in terms of measuring the fitness of the individuals, but also in finding ways out of local optima in solution space.

Consider, for instance, the 6-exactly-one-on function with a total of 64 transition states. In this case, only six cases are of the type "1", whereas 58 belong to the class "0". So, the possibilities are immense for candidate solutions that are good at realizing all this excess of "0" cases, but unfortunately that won't help our cause. Indeed, fitness functions based on the number of hits are unable to get populations out of local optima, but a fitness function based on the sensitivity/specificity indicators can, as it tends to minimize the number of false negatives. Indeed, such fitness functions are the key to dealing with such unbalanced sets of fitness cases, be it in logic synthesis or classification problems. So, for this problem, the fitness function will be therefore evaluated by equation (3.14).

Suppose we were again interested in finding a parsimonious solution to a higher order exactly-one-on function (say, 11-exactly-one-on function) in terms of ANDs, ORs, and NOTs. Again, we could try the direct approach, but as for the odd-parity functions, we would soon find out that these functions are also hard to design (Table 4.16). In fact, I was unable to find a perfect solution to the 9-exactly-one-on function using this approach. And indeed, not even the use of XOR in the function set could help me in finding a perfect solution to the problem at hand. The only alternative then consists of using UDFs as a way of identifying some underlying pattern in their design (Table 4.17).

And as you can see in Table 4.17, the use of lower order exactly-one-on functions as UDFs in the design of higher order exactly-one-on functions results in a much better performance (compare Tables 4.16 and 4.17), which

Table 4.16
Parameters for the *n*-exactly-one-on problem without UDFs.

	2-1On	3-1On	4-1On	5-1On	6-1On
Number of runs	100	100	100	100	100
Number of generations	50	1,000	10,000	10,000	100,000
Population size	20	30	30	30	30
Number of fitness cases	4	8	16	32	64
Function set	A O N	A O N	A O N	A O N	A O N
Terminal set	a b	a b c	a b c d	a b c d e	a b c d e f
Head length	10	10	10	10	10
Gene length	21	21	21	21	21
Number of genes	2	2	2	4	4
Linking function	A	A	A	A	A
Chromosome length	42	42	42	84	84
Mutation rate	0.044	0.044	0.044	0.044	0.044
Inversion rate	0.1	0.1	0.1	0.1	0.1
IS transposition rate	0.1	0.1	0.1	0.1	0.1
RIS transposition rate	0.1	0.1	0.1	0.1	0.1
One-point recombination rate	0.3	0.3	0.3	0.3	0.3
Two-point recombination rate	0.3	0.3	0.3	0.3	0.3
Gene recombination rate	0.3	0.3	0.3	0.3	0.3
Gene transposition rate	0.1	0.1	0.1	0.1	0.1
Fitness function	Eq. (3.14)	Eq. (3.14)	Eq. (3.14)	Eq. (3.14)	Eq. (3.14)
Success rate	98%	85%	50%	38%	39%

Table 4.17
Parameters for the n-exactly-one-on problem with UDFs.

	3-1On	4-1On	5-1On	6-1On
Number of runs	100	100	100	100
Number of generations	1,000	10,000	10,000	10,000
Population size	30	30	30	30
Number of fitness cases	8	16	32	64
Function set	A O N	A O N	A O N	A O N
User defined functions	2-1On	3-1On	4-1On	5-1On
Terminal set	a b c	a b c d	a b c d e	a b c d e f
Head length	10	10	10	10
Gene length	21	21	21	21
Number of genes	2	2	2	2
Linking function	A	A	A	A
Chromosome length	42	42	42	42
Mutation rate	0.044	0.044	0.044	0.044
Inversion rate	0.1	0.1	0.1	0.1
IS transposition rate	0.1	0.1	0.1	0.1
RIS transposition rate	0.1	0.1	0.1	0.1
One-point recombination rate	0.3	0.3	0.3	0.3
Two-point recombination rate	0.3	0.3	0.3	0.3
Gene recombination rate	0.3	0.3	0.3	0.3
Gene transposition rate	0.1	0.1	0.1	0.1
Fitness function	Eq. (3.14)	Eq. (3.14)	Eq. (3.14)	Eq. (3.14)
Success rate	98%	83%	85%	53%

means that there is indeed a pattern in the structure of these functions. Let's then see if we can find it.

Compared to the odd-parity functions, the structural pattern governing the design of the exactly-one-on functions is much harder to find, and being able to find the most parsimonious representations is therefore crucial in order to discern this pattern. For this task, we are going to use the already familiar strategy of letting the system evolve freely without constraints concerning the size of the evolving solutions, but as soon as a perfect solution is found, parsimony pressure is applied in order to create perfect solutions that are as compact as possible. Thus, before applying the parsimony pressure the fitness is evaluated by equation (3.14), whereas during the simplification process the fitness is evaluated by equation (4.21), where rf_i corresponds obviously to the sensitivity/specificity raw fitness.

Figure 4.10 shows the most parsimonious solutions designed using this strategy. And as you can see, there is indeed a pattern in the design of this class of exactly-one-on functions and this pattern can be easily exploited to build extremely parsimonious solutions to any higher order exactly-one-on function, such as the hard 11-exactly-one-on function.

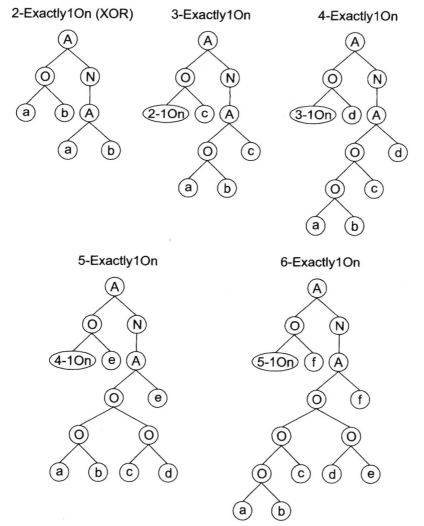

Figure 4.10. Finding patterns to design parsimonious solutions to *n*-exactly-one-on functions. Note that the simple structure that emerges here can be used to design parsimonious solutions to any exactly-one-on function.

4.4 Evolving Cellular Automata Rules for the Density-classification Problem

The density-classification task is a challenging problem, and different adaptive algorithms, namely, GAs and GP, have been used to try to evolve better than human-written rules. Gene expression programming was also used to evolve better rules for this task (Ferreira 2001), and two new rules were discovered which are not only better than any human-written rule but also better than the rules discovered both by the GA or GP. In this section, I describe these two GEP rules and how they were discovered.

The starting point for the density-classification task was the Gacs-Kurdyumov-Levin (GKL) rule, designed in 1978 by hand to study reliable computation and phase transitions in one-dimensional spatially extended systems (Mitchell 1996). Although not especially designed for the density-classification task, the GKL rule was for some time the rule with the best performance on this task. Its unbiased performance is shown in Table 4.18. In 1993, Lawrence Davis obtained a new rule by modifying by hand the GKL rule (Koza et al. 1999). This new rule, Davis rule, achieved an accuracy slightly better than the GKL rule (Table 4.18). Similarly, Rajarshi Das cleverly modified rules discovered by the GA, obtaining a new rule (Das rule) that performed slightly better than the GKL rule and the Davis rule (Koza et al. 1999). Genetic programming also discovered a new rule (GP rule) slightly better than the previous rules (Koza et al. 1999). And gene expression programming discovered two new rules (GEP$_1$ and GEP$_2$ rules) better than all the previous rules (Ferreira 2001). And finally, Juillé and Pollack (1998), using coevolutionary learning, discovered two new rules

Table 4.18
Accuracies measured at $N = 149$.

GKL rule	0.815
Davis rule	0.818
Das rule	0.823
GP rule	0.824
GEP1 rule	0.825
GEP2 rule	0.826
Coevolution1 rule	0.851
Coevolution2 rule	0.860

(Coevolution$_1$ and Coevolution$_2$) significantly better than all the previous rules. Their performances are also shown in Table 4.18.

Cellular automata (CA) have been studied widely as they are idealized versions of massively parallel, decentralized computing systems capable of emergent behaviors (for overviews of CA theory and applications see, e.g., Toffoli and Margolus 1987 and Wolfram 1986). These complex behaviors result from the simultaneous execution of simple rules at multiple local sites. In the density-classification task, a simple rule involving a small neighborhood and operating simultaneously in all the cells of a one-dimensional cellular automaton, should be capable of making the CA converge into a state of all 1's if the initial configuration (IC) has a higher density of 1's, or into a state of all 0's if the IC has a higher density of 0's.

4.4.1 The Density-classification Task

The simplest CA is a wrap-around array of N binary-state cells, where each cell is connected to r neighbors from both sides. The state of each cell is updated by a defined rule. The rule is applied simultaneously in all the cells, and the process is iterated for t time steps.

In the most frequently studied version of this problem, $N=149$ and the neighborhood is 7 (the central cell is represented by "u"; the $r = 3$ cells to the left are represented by "c", "b", and "a"; the $r = 3$ cells to the right are represented by "1", "2", and "3"). Figure 4.11 shows a CA with $N = 11$ in which the updated state for the cellular automaton "u" is shown.

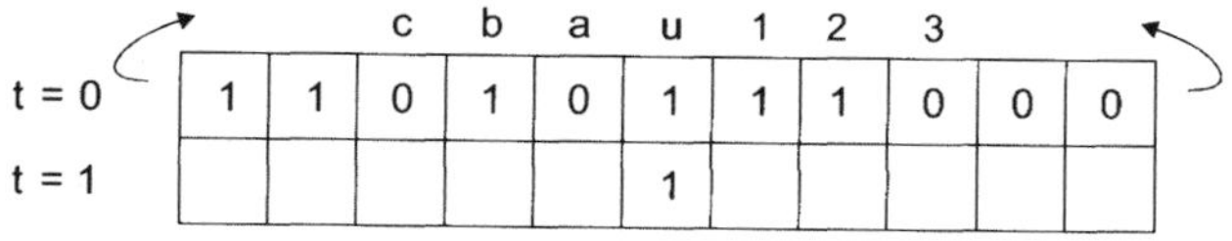

		c	b	a	u	1	2	3			
t = 0	1	1	0	1	0	1	1	1	0	0	0
t = 1					1						

Figure 4.11. A one-dimensional, binary-state, $r = 3$ cellular automaton with $N = 11$. The arrows represent the periodic boundary conditions. The updated state is shown only for the central cell.

The task of density-classification consists of correctly determining whether ICs contain a majority of 1's or a majority of 0's, by making the system converge, respectively, to an all 1's state (black or "on" cells in a space-time diagram), or to a state of all 0's (white or "off" cells). As the density of an IC is a function of N arguments, the actions of local cells with limited information and communication must be coordinated with one another in order to classify correctly the ICs. Indeed, to find, by hand, in a search space of 2^{128} transition states, rules that perform well, is an almost impossible task and, therefore, several algorithms were used to evolve better than human-written rules (Das et al. 1994; Juillé and Pollack 1998; Koza et al. 1999; Ferreira 2001). The best rules with performances of 86.0% (Coevolution$_2$) and 85.1% (Coevolution$_1$) were discovered using a coevolutionary approach between ICs and rules evolved by a GA (Juillé and Pollack 1998). The rules discovered by gene expression programming are better than all the human-written rules and better than the GP rule or the rules evolved by the GA (Mitchell et al. 1993 and Mitchell et al. 1994), and were discovered using computational resources that are more than 60,000 times smaller than those used by the GP technique.

4.4.2 Two Rules Discovered by GEP

In one experiment, F = {A, O, N, I} ("I" represents the already familiar IF function of three arguments) and T = {c, b, a, u, 1, 2, 3}. The parameters used per run are shown in the first column of Table 4.19.

Table 4.19
Parameters for the density-classification task.

	GEP1	GEP2
Number of generations	50	50
Population size	30	50
Number of ICs	25	100
Function set	A O N I	I M
Terminal set	c b a u 1 2 3	c b a u 1 2 3
Head length	17	4
Gene length	52	13
Number of genes	1	3
Linking function	--	I
Chromosome length	52	39
Mutation rate	0.038	0.051
One-point recombination rate	0.5	0.7
IS transposition rate	0.2	--
RIS transposition rate	0.1	--

The fitness was evaluated against a set of 25 unbiased ICs (i.e., ICs with equal probability of having a one or a zero at each cell). For this task, the fitness f_i of an individual program i was a function of the number of ICs n for which the system stabilized correctly to a configuration of all 0's or 1's after $2 \times N$ time steps, and it was designed in order to privilege individuals capable of correctly classifying ICs both with a majority of 1's and 0's. Thus, if the system converged, in all cases, indiscriminately to a configuration of 1's or 0's, only one fitness point was attributed; if, in some cases, the system correctly converged either to a configuration of 0's or 1's, $f_i = 2$; in addition, rules converging to an alternated pattern of all 1's and all 0's were eliminated, as they are easily discovered and disseminate very quickly through the populations halting the discovery of good rules; and finally, when an individual program could correctly classify ICs both with majorities of 1's and 0's, a bonus equal to the number of ICs C was added to the number of correctly classified ICs, being in this case $f_i = n + C$. For instance, if a program correctly classified two ICs, one with a majority of 1's and another with a majority of 0's, it received $2 + 25 = 27$ fitness points.

In this experiment a total of seven runs were made. In generation 27 of run 5, the following rule was discovered (only the K-expression is shown):

```
0123456789012345678901234 5678
OAIIAucONObAbIANIb1u23u3a12aa
```
(4.27)

This program (GEP$_1$ rule) has an accuracy of 0.82513 tested over 100,000 unbiased ICs in a 149×298 lattice, thus better than the 0.824 of the GP rule tested in a 149×320 lattice (Juillé and Pollack 1998, Koza et al. 1999). Its rule table is shown in Table 4.20. Figure 4.12 shows two space-time diagrams obtained with this new rule.

As a comparison, genetic programming used populations of 51,200 individuals and 1,000 ICs for 51 generations (Koza et al. 1999), thus a total of $51,200 \times 1,000 \times 51 = 2,611,200,000$ fitness evaluations were made, whereas gene expression programming only made $30 \times 25 \times 50 = 37,500$ fitness evaluations. Thus, at this task, gene expression programming not only discovered more and better rules than the GP technique, but also managed to do so using resources that surpassed the GP technique by a factor of 69,632.

In another experiment a rule with an accuracy of 0.8255, thus slightly better than the GEP$_1$ rule, was discovered. Again, its performance was evaluated over 100,000 unbiased ICs in a 149×298 lattice. In this case $F = \{I, M\}$ ("I" represents again the familiar IF function with three arguments and "M"

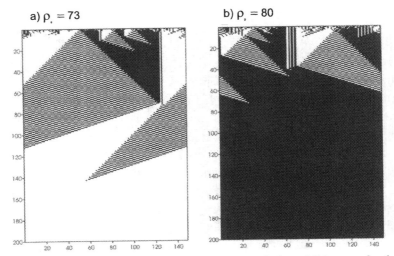

a) $\rho_t = 73$ b) $\rho_t = 80$

Figure 4.12. Space-time diagrams describing the evolution of CA states for the GEP$_1$ rule. The number of 1's in the IC (ρ_0) is shown above each diagram. In both cases the CA converged correctly to a uniform pattern.

represents the already familiar majority function with three arguments), and the set of terminals was obviously the same. In this case, a total of 100 unbiased ICs and three-genic chromosomes with sub-ETs linked by IF were used. The parameters used per run are shown in the second column of Table 4.19.

In this experiment, the fitness function was slightly modified by introducing a ranking system, in which individuals capable of correctly classifying between 2 ICs and 3/4 of the ICs received one bonus equal to C; if correctly

Table 4.20
Description of two rules discovered by GEP for the density-classification task. The output bits are given in lexicographic order starting with 0000000 and finishing with 1111111.

GEP1	00010001 00000000 01010101 00000000
	00010001 00001111 01010101 00001111
	00010001 11111111 01010101 11111111
	00010001 11111111 01010101 11111111
GEP2	00000000 01010101 00000000 01110111
	00000000 01010101 00000000 01110111
	00001111 01010101 00001111 01110111
	11111111 01010101 11111111 01110111

classified between 3/4 and 17/20 of the ICs received two bonus; and if correctly classified more than 17/20 of the ICs received three bonus. Also, in
this experiment, individuals capable of correctly classifying only one kind of
situation, although not indiscriminately, were differentiated and had a fitness equal to n.

By generation 43 of run 10, the following rule (GEP$_2$ rule) was discovered
(the sub-ETs are linked with IF):

```
0123456789012012345678901201234356789012
MIuua1113b21cMIM3au3b2233bM1MIacc1cb1aa                    (4.28)
```

This program (GEP$_2$ rule) has an accuracy of 0.8255 tested over 100,000
unbiased ICs in a 149×298 lattice, thus is better than the GEP$_1$ rule and also
better than the GP rule. Its rule table is shown in Table 4.20. Figure 4.13
shows two space-time diagrams obtained with this new rule.

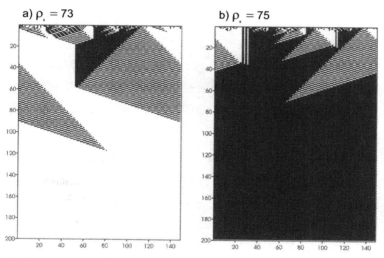

Figure 4.13. Space-time diagrams describing the evolution of CA states for the
GEP$_2$ rule. The number of 1's in the IC (ρ_0) is shown above each diagram. In both
cases the CA converged correctly to a uniform pattern.

In this chapter we have learned how to apply the basic gene expression
algorithm to very different problem domains, introducing, along the way, a
varied set of new tools that significantly enlarged the scope of the algorithm.
These new tools include new fitness functions for dealing with classification
problems with multiple outputs; new fitness functions with parsimony

pressure to create simpler and more intelligible programs; a new class of comparison functions; and a facility for handling user defined functions to create complex modular programs composed of simpler building blocks. In the next chapter we are going to enlarge the scope of gene expression programming by introducing the elegant structure developed to manipulate random numerical constants. This elegant structure is indeed the cornerstone of several new algorithms, including algorithms for polynomial induction (chapter 7), parameter optimization (chapter 8), decision tree induction (chapter 9), and total neural network induction (chapter 10).

5 Numerical Constants and the GEP-RNC Algorithm

Numerical constants are an integral part of most mathematical models and, therefore, it is important to allow their integration in the models designed by evolutionary techniques.

We know already that evolutionary algorithms handle functional building blocks somewhat unconventionally. So, it is not surprising that the simplest of all mathematical building blocks – numerical constants – can also be handled somewhat unconventionally.

In this chapter we are going to discuss different methods of handling numerical constants in evolutionary computation, using gene expression programming as a framework. We will start by introducing the simpler forms of handling numerical constants, followed by a detailed description of the GEP-RNC algorithm (GEP with the facility for handling random numerical constants) and all its components. The chapter finishes with a case study in which three different approaches to the problem of constants' creation are compared on four different problems: two simple computer generated problems and two real-world problems from two different fields (disease diagnosis and analog circuit design).

5.1 Handling Constants in Automatic Programming

It is assumed that the creation of floating-point constants is necessary to design mathematical models using evolutionary techniques (see, e.g., Koza 1992 and Banzhaf 1994). Most of these techniques, however, use the simple approach of using a fixed set of numerical constants that are then added to the terminal set, as no special facilities for handling random numerical constants were developed. But Koza also implemented a special facility for handling random numerical constants in genetic programming (Koza 1992). For

Cândida Ferreira: *Gene Expression Programming*, Studies in Computational Intelligence (SCI) **21**, 181–232 (2006)
www.springerlink.com

that, a special terminal named "ephemeral random constant" was introduced. For each ephemeral random constant used in the parse trees of the initial population, a random number of a special data type in a specified range is generated. Then these random numerical constants are moved around from tree to tree by the crossover operator.

Gene expression programming solves the problem of constants' creation differently (Ferreira 2001). This technique uses an extra terminal "?" and an extra domain Dc composed of the symbols chosen to represent the random numerical constants. For each gene, the RNCs are generated during the inception of the initial population and kept in an array. The values of each random constant are only assigned during gene expression. Furthermore, a special operator is used to introduce genetic variation in the available pool of random constants by mutating the random constants directly. In addition, the usual GEP operators (mutation, inversion, transposition, and recombination) plus a Dc-specific inversion and a Dc-specific transposition guarantee the effective circulation of the random constants in the population. Indeed, with this scheme of constants' manipulation, the appropriate diversity of random constants can be generated in the beginning of a run and maintained easily afterwards by the genetic operators.

But numerical constants can be created rather more simply. We have already seen in section 3.4, Solving a Simple Problem with GEP, how the basic gene expression algorithm creates integer constants by performing simple mathematical operations, such as x/x to create one or $(x-x)$ to create zero. Indeed, the simple problem of section 3.4 required the discovery of constants 2 and 3 in order to design a perfect solution and, as you would recall, the basic GEA had no problems in finding a perfect match for the target function (3.19). In fact, most problems can be solved similarly by creating numerical constants from scratch. But there are problems, however, for which the required numerical constants are not that easily created, and special facilities for handling them are absolutely necessary. We will analyze one such problem later in this chapter, the analog circuit design problem, where numerical constants play a crucial role in the design of good solutions.

Yet another simple and very popular – albeit not very efficient – form of creating numerical constants consists of adding a small number of numerical constants to the terminal set and then deal with them as extra terminals. For instance, the simple problem of section 3.4 could also be solved by using the terminal set T = {a, 0, 1, 2, 3} so that numerical constants could be directly integrated in the evolving solutions. Let's now analyze how this method compares with the simpler approach in which numerical constants are created from scratch by the algorithm itself.

In order to do so, we are going to conduct two experiments with 100 runs each and compare the performance of the two approaches by evaluating the success rate on the simple problem of section 3.4. We are going to use again the fitness function based on the selection range and precision (using an absolute error of 100 for the selection range and 0.01 for the precision) as this guarantees that all the solutions with maximum fitness match perfectly the target function (3.19). The performance of both experiments and the parameters used per run are summarized in Table 5.1. And as you can see, when the algorithm itself is used to create small integer constants from scratch, the performance is considerably higher than in the case where the constants

Table 5.1
Performance and settings used in the simple function problem without using numerical constants (**GEP**) and with a fixed set of NCs (**GEP-NC**).

	GEP	GEP-NC
Number of runs	100	100
Number of generations	50	50
Population size	30	30
Number of fitness cases	10 (Table 3.2)	10 (Table 3.2)
Function set	+ - * /	+ - * /
Terminal set	a	a 0 1 2 3
Head length	7	7
Gene length	15	15
Number of genes	3	3
Linking function	+	+
Chromosome length	45	45
Mutation rate	0.044	0.044
Inversion rate	0.1	0.1
IS transposition rate	0.1	0.1
RIS transposition rate	0.1	0.1
One-point recombination rate	0.3	0.3
Two-point recombination rate	0.3	0.3
Gene recombination rate	0.3	0.3
Gene transposition rate	0.1	0.1
Fitness function	Equation (3.3a)	Equation (3.3a)
Selection range	100	100
Precision	0.01	0.01
Success rate	68%	10%

are ready-made and easily available to integrate in the evolving models (68% success rate as opposed to 10%).

How can we explain these results? First of all, we already know that the creation of small integer constants can be easily achieved by performing simple arithmetic operations and, therefore, the GEP algorithm has no problems in creating them. And second, by including numerical constants in the terminal set, we are increasing, perhaps unnecessarily, the complexity of the problem, and this might be responsible for the observed decrease in performance in the GEP-NC approach.

The comparison of the structures of the sub-ETs created with the simpler approach (that is, without the explicit use of numerical constants) with the structures of the sub-ETs created by using integer constants in the terminal set can give us some pointers as to why the latter structures are more complicated (or perhaps less flexible) and, consequently, trickier to handle by the algorithm. Below are listed the first 10 perfect solutions discovered in the experiment summarized in the first column of Table 5.1 and all the perfect solutions (10 in total) obtained in the second experiment (the numbers in square brackets indicate, respectively, the run number and the generation by which they were discovered and the numbers in parentheses correspond to the total number of nodes in the sub-ETs):

```
GEP without NCs:
012345678901234012345678901234012345678901234
/+aa-a+aaaaaaaa/*+*a+-aaaaaaaa+a+/+-aaaaaaaa-[00,44] = 1000 (33)
-aaa---aaaaaaaa/+++*//aaaaaaaa+aa-/--aaaaaaaa-[01,39] = 1000 (21)
*/+a+*+aaaaaaaa/a++*/aaaaaaaaaa/+*/+aaaaaaaa-[02,41] = 1000 (15)
-a**-*aaaaaaaa+/a*aaaaaaaaaaa/*/aa+aaaaaaaa-[04,10] = 1000 (27)
*a/a+//aaaaaaaa/-+a-/-aaaaaaaa-+*-+-*aaaaaaaa-[05,16] = 1000 (39)
+aa/aa+aaaaaaaa/a-a/+aaaaaaaa*+/*-a+aaaaaaaa-[06,27] = 1000 (17)
+-+a/-+aaaaaaaa+a*//**aaaaaaaa-*a/*a+aaaaaaaa-[07,31] = 1000 (37)
+aa/---+aaaaaaaaaaaaaa/aaaaaaaa*/a*+aaaaaaaaaa-[10,23] = 1000 (13)
aaa**/+aaaaaaaa**//a*+aaaaaaaa+aa+/a*aaaaaaaa-[11,35] = 1000 (17)
*//a+*/aaaaaaaa*+a//**aaaaaaaa*//a///aaaaaaaa-[12,18] = 1000 (39)


GEP with NCs:
012345678901234012345678901234012345678901234
+0a223a0233aa12-33-2+-033aa312*a+1/+2a23111aa-[10,24] = 1000 (15)
+a0//a1a0aa0200*+a+2/0a2221011-3*31/003a000a3-[19,06] = 1000 (17)
-00++aaa001a03a*a+/*a221233302a+2021/3312a03a-[31,13] = 1000 (13)
/+/*a2a2311aa0001*a1-/13321a13*10a0003013a201-[37,12] = 1000 (13)
/*-/--3a2032010/*23a-3320a11a0*1*a/a223130a30-[59,22] = 1000 (25)
+-+a211322111aa*a++/2-a21112a2013-a/21a113022-[63,10] = 1000 (19)
```

```
*++//2a0aa22a33/0a231213122123*2aa1-103aa3a31-[72,11] = 1000 (17)
-*//-3-a2a02123-+/-1-3320331a0*3a-1*1a2a3201a-[81,45] = 1000 (27)
*a2a--aa30a333a*+//1a/a2aa032a/*a00*aa3a0331a-[84,38] = 1000 (19)
*a+3/a*2130a1200-2212100101aa0/*31*-0a310a2aa-[96,33] = 1000 (19)
```

And a remarkable thing can be straightaway observed just by computing the average program size of these perfect solutions in both experiments: in the first experiment, we get an average size of 25.8 against 18.4 in the second, which corresponds to a decrease of 28.7%. And this means that the solutions created in the second experiment are much more compact and, therefore, have much less room for neutral or redundant motifs to appear, which is indeed a serious handicap in evolutionary terms (see a discussion of The Role of Neutrality in Evolution in chapter 12). Consequently, in the second experiment, the algorithm has much more difficulties in creating and manipulating small building blocks and combining them in different arrangements to test how they work together. In other words, the GEP-NC system is much less flexible and consequently performs considerably worse in this kind of situation.

However, when the constants required to solve a problem are small floating-point constants, then the inclusion of numerical constants in the terminal set is advantageous and the algorithm performs considerably better when numerical constants are explicitly available. To demonstrate this, a simple polynomial function with rational coefficients was chosen:

$$y = 2.718x^2 + 3.141636x \tag{5.1}$$

For this study, a set of 10 random fitness cases chosen from the real interval [-10, 10] was used (Table 5.2). For both experiments, the same function set was used and consisted of F = {+, -, *, /}. For the basic GEA without numerical constants, the terminal set consisted obviously of the independent variable alone, giving T = {x}, whereas for the GEP-NC algorithm, a small set of five numerical constants chosen randomly from the real interval [0, 2] was used, giving T = {x, a, b, c, d, e}, where x corresponds to the independent variable, $a = 0.298$, $b = 1.083$, $c = 1.466$, $d = 0.912$, and $e = 1.782$. The fitness function was evaluated by equation (3.3b), with a selection range of 100% and maximum precision (0% error), giving $f_{max} = 1000$. This experiment, with its two approaches, is summarized in Table 5.3.

And as you can see in Table 5.3, the GEP-NC system performs considerably better than the simpler approach, with an average best-of-run fitness of 987.384 and an average best-of-run R-square of 0.999980229 (as opposed to

Table 5.2
Set of 10 random computer generated fitness cases used in
the polynomial function problem with rational coefficients.

x	$f(x)$
-8.5013	169.7278883
-0.8696	-0.676572453
3.7181	49.25514232
5.0878	86.34118911
-4.313	37.01043094
1.9775	16.84126999
-8.767	181.3638583
-5.5617	66.60191701
-1.4234	1.035098188
6.9014	151.1379353

an average best-of-run fitness of 934.534 and an average best-of-run R-square of 0.999919082).

It is also interesting to take a look at the best-of-experiment solutions discovered with both systems (see Figures 5.1 and 5.2). The one designed using the simpler approach was found in generation 395 of the second run, and has an R-square of 0.999994826 and a fitness of 896.542. Its chromosome is shown below (the sub-ETs are linked by addition):

```
01234567890120123456789012
/x-/-+xxxxxxx*++x/+xxxxxxx
```
                                                                    (5.2a)

And as you can see in Figure 5.1, it codes for the following function:

$$y = 3x^2 + 3.5x$$ (5.2b)

which is a slightly simplistic approximation to the target function (5.1). Curiously enough, this solution was found in 71 out of 100 runs, which suggests that this solution is a strong attractor in the particular fitness landscape explored in this experiment. This is perhaps the reason why the algorithm performed so poorly, and therefore a different fitness function or a different function set or a different chromosome structure would most probably be more appropriate to solve this particular problem.

The best-of-experiment solution designed with the GEP-NC algorithm was found in generation 63 of run 25 and has an R-square of 0.9999999999082

Table 5.3
Performance and settings used in the polynomial function problem without numerical constants (**GEP**) and with a fixed set of NCs (**GEP-NC**).

	GEP	GEP-NC
Number of runs	100	100
Number of generations	1000	1000
Population size	50	50
Number of fitness cases	10 (Table 5.2)	10 (Table 5.2)
Function set	+ - * /	+ - * /
Terminal set	x	x a b c d e
Head length	6	6
Gene length	13	13
Number of genes	2	2
Linking function	+	+
Chromosome length	26	26
Mutation rate	0.044	0.044
Inversion rate	0.1	0.1
IS transposition rate	0.1	0.1
RIS transposition rate	0.1	0.1
One-point recombination rate	0.3	0.3
Two-point recombination rate	0.3	0.3
Gene recombination rate	0.3	0.3
Gene transposition rate	0.1	0.1
Fitness function	Equation (3.3b)	Equation (3.3b)
Selection range	100%	100%
Precision	0.0%	0.0%
Average best-of-run fitness	934.534	987.384
Average best-of-run R-square	0.9999190824	0.9999802294

and a fitness of 983.813. Its chromosome is shown below (the sub-ETs are linked by addition):

```
0123456789012 0123456789012
+x**x+xabbcae**+x+*xabbcae
```
(5.3a)

And as you can see in Figure 5.2, it codes for the following function:

$$y = 2.762x^2 + 3.19258x \qquad (5.3b)$$

which is a pretty good approximation to the target function (5.1).

It is worth pointing out, however, that in real-world problems one never knows either the type or the range of the numerical constants that are needed

a.   `012345678901201234567890 12`
      `/x-/-+xxxxxxx*++x/+xxxxxxx`

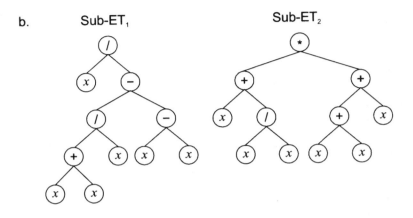

b.   Sub-ET₁   Sub-ET₂

c. $$y = \left(\frac{x}{2}\right) + (3x^2 + 3x) = 3x^2 + 3.5x$$

Figure 5.1. Best solution created in the experiment summarized in the first column of Table 5.3. This program has an R-square of 0.999994826 and was found in generation 395 of run 2. **a)** The chromosome of the individual. **b)** The sub-ETs codified by each gene. **c)** The corresponding mathematical expression after linking with addition (the contribution of each sub-ET is shown in brackets). Note that this model is a rather simplistic approximation to the target function (5.1).

to solve a problem, and choosing the right ones for a problem might be really complicated. So, a more flexible approach is required to handle a wider diversity of random numerical constants, and an elegant and efficient way of doing so is presented below.

5.2 Genes with Multiple Domains to Encode RNCs

A facility for handling random numerical constants can be easily implemented in gene expression programming. We have already met two different domains in GEP genes: the head and the tail. And now another one – the Dc domain – will be introduced. This domain was especially designed to handle

a. 012345678901201234567890 12
+x**x+xabbcae**+x+*xabbcae

b.

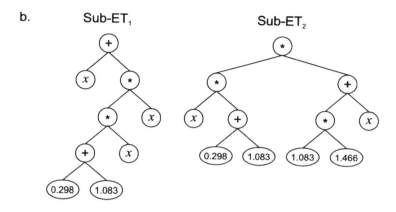

c. $y = (1.381x^2 + x) + (1.381x^2 + 2.19258x) = 2.762x^2 + 3.19258x$

Figure 5.2. Best solution created in the experiment summarized in the second column of Table 5.3. This program has an R-square of 0.9999999999082 and was found in generation 63 of run 25. **a)** The chromosome of the individual. **b)** The sub-ETs codified by each gene (the actual values of the constants a, b, and c are shown in the sub-ETs). **c)** The corresponding mathematical expression after linking with addition (the contribution of each sub-ET is shown in brackets). Note that this program is a very good approximation to the target function (5.1).

random numerical constants and consists of an extremely elegant, efficient, and original way of dealing with them.

Structurally, the Dc comes after the tail, has a length equal to t, and is composed of the symbols used to represent the random numerical constants. Therefore, another region with defined boundaries and its own alphabet is created in the gene.

For each gene the numerical constants are randomly generated at the beginning of a run, but their circulation is guaranteed by the usual genetic operators of mutation, inversion, transposition, and recombination. Furthermore, special Dc-specific operators such as mutation, inversion, and IS transposition, guarantee a more generalized shuffling of the numerical constants. And there is also a special mutation operator that allows the permanent introduction of variation in the set of random numerical constants throughout the learning process.

Consider, for instance, the simple chromosome below composed of just one gene with an $h = 7$ (the Dc is shown in bold):

```
0123456789012356789012
+?*+?**aaa??aaa68083295
```
(5.4)

where the terminal "?" represents the random numerical constants. The expression of this kind of chromosome is done exactly as before, obtaining:

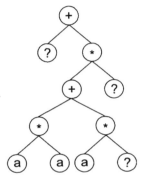

Then the ?'s in the ET are replaced from left to right and from top to bottom by the symbols (numerals, for simplicity) in the Dc, obtaining:

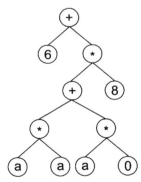

The values corresponding to these symbols are kept in an array. For simplicity, the number represented by the numeral indicates the order in the array. For instance, for the following 10 elements array of RNCs:

$$C = \{0.611, 1.184, 2.449, 2.98, 0.496, 2.286, 0.93, 2.305, 2.737, 0.755\}$$

the chromosome (5.4) above gives:

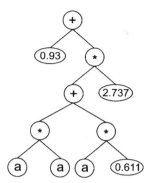

We will see in chapters 7 and 8 that genes encoding this type of domain can be used to great advantage in polynomial induction and in parameter optimization, as both these problems require large quantities or numerical constants. But this elegant structure can also be used to design decision trees with numeric attributes (see chapter 9) and to fine-tune the weights and thresholds of evolvable neural networks (see chapter 10).

5.3 Multigenic Systems with RNCs

The creation of a multigenic system in which the basic units are the complex genes of the previous section is very easy indeed, and can be achieved exactly as we did for the basic gene expression algorithm (see section 2.1.3, Multigenic Chromosomes). This means that each gene encodes a different sub-ET and that the sub-ETs are posttranslationally linked by a linking function, creating a more complex program composed of multiple sub-ETs.

Consider, for example, the following chromosome with length 34, composed of two genes, each with $h = 5$ and $n_{max} = 2$ and, therefore, with a gene length of 17 (the Dc's are shown in bold):

```
01234567890123456012345678 90123456
-//--?a?aa?313500/-*a-???a?a185516
```
(5.5)

and its arrays of random numerical constants:

C_1 = {0.699, 0.887, -0.971, -1.785, 1.432, 0.287, -1.553, -1.135, 0.379, 0.229}
C_2 = {1.446, -0.842, -1.054, -1.168, 1.085, 1.470, -0.241, -0.496, 0.194, 0.302}

where C_1 represents the set of RNCs of gene 1 and C_2 represents the RNCs of gene 2. This complex structure codes for two different sub-ETs, each incorporating a particular set of random numerical constants (Figure 5.3). Then these sub-ETs are as usual linked by a certain linking function, such as addition or multiplication, forming a much more complex program composed of smaller sub-programs (Figure 5.3 c).

It is worth emphasizing that this system – the GEP-RNC system – is considerably more complex than the basic gene expression algorithm. Notwithstanding, this system still performs with great efficiency, so much so that it

a. 012345678901234560123456789012345 6
    -//--?a?aa?**313500**/-*a-???a?a**185516**

    C_1 = {0.699, 0.887, -0.971, -1.785, 1.432, 0.287, -1.553, -1.135, 0.379, 0.229}
    C_2 = {1.446, -0.842, -1.054, -1.168, 1.085, 1.470, -0.241, -0.496, 0.194, 0.302}

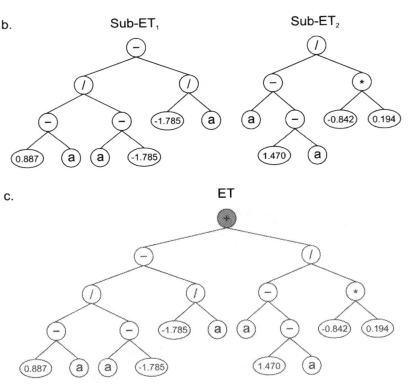

Figure 5.3. Expression of multigenic chromosomes encoding sub-ETs with random numerical constants. **a)** A two-genic chromosome with its arrays of random numerical constants. **b)** The sub-ETs codified by each gene. **c)** The result of posttranslational linking with addition (the linking function is shown in gray).

can indeed be used as the basis for creating yet more complex systems that require the swift handling of huge quantities of random numerical constants. We will describe these systems in chapters 7, 8, 9, and 10, but for now let's see how this complex system manages to fine-tune the random numerical constants in order to find good solutions to the problems in hand.

5.4 Special Search Operators for Fine-tuning the RNCs

The efficient evolution of such complex entities composed of different domains and different components requires a special set of genetic operators. The operators of the basic gene expression algorithm (mutation, inversion, transposition, and recombination) are easily transposed to the GEP-RNC system, but obviously the boundaries of each domain must be maintained and the different alphabets must be used within the confines of the corresponding domain. Mutation was as expected extended to the Dc domain, although I decided to keep it separated from the mutation operator that controls the rate of mutation in the head/tail domains in order to allow not only a better control but also a better understanding of this important operator. The inversion operator was also transposed to the GEP-RNC algorithm, with the primary inversion operator being restricted to the heads of genes and a new Dc-specific inversion restricted obviously to the Dc-domain. Both IS and RIS transposition were transposed to the GEP-RNC algorithm and their action continues to be obviously restricted to the heads and tails of genes. However, a special transposition operator was created that operates within the Dc alone and also helps with the circulation of the RNCs in the population. Furthermore, a special mutation operator – direct mutation of random numerical constants – was also created in order to directly introduce genetic variation in the sets of RNCs; this way a constant flux of new numerical constants is maintained throughout the adaptive process.

The extension of all forms of recombination to the GEP-RNC algorithm is straightforward as their actions never result in mixed domains or alphabets, and they are also very effective not only at reshaping the expression trees but also contribute to the fine-tuning of the numerical constants. Gene transposition was also extended to the GEP-RNC algorithm and, like recombination, its implementation is straightforward as it never gives rise to mixed domains or alphabets. However, for it to be productive, the set of RNCs attached to the transposing gene must also go with it.

5.4.1 Dc-specific Mutation

By controlling separately the point mutations that take place in the head/tail domains from the point mutations that occur in the Dc domain, we can better understand the importance of this operator in the fine-tuning of the numerical constants. But in practice, when all the genetic operators are working together, I usually use the same mutation rate both for the head/tail domains and the Dc domain. Thus, the fact that they are controlled separately is almost irrelevant and one could choose to just simply extend the basic mutation operator to the Dc domain. Let's now analyze how this operator works and how it contributes to the fine-tuning of the random numerical constants.

Consider, for instance, the two-genic chromosome below with $h = 5$ and its arrays of random numerical constants:

```
0123456789012345601234567890123456
+/?**aaaa?a086601-?++???a??a435968                    (5.6)
```

C_1 = {0.879, -0.244, -1.204, 1.212, -0.524, -1.178, 1.919, -0.763, 1.970, 1.369}
C_2 = {-1.716, -0.642, -1.298, -1.155, 0.205, -1.022, 0.615, -1.441, 0.872, -1.396}

Now suppose that two point-mutations occurred in the Dc domain, changing the "6" at position 14 in gene one into "2" and the "5" at position 13 in gene two into "7", giving:

```
0123456789012345601234567890123456
+/?**aaaa?a086201-?++???a??a437968                    (5.7)
```

As you can see in Figure 5.4, this new chromosome codes for a new mathematical expression, which is slightly different from the one encoded in the parental chromosome because a different numerical constant is used in the new sub-ET$_2$. It is worth pointing out, however, that a different constant is used not because the RNCs themselves have changed but because the symbol "5" at position 13 in gene 2 changed into "7", pointing to a different address in the array of random numerical constants. More specifically, in the parental sub-ET$_2$ the instruction says to use the numerical constant at position 5 (the constant -1.022), whereas in the daughter sub-ET the instruction says to use the numerical constant at position 7 (the constant -1.441).

It is worth noticing that the first mutation in gene one is an example of a neutral mutation, as it has no effect whatsoever in the expression tree. In fact, in this particular sub-ET, only the first position of the Dc is being expressed, with all the rest of this domain being part of a relatively large

a.
```
0123456789012345601234567890123456
+/?**aaaa?a086601-?++???a??a435968
+/?**aaaa?a086201-?++???a??a437968
```

C_1 = {0.879, -0.244, -1.204, 1.212, -0.524, -1.178, 1.919, -0.763, 1.970, 1.369}
C_2 = {-1.716, -0.642, -1.298, -1.155, 0.205, -1.022, 0.615, -1.441, 0.872, -1.396}

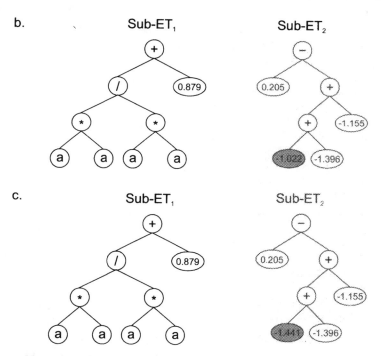

Figure 5.4. Illustration of Dc-specific mutation. **a)** The mother and daughter chromosomes and their random numerical constants (the RNCs are exactly the same for both mother and daughter and therefore are shown just once). **b)** The sub-ETs encoded by the mother chromosome (before mutation). **c)** The sub-ET encoded by the daughter chromosome (after mutation). The nodes affected by mutation are highlighted. Note that the first mutation is an example of a neutral mutation as it has no expression in the sub-ET.

noncoding region. So, similarly to what we have seen on the head/tail domain, the Dc domain also has the potential to have noncoding sequences at its end. And we already know that this kind of noncoding sequences in the genome plays an important role in evolution as these regions are ideal places for the accumulation of neutral mutations (for a discussion of The Role of Neutrality in Evolution see chapter 12).

5.4.2 Dc-specific Inversion

The Dc-specific inversion is similar to the inversion operator that works in the heads of genes, with the difference that it is restricted to the Dc domain. Thus, this operator randomly chooses the chromosome, the gene with its respective Dc to be modified, and the start and termination points of the sequence to be inverted. Typically, a small inversion rate of 0.1 is used as this operator is seldom used as the only source of genetic variation.

Consider, for instance, the two-genic chromosome below with $h = 5$:

```
0123456789012345601234567890123456
*+/+/a?a??a407309+++*+??a???737256
```
(5.8)

and its arrays of random numerical constants:

C_1 = {1.227, 1.361, 0.512, 1.467, 1.666, 1.901, -1.583, 1.338, 0.620, -1.180}
C_2 = {-0.286, -1.551, -1.165, 0.783, -1.872, 0.064, 1.152, -1.066, 1.343, 1.851}

Suppose now that the sequence "4073" in gene one was chosen to be inverted, giving:

```
0123456789012345601234567890123456
*+/+/a?a??a370409+++*+??a???737256
```
(5.9)

As you can see in Figure 5.5, the program encoded in this new chromosome is slightly different from the one encoded in the parent chromosome. Note that, as a result of this event of inversion, not only a new numerical constant (the constant 1.467) is incorporated in the daughter tree but also a different combination of old numerical constants is tested (constants 1.338 and 1.277).

5.4.3 Dc-specific Transposition

The Dc-specific transposition is similar to the IS transposition that operates in the head/tail domain of genes, with the difference that it is restricted to the Dc domain. Thus, this operator randomly chooses the chromosome, the gene with its respective Dc to be subjected to transposition, the start and termination points of the transposon, and the target site; then it moves the transposon from the place of origin to the target site. Typically, a small transposition rate of 0.1 is used as this operator is usually used together with other, more powerful genetic operators.

a.   `0123456789012345601234567890123456`
      `*+/+/a?a??a407309+++*+??a???737256`
      `*+/+/a?a??a370409+++*+??a???737256`

C_1 = {1.227, 1.361, 0.512, 1.467, 1.666, 1.901, -1.583, 1.338, 0.620, -1.180}
C_2 = {-0.286, -1.551, -1.165, 0.783, -1.872, 0.064, 1.152, -1.066, 1.343, 1.851}

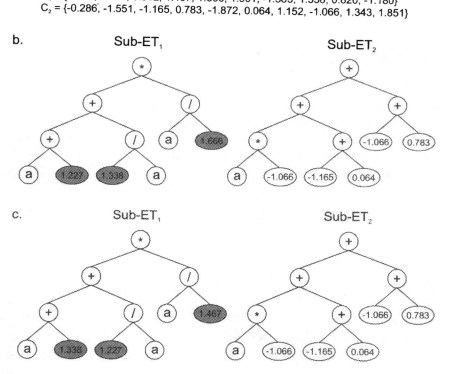

Figure 5.5. Illustration of Dc-specific inversion. **a)** The mother and daughter chromosomes with their random numerical constants (the RNCs are exactly the same for both mother and daughter and therefore are shown just once). **b)** The sub-ETs encoded by the mother chromosome (before inversion). **c)** The sub-ETs encoded by the daughter chromosome (after inversion). The nodes affected by inversion are highlighted. Note that not only a new constant is expressed in the daughter tree but also different interactions are tested for two of the old constants.

Consider, for instance, the two-genic chromosome below with $h = 5$:

`0123456789012345601234567890123456`
`+--+?aa?aaa655501-a+?-?aaa??577250`   (5.10)

and its arrays of random numerical constants:

$C_1 = \{-0.273, -0.376, -1.089, -1.544, -0.668, 1.969, -1.978, -1.192, -1.930, 0.053\}$
$C_2 = \{1.499, -0.346, 0.391, -0.232, 0.307, 1.586, -0.300, 0.729, -0.885, -0.1897\}$

Suppose now that the sequence "250" in gene 2 was chosen as a transposon and that the insertion site was bond 1 in Dc (between positions 10 and 11). Then the following chromosome is obtained:

```
0123456789012345601234567890123456
+--+?aa?aaa655501-a+?-?aaa??250772
```
(5.11)

Note that, before moving to the insertion site, the transposon is deleted at the place of origin, thus maintaining the Dc length.

As you can see in Figure 5.6, these two chromosomes encode slightly different solutions because the random numerical constants were moved around by the transposition operator and, as a result, not only a new constant was tested (the constant 0.391) but also different interactions were tested for old constants; in this case, the constant 1.586 was moved around in sub-ET_2.

5.4.4 Direct Mutation of RNCs

We have already seen that all the genetic operators contribute, directly or indirectly, to move the random numerical constants around. And, in fact, this permanent shuffling of RNCs is more than sufficient to allow an efficient evolution of good solutions as the appropriate number of RNCs can be generated at the beginning of each run. Notwithstanding, we can also implement a special mutation operator that replaces the value of a particular numerical constant by another.

This operator, called direct mutation of RNCs, randomly selects particular targets in the arrays where the RNCs are kept, and randomly generates a new numerical constant.

Consider, for instance, the two-genic chromosome below with $h = 5$ and its arrays of random numerical constants:

```
0123456789012345601234567890123456
**a//aa?a??461226*a+*?a?aaaa406961
```
(5.12)

$C_1 = \{0.139, -0.299, -1.024, -0.330, \mathbf{0.510}, -1.864, 1.008, -0.712, -1.740, 1.552\}$
$C_2 = \{-0.986, -0.147, -1.113, -1.577, 0.210, \mathbf{0.218}, 1.705, \mathbf{-0.770}, 1.845, 1.954\}$

Now suppose that a mutation occurred at position 4 in C_1, changing the constant 0.510 occupying that position by -0.256, and that two other mutations

a.
```
01234567890123456012345678 90123456
+--+?aa?aaa655501-a+?-?aaa??577250
+--+?aa?aaa655501-a+?-?aaa??250772
```

C_1 = {-0.273, -0.376, -1.089, -1.544, -0.668, 1.969, -1.978, -1.192, -1.930, 0.053}
C_2 = {1.499, -0.346, 0.391, -0.232, 0.307, 1.586, -0.300, 0.729, -0.885, -0.1897}

b.

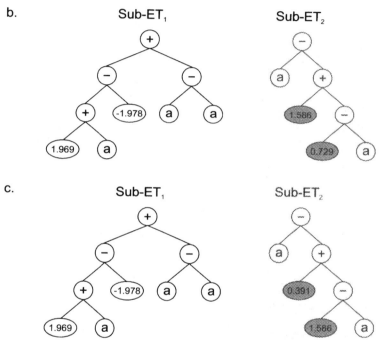

Figure 5.6. Illustration of Dc-specific transposition. a) The mother and daughter chromosomes with their random numerical constants (the RNCs are exactly the same for both mother and daughter and therefore are shown just once). b) The sub-ETs encoded by the mother chromosome (before transposition). c) The sub-ETs encoded by the daughter chromosome (after transposition). The nodes affected by transposition are highlighted. Note that not only a new constant is expressed in the daughter tree but also a different interaction is being tested for constant 1.586.

occurred in C_2, changing the number 0.218 at position 5 by 0.853 and –0.770 at position 7 by –0.256, giving:

```
01234567890123456012345678 90123456
**a//aa?a??461226*a+*?a?aaaa406961
```
(5.13)

C_1 = {0.139, -0.299, -1.024, -0.330, **-0.256**, -1.864, 1.008, -0.712, -1.740, 1.552}
C_2 = {-0.986, -0.147, -1.113, -1.577, 0.210, **0.853**, 1.705, **-0.256**, 1.845, 1.954}

As you can see in Figure 5.7, these two individuals represent slightly different solutions because one of the mutated constants happened to be part of
sub-ET$_1$ and, as a result, a slightly different program is generated. Note that
the two mutations that occurred in C$_2$ have no effect whatsoever in the

a.   0123456789012345601234567890123456
       **a//aa?a??461226*a+*?a?aaaa406961-[m]

C$_{m1}$ = {0.139, -0.299, -1.024, -0.330, **0.510**, -1.864, 1.008, -0.712, -1.740, 1.552}
C$_{m2}$ = {-0.986, -0.147, -1.113, -1.577, 0.210, **0.218**, 1.705, **-0.770**, 1.845, 1.954}

       0123456789012345601234567890123456
       **a//aa?a??461226*a+*?a?aaaa406961-[d]

C$_{d1}$ = {0.139, -0.299, -1.024, -0.330, **-0.256**, -1.864, 1.008, -0.712, -1.740, 1.552}
C$_{d2}$ = {-0.986, -0.147, -1.113, -1.577, 0.210, **0.853**, 1.705, **-0.256**, 1.845, 1.954}

b.   Sub-ET$_1$ Sub-ET$_2$

c.   Sub-ET$_1$ Sub-ET$_2$

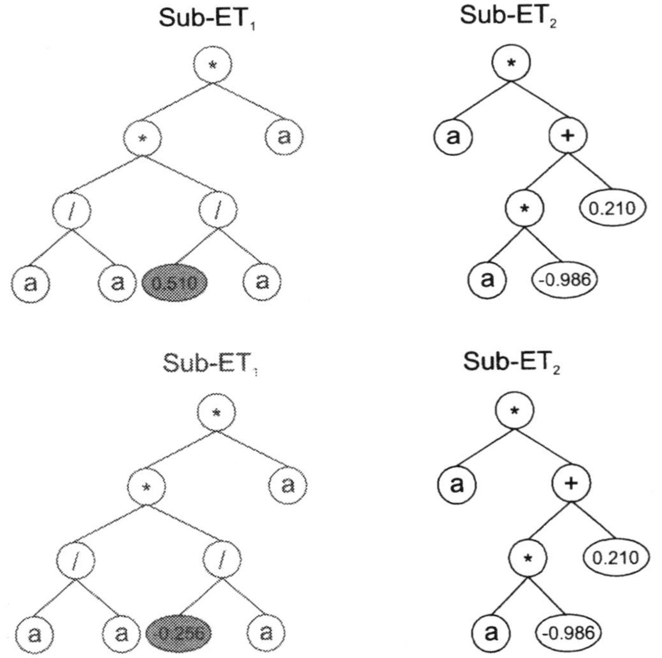

Figure 5.7. Illustration of direct mutation of random numerical constants. **a)** The
mother and daughter chromosomes with their arrays of RNCs. **b)** The sub-ETs
encoded by the mother chromosome (before mutation). **c)** The sub-ETs encoded by
the daughter chromosome (after mutation). The nodes affected by this event of
mutation are highlighted. Note that, of the three point mutations, just one of them
(the substitution of 0.510 by -0.256) left its mark in the expression tree; the other
two are neutral as they are not being used to build the sub-ETs.

expression of sub-ET$_2$, as these constants are not used to build this sub-ET. This is yet another example of how the GEP system explores the idea of neutrality not only to build the core structure of the expression trees but also to fine-tune this same structure and all its elements.

On the other hand, it is also worth pointing out that the direct mutation of random numerical constants might also have a great impact in the overall solution. This might occur whenever the mutated constant happens to be used more than once in a particular sub-ET.

Interestingly, the direct mutation of RNCs seems to have a very limited impact on the creation of good models and better results are indeed obtained when this operator is switched off. Therefore, we can conclude that a well-dimensioned initial diversity of RNCs is more than sufficient to allow their evolutionary tuning, as they are constantly being moved around by all the genetic operators. Typically, per gene, I use an array length of 10 RNCs for Dc lengths equal to or less than 20. For larger Dc domains we could increase the number of elements but, even for bigger structures, an array of length 10 seems to be more than enough.

5.5 Solving a Simple Problem with GEP-RNC

In this section we are going to analyze a successful run in its entirety in order to understand how populations of such complex entities adapt, not only by changing the size and shape of their structures but also by fine-tuning the numerical constants that integrate them.

The simple problem we are going to solve using the GEP-RNC algorithm is the same simple problem we solved in section 3.4 with the basic gene expression algorithm. In this case, however, we are going to use just one gene per chromosome so that both the chromosomal structure and the corresponding array of random numerical constants could fit side-by-side to allow a better visualization of the evolutionary process.

For this relatively simple function requiring integer constants, we are going to use a set of 10 integer random constants and represent them by the numerals 0-9, that is, R = {0, 1, 2, 3, 4, 5, 6, 7, 8, 9}, so that the numeral corresponds to the position each random constant occupies in the array of RNCs. And the ephemeral random constants "?" will be drawn from the integer interval [0, 9]. The fitness will be evaluated by equation (3.3a), using an absolute error of 100 for the selection range and a precision for the error equal to 0.01. Thus, for the 10 fitness cases used in this problem (again the training samples shown in Table 3.2), f_{max} = 1000. The complete list of the parameters used per run is shown in Table 5.4.

The evolutionary dynamics of the successful run we are going to analyze is shown in Figure 5.8. And as you can see, in this run, a perfect solution was found in generation 7.

The initial population of this run and the fitness of each individual in the particular environment of Table 3.2, is shown below (the best of generation is shown in bold):

```
Generation N: 0
01234567890123456789012
//*++++aa??a???82565470-{7,9,3,8,7,8,0,2,7,5}-[ 0] = 820.4595
*//?a++?a?a???a28979219-{8,7,5,9,3,8,5,7,0,2}-[ 1] = 824.7692
```

Table 5.4
Parameters for the simple symbolic regression problem.

Number of generations	50
Population size	20
Number of fitness cases	10 (Table 3.2)
Function set	+ - * /
Terminal set	a ?
Random constants array length	10
Random constants range	Integer interval [0, 9]
Head length	7
Gene length	23
Number of genes	1
Chromosome length	23
Mutation rate	0.044
Inversion rate	0.1
IS transposition rate	0.1
RIS transposition rate	0.1
One-point recombination rate	0.3
Two-point recombination rate	0.3
Dc-specific mutation rate	0.044
Dc-specific inversion rate	0.1
Dc-specific transposition rate	0.1
Random constants mutation rate	0.01
Fitness function	Equation (3.3a)
Selection range	100
Precision	0.01

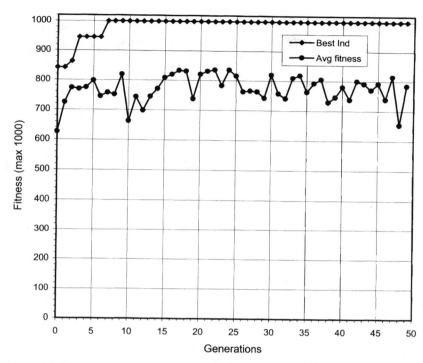

Figure 5.8. Progression of average fitness of the population and the fitness of the best individual for a successful run of the experiment summarized in Table 5.4.

```
-/?a/a/?aaaa?a?93035855-{2,8,0,8,1,7,6,8,5,8}-[ 2] = 707.3825
-/?+?*??aaaaa?a29276817-{3,2,5,6,0,4,3,6,6,9}-[ 3] = 781.2165
+*a-a/a?a??????73341268-{8,2,1,8,5,6,6,4,5,0}-[ 4] = 505.3105
+*++++/????a?aa00881894-{2,4,5,4,0,6,3,4,2,9}-[ 5] = 828.4936
-+/+++*aa???aaa80314936-{3,6,3,5,6,7,0,2,0,2}-[ 6] = 719.5594
//*//?/a???a??a60017613-{2,9,2,5,0,1,9,0,9,9}-[ 7] = 0
*/a?*++aa?a?aa?42294067-{7,2,2,9,3,2,8,6,1,6}-[ 8] = 823.6289
-a-+a+/??a?????76298538-{6,5,4,6,8,9,1,5,3,2}-[ 9] = 748.9699
***/-/-?aa?a?a?63343489-{8,2,6,4,2,6,5,1,2,3}-[10] = 501.2423
*+*/+a*?aa??aa?19397639-{3,3,1,6,3,1,6,9,1,2}-[11] = 188.0554
/++-/++??????a?47231525-{6,0,0,7,0,7,0,4,8,4}-[12] = 831.1588
-**/+*a??aaa??a22306510-{5,2,9,8,2,2,2,9,5,1}-[13] = 286.3837
+?++/--?aaa??a?80472239-{2,2,5,9,5,8,9,8,9,2}-[14] = 838.3493
--**/a-aaaaaaa?10288804-{0,9,3,6,3,7,6,7,0,2}-[15] = 735.4132
*++*a/*aaaaa???08586041-{6,1,8,2,1,0,1,4,2,2}-[16] = 229.4964
//-?++*?a?aaa?a47461780-{2,6,1,4,9,7,8,8,6,6}-[17] = 821.8963
*--?++aaaa??a?a73345287-{2,7,5,3,8,8,9,8,1,2}-[18] = 500.0852
+/?aa*-??????aa70225088-{6,4,2,7,3,1,6,7,1,0}-[19] = 842.5013
```

The expression of the best individual of this generation (chromosome 19) and the corresponding mathematical expression are shown in Figure 5.9. Note how simplistic this individual is, not only structurally but also functionally: indeed, it encodes the constant expression $y = 8$ and I doubt the perfect solution found later in the run will be a direct descendant of this best-of-generation solution (indeed, the next best-of-generation has nothing in common with this solution as the lack of homology both in the chromosome sequence and the array of RNCs indicates). Notwithstanding, this individual is the best of this initial population and therefore will be cloned unchanged into the next generation.

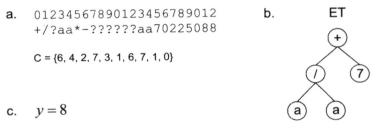

a.   01234567890123456789012
      +/?aa*-??????aa70225088

      C = {6, 4, 2, 7, 3, 1, 6, 7, 1, 0}

c. $y = 8$

b.   ET

Figure 5.9. Best individual of generation 0 (chromosome 19). It has a fitness of 842.5013 evaluated against the set of fitness cases presented in Table 3.2. **a)** The chromosome of the individual with its random numerical constants. **b)** The fully expressed individual, which means that the constants themselves and not their symbols are shown in the expression tree; coincidentally, in this case, the numeral "7" in the Dc corresponds to the numerical constant 7. **c)** The corresponding mathematical expression.

In the next generation there was no improvement in best fitness and the best two individuals of this generation are the clone of the best individual of the initial population and a major variant of this individual (both these chromosomes are shown in bold):

```
Generation N: 1
01234567890123456789012
+/?aa*-??????aa70225088-{6,4,2,7,3,1,6,7,1,0}-[ 0] = 842.5013
+*++++/????a?aa00881894-{2,4,5,4,0,6,3,4,2,9}-[ 1] = 828.4936
//*?+++aa??a???70216547-{7,9,3,8,7,8,0,2,7,5}-[ 2] = 821.6634
//*++++aa??a???82565470-{7,9,3,8,7,8,0,2,7,5}-[ 3] = 820.4595
-a-+a+??aaaaa??76298438-{6,5,4,6,8,9,1,5,3,2}-[ 4] = 760.5252
+*++++/??a?????76298538-{2,4,5,4,0,6,3,4,2,9}-[ 5] = 623.6609
-/?+?*/??a????a29276817-{6,0,0,7,0,7,0,4,8,4}-[ 6] = 0
*--?++aaaa???aa70225088-{2,7,5,3,8,8,9,8,1,2}-[ 7] = 634.9871
```

```
+?++/--?aaa??a?80472239-{2,2,5,9,1,8,9,8,9,2}-[ 8]  =  822.3493
/++-/++??????a?45251327-{6,0,0,7,0,7,0,4,8,4}-[ 9]  =  833.5443
-//+/a+??????a?47231525-{3,2,5,6,0,4,3,6,6,9}-[10]  =  840.076
+++/???????a?aa00881894-{2,4,5,4,0,6,3,2,2,9}-[11]  =  840.8194
*--?++aaaa??a?a75347287-{2,7,5,3,8,8,9,8,1,2}-[12]  =  174.7606
***/-/-?aa?a?aa63433489-{8,2,6,4,2,6,5,1,2,3}-[13]  =  821.5259
-/?a/a/?aa?a?a93035855-{2,8,0,8,1,7,6,8,5,8}-[14]  =  707.3825
+/?aa*-?a????aa87022508-{6,4,2,7,3,1,6,7,1,0}-[15]  =  829.5259
+/???aa?????a?a73345287-{6,4,2,7,3,1,6,7,1,0}-[16]  =  842.5013
++-++?/????a?a?00881094-{2,4,5,4,0,6,3,4,2,9}-[17]  =  829.926
+?-/?a??aaaaa?a29276817-{3,2,5,6,0,4,3,6,6,9}-[18]  =  792.8509
*//?a++?a?a???a28979219-{8,7,5,9,3,8,5,7,0,2}-[19]  =  824.7692
```

Note that chromosome 16, one of the best of this generation, encodes exactly the same expression as the clone created by elitism (chromosome 0). However, these chromosomes share little homology in terms of gene sequence, probably due to a major event of recombination coupled with other modifications that occurred during reproduction.

In the next generation a new individual was created, chromosome 5, considerably better than the best individuals of the previous generations. Its expression is shown in Figure 5.10. The complete population is shown below (the best of generation is shown in bold):

```
Generation N: 2
012345678901234567890012
+/???aa?????a?a73345287-{6,4,2,7,3,1,6,7,1,0}-[ 0]  =  842.5013
+++/???????a?aa18088194-{2,4,5,4,0,6,3,2,2,9}-[ 1]  =  842.2943
-//+/a+??????a?00881094-{3,2,5,6,0,4,3,6,6,9}-[ 2]  =  839.591
+*//?a+?a?a??a?45251379-{8,7,5,9,3,8,5,7,0,2}-[ 3]  =  664.1066
+?++/--?aaa??a?80472237-{2,2,5,9,1,8,9,8,9,2}-[ 4]  =  822.3493
/+?++*+??aa????76298538-{6,0,0,7,0,7,0,4,8,4}-[ 5]  =  863.8036
++-++?/????a?a?93181094-{2,4,5,4,0,6,3,4,2,9}-[ 6]  =  765.1267
+++-/++??????a?45251327-{2,4,5,4,0,6,3,4,2,9}-[ 7]  =  840.079
+/?aa*-?a????aa87022507-{6,9,2,7,3,1,6,7,1,0}-[ 8]  =  829.5259
++-++?/????a?a?00881094-{2,4,5,4,0,6,3,4,2,9}-[ 9]  =  829.926
-/?a/*/?aa?a?a93035855-{2,8,0,8,1,7,6,8,5,8}-[10]  =  754.9694
*//?a++?a?a???a28979219-{8,7,5,9,3,8,5,7,0,2}-[11]  =  824.7692
/++-/++???a??a?45251327-{6,0,0,7,0,7,0,4,8,4}-[12]  =  817.4693
++?+?a/????a?aa00886894-{2,4,5,4,0,6,3,2,2,9}-[13]  =  836.6383
+?a???*??a?????76298538-{2,4,5,4,0,6,3,4,2,9}-[14]  =  834.6383
++-++*/????a?a?23152235-{2,4,5,4,0,6,3,4,2,9}-[15]  =  797.9468
//*++++aa??a???82565470-{7,9,3,8,7,8,0,2,7,5}-[16]  =  820.4595
+/???aa?????a?a73324287-{6,4,2,7,3,1,6,7,1,0}-[17]  =  842.5013
-/?a/a/?aa?a?a?00835855-{2,8,0,8,1,7,6,8,5,8}-[18]  =  800.2763
/++-/++???????a28912328-{6,0,0,7,0,7,0,4,8,4}-[19]  =  0
```

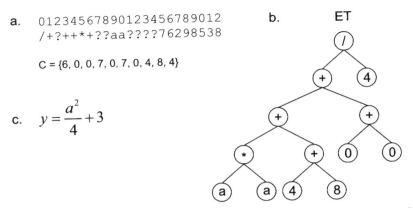

a. 01234567890123456789012
    /+?++*+??aa????76298538

    C = {6, 0, 0, 7, 0, 7, 0, 4, 8, 4}

c. $y = \dfrac{a^2}{4} + 3$

b.   ET

Figure 5.10. Best individual of generation 2 (chromosome 5). It has a fitness of 863.8036 evaluated against the set of fitness cases presented in Table 3.2. **a)** The chromosome of the individual with its random numerical constants. **b)** The fully expressed individual, meaning that the constants themselves and not their symbols are shown in the ET. **c)** The corresponding mathematical expression.

It is worth pointing out that the structure of this best-of-generation solution is totally unrelated to the best-of-generation of the previous populations (compare Figures 5.9 and 5.10). Indeed, as we had predicted, the best of the initial population had no potential to evolve into the perfect solution and, as expected, the best of generation 2 comes from a different lineage, this one with real potential to evolve into the perfect solution.

In the next generation a new individual was created, chromosome 17, considerably better than the best individual of the previous generation. Its expression is shown in Figure 5.11. The complete population is shown below (the best of generation is shown in bold):

```
Generation N: 3
01234567890123456789012
/+?++*+??aa????76298538-{6,0,0,7,0,7,0,4,8,4}-[ 0] = 863.8036
-/?a/*/?a??a?aa87022507-{2,8,0,8,1,7,6,8,5,8}-[ 1] = 0
+-/++/a??????a?20881094-{3,2,5,6,0,4,3,6,6,9}-[ 2] = 791.4108
++-++a/????a?a?52235152-{2,4,5,4,0,6,3,4,2,9}-[ 3] = 828.7938
-+-+++/????a?a?23152235-{2,4,5,4,0,6,3,4,2,9}-[ 4] = 804.6464
+/??aaa?a????aa87092507-{6,9,2,7,3,1,6,7,1,0}-[ 5] = 782.5773
//*++++aa??a???81688194-{7,9,3,8,7,8,0,2,7,5}-[ 6] = 821.5616
+++/?+?????a?aa13052470-{2,4,5,4,0,6,7,2,2,9}-[ 7] = 833.6277
//*++++aa??a??a73345257-{7,9,3,8,7,8,0,2,7,5}-[ 8] = 821.8415
```

```
/+?+/*+?a?a?????76296538-{6,0,0,7,0,7,0,4,8,4}-[ 9]  = 825.5259
++-++?/????a?a?93181094-{2,4,5,4,0,6,3,4,2,9}-[10]  = 765.1267
+/???aa?????a??82565470-{6,4,2,7,3,1,6,7,1,0}-[11]  = 833.5259
//*++*+aa??a???82565470-{7,9,3,8,7,8,0,2,7,5}-[12]  = 821.5384
+*//?a+?a?a??a?45251379-{8,7,5,9,3,7,5,7,0,2}-[13]  = 669.5308
*//?a++??aa???a28979219-{8,7,5,9,3,8,5,7,0,2}-[14]  = 722.7797
*//?a++?a?a???a79219799-{8,7,5,9,3,8,5,7,0,2}-[15]  = 763.8067
+++-/++????a?a?00881327-{2,4,5,4,0,6,3,4,2,9}-[16]  = 834.5455
*+/+a++??aa?????76298530-{6,0,0,7,0,7,0,4,8,4}-[17]  = 945.5585
+/?aa*-?aa???a?93035855-{6,9,2,7,3,1,6,7,1,0}-[18]  = 825.5259
+a?/a*-?a????aa87027052-{6,9,2,7,3,1,6,7,1,0}-[19]  = 822.5909
```

Note that the best of this generation and the best of the one before, despite coding for very different expressions, not only share some sequence homology (for instance, they differ just at one position in the Dc) but also share the same set of random numerical constants. This most probably means that the best program of this generation is a daughter of the best individual of the previous generation.

a.   0123456789 0123456789012
      *+/+a++??aa?????76298530

   C = {6, 0, 0, 7, 0, 7, 0, 4, 8, 4}

c. $y = \dfrac{a^2}{2} + 2a$

b.   ET

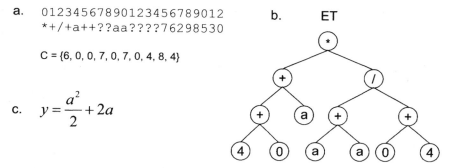

Figure 5.11. Best individual of generation 3 (chromosome 17). It has a fitness of 945.5585 evaluated against the set of fitness cases presented in Table 3.2. **a)** The chromosome of the individual with its random numerical constants. **b)** The fully expressed individual, meaning that the constants themselves and not their symbols are shown in the ET. **c)** The corresponding mathematical expression.

In the next three generations there was no improvement in best fitness, and the best individuals of all the three populations are all of them clones of the best individual of generation 3; in fact, all of them were generated by elitism (these chromosomes are highlighted in all the populations):

Generation N: 4
012345678901234567890012
+/+a++??aa????76298530-{6,0,0,7,0,7,0,4,8,4}-[0] = 945.5585
+-/??*+aa???a??82565470-{3,2,5,6,0,4,3,6,6,9}-[1] = 870.9714
*//*a++??aa???a91297982-{8,7,5,9,3,8,5,7,0,2}-[2] = 835.8194
a??/a++????a?a?00881327-{6,9,2,7,3,1,6,7,1,0}-[3] = 816.5909
+/?aa*-?aa???a?93035420-{6,9,2,7,3,1,6,7,1,0}-[4] = 825.5259
+++-+aa?a????a?20925007-{2,4,5,4,0,6,3,4,2,9}-[5] = 838.4936
+-??/++??????aa70881094-{3,2,5,6,0,4,3,6,6,9}-[6] = 841.0194
++-++a/????a?a?53235152-{2,4,5,4,0,6,3,4,2,9}-[7] = 830.7938
//*++++aa??a???81988194-{7,9,3,8,7,8,0,2,7,5}-[8] = 821.1328
*//?a++?a?a???a79219799-{8,7,5,9,3,8,5,7,6,2}-[9] = 763.8067
//*++aa????a???82565470-{7,9,3,8,7,8,0,2,7,5}-[10] = 762.2184
+?++*+/??aa????76296538-{6,0,0,7,0,7,0,4,8,4}-[11] = 0
-++//++????a?a?00881327-{2,4,5,4,0,6,3,4,2,9}-[12] = 780.873
+a?/**-?a????aa87027052-{6,9,2,7,3,1,6,6,1,0}-[13] = 822.5909
+++/?+?????a?aa13057825-{2,4,5,4,0,6,7,2,2,9}-[14] = 835.6277
++-++?/????a?a?93181094-{2,4,5,4,0,6,3,4,2,9}-[15] = 765.1267
+*//?a+?a?a??a?45651379-{8,7,5,9,3,7,5,7,0,2}-[16] = 669.5308
a++++-/????a?a?20881327-{2,4,5,4,0,6,3,4,2,9}-[17] = 816.5909
+/?aa*-?aa???a?93035855-{6,9,2,7,3,1,6,7,1,0}-[18] = 825.5259
/a?*+++aa??a??a73345257-{7,9,3,8,7,8,0,2,7,5}-[19] = 821.3651

Generation N: 5
012345678901234567890012
+/+a++??aa????76298530-{6,0,0,7,0,7,0,4,8,4}-[0] = 945.5585
+/?aa*-?a??a???48198819-{6,9,2,7,3,1,6,7,1,0}-[1] = 835.8194
a??/a++?????a??82565470-{6,9,2,7,3,1,6,7,1,0}-[2] = 816.5909
*//++aa???aa?a?20881327-{7,9,3,8,7,8,0,2,7,5}-[3] = 796.6234
+/?aa*-?aa???a?93335420-{6,9,2,7,3,1,6,7,1,0}-[4] = 825.5259
+-/??*+???a???a91297987-{3,2,5,6,0,4,3,6,6,9}-[5] = 718.8821
a??/a++???a?a?00881327-{6,9,2,7,3,1,6,7,1,0}-[6] = 816.5909
+/?aa?aa???a?93065855-{6,9,2,7,3,1,6,7,1,0}-[7] = 821.6173
+-??/++????a??a70781094-{3,2,5,6,0,4,3,6,6,9}-[8] = 841.0194
//++*++aa??a?a?81488194-{7,9,3,8,7,8,0,2,7,5}-[9] = 821.8661
a++++-/??????a?82765470-{2,4,5,4,0,6,3,4,2,9}-[10] = 816.5909
?+/+/++??a?a???76298530-{6,0,0,7,0,7,0,4,8,4}-[11] = 835.8194
//*++++aa??a???81988194-{7,9,3,8,7,8,0,2,7,5}-[12] = 821.1328
++-++a/????a?a?75232351-{2,4,5,4,0,6,3,4,2,9}-[13] = 830.4538
+*//?a+?a?a??a?45156379-{8,7,5,9,3,7,5,7,0,2}-[14] = 727.4235
//*++++?aa???a?93035420-{7,9,3,8,7,8,0,2,7,5}-[15] = 810.5616
++-/??*aa?a?a?00881327-{3,2,5,6,0,4,3,6,6,9}-[16] = 548.4261
*//*a++aa???a??82565402-{8,1,5,9,3,8,5,7,0,9}-[17] = 738.0844
+++-+aa?a?a??a?20925007-{2,4,5,4,0,6,3,4,2,9}-[18] = 832.7814
+/+aa*-?aa???a?93035420-{6,9,2,7,3,1,6,7,1,0}-[19] = 760.5252

```
Generation N:  6
012345678901234567890

*+/+a++??aa????76298530-{6,0,0,7,0,7,0,4,8,4}-[ 0] = 945.5585
*+/+a++a?aa????20925007-{6,0,0,7,0,7,0,5,8,4}-[ 1] = 849.9752
/+/+?++??a?a????76298530-{6,0,0,7,0,7,0,4,8,4}-[ 2] = 808.5244
-++++a/??????a?82765470-{2,4,5,4,0,6,3,4,2,9}-[ 3] = 0
+++-+aa?a?a??a?20925007-{2,4,5,4,0,6,3,4,2,9}-[ 4] = 832.7814
//*++++aa?aa???81025420-{7,9,3,8,7,8,0,2,7,5}-[ 5] = 820.239
a??/a++???a??a?00881327-{6,9,2,7,3,1,6,7,1,0}-[ 6] = 816.5909
+/*--???aa???aa97628530-{6,9,2,7,3,1,6,7,1,0}-[ 7] = 809.5608
/++-++a???a?a?75232351-{2,4,5,4,0,6,3,4,2,9}-[ 8] = 818.898
*/*++++?aa???a?90988194-{7,9,3,8,7,8,0,2,7,5}-[ 9] = 382.0706
/+-++a/????a?a?75235321-{7,9,3,8,7,8,0,2,7,5}-[10] = 706.7844
a??/a++?????a??82565471-{6,9,2,7,3,1,6,7,1,0}-[11] = 816.5909
//++*++aa??a?a?81488194-{7,9,3,8,7,8,0,2,7,5}-[12] = 821.8661
//++*++aa??a?a?81488194-{7,9,3,8,7,8,0,2,7,5}-[13] = 821.8661
*a??a*/aa????a?32565402-{8,1,5,9,3,8,5,7,0,9}-[14] = 611.6435
+/?aa*-?aa???a?94533320-{6,9,2,7,3,1,6,7,1,0}-[15] = 825.5259
//++*++?a??a?a?81488174-{7,9,3,8,7,8,0,2,7,5}-[16] = 821.5576
?/??a*-?aa???a?93335420-{6,9,2,7,3,1,6,7,1,0}-[17] = 821.5259
+/*++++?a????a?40354220-{2,4,5,4,0,6,3,4,2,9}-[18] = 725.656
+++-+aa?a?a??a?93335420-{2,4,5,4,0,6,3,6,2,9}-[19] = 834.5455
```

Finally, by generation 7 an individual with maximum fitness (chromosome 1), matching exactly the target function (3.19), was created. Its expression is shown in Figure 5.12. The complete population is shown below (the best of generation is shown in bold):

```
Generation N:  7
012345678901234567890

*+/+a++??aa????76298530-{6,0,0,7,0,7,0,4,8,4}-[ 0] = 945.5585
*+/+a++??aa????20925007-{6,0,0,7,0,7,0,5,8,4}-[ 1] = 1000
+/?aa*-?aa???a?94533394-{6,9,2,7,3,1,6,7,1,0}-[ 2] = 825.5259
*+/+a++a?aa????76298530-{6,0,0,7,0,7,0,4,8,4}-[ 3] = 0
+aa???/?aa???aa67928530-{6,9,2,7,3,1,6,7,1,0}-[ 4] = 804.8522
*+/+a*+??aa????76298530-{6,0,0,7,0,7,0,4,8,4}-[ 5] = 530.2093
*+/+a+a??aa????76458530-{6,0,0,7,0,7,0,4,8,4}-[ 6] = 834.5455
//++*++?a??a?a?81488174-{2,9,3,8,7,8,0,2,7,5}-[ 7] = 821.5576
?a???a/?aa???a?93335820-{6,9,2,7,3,1,6,7,1,0}-[ 8] = 821.5259
//*++++aa?aa???81025074-{7,9,3,8,7,8,0,2,7,5}-[ 9] = 820.239
*/*++++?aa???a?90988120-{7,9,3,8,7,8,0,2,7,5}-[10] = 382.0706
+++/+aa?aaaa???81025420-{2,4,5,4,0,6,3,6,2,9}-[11] = 755.3276
++/+/+???a?a????76298530-{6,0,0,7,0,7,0,4,8,4}-[12] = 829.846
+/++++a?a??????40154220-{2,4,5,4,0,6,3,4,2,9}-[13] = 838.4415
```

```
?/?aa*-?aa???a?29453350-{6,9,2,7,3,1,6,7,1,0}-[14]  = 829.5259
//++*++aaa?a?a?81488174-{7,9,3,8,7,8,4,2,7,5}-[15]  = 822.359
+a++*+/??aa????76298530-{6,0,0,7,0,7,0,4,8,4}-[16]  = 789.711
/++?++a????a?a?32322331-{2,4,5,4,0,6,3,4,2,9}-[17]  = 836.1209
//*++++aa?a??aa93330420-{7,9,3,8,7,8,0,2,7,5}-[18]  = 822.1262
+++-+aa?a?a??a?93335420-{2,4,5,4,0,6,3,6,2,9}-[19]  = 834.5455
```

Note that the best individual of this generation (chromosome 1) encodes exactly the same structure as the best of the previous generation (compare Figures 5.11 and 5.12); indeed, their sequences in the head/tail domain match perfectly. However, the Dc domain came from another individual, most probably through recombination, as they are completely different. And although the set of RNCs was inherited (with just one mutation) from the best of the previous generation, the new Dc allowed the fine-tuning of the numerical constants by enabling a different set of constants to get expressed. And thanks to this modification, a new individual with maximum fitness was created, matching perfectly the target function (3.19).

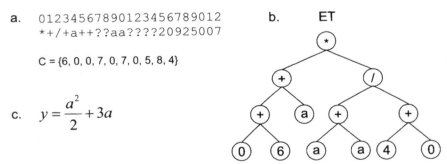

a. 012345678901234567890012
    *+/+a++??aa????20925007

    C = {6, 0, 0, 7, 0, 7, 0, 5, 8, 4}

c. $y = \dfrac{a^2}{2} + 3a$

b.   ET

Figure 5.12. Perfect solution to the simple symbolic regression problem. This program was found in generation 7 and has maximum fitness. **a)** The chromosome of the individual with its random numerical constants. **b)** The fully expressed individual. **c)** The corresponding mathematical expression. Note that it matches perfectly the target function (3.19).

5.6 Three Approaches to the Creation of NCs

In this section we are going to analyze three different approaches to the problem of constants' creation in evolutionary computation by comparing the performance of three different algorithms. The first, called GEA-B (from basic GEA), creates numerical constants from scratch if necessary, but no explicit constants are used as terminals; the second, called GEP-NC, uses a fixed set of numerical constants and handles them explicitly either to create new ones or to incorporate them directly in the evolving programs; the last one, called GEP-RNC, uses the facility to manipulate directly random numerical constants. The comparison between the three approaches will be made on four different problems. The first is an artificial problem of sequence induction requiring integer constants; the second is a computer generated problem of function finding requiring floating-point constants; and the last two are complex real-world problems from two different fields: diagnosis of breast cancer and analog circuit design.

5.6.1 Problems and Settings

For the sequence induction task, the following test sequence was chosen:

$$a_n = 4n^4 + 3n^3 + 2n^2 + n \qquad (5.14)$$

where n consists of the nonnegative integers. This sequence was chosen because it can be exactly solved by all the three algorithms and therefore can provide an accurate measure of their performance in terms of success rate. For this problem, the first 10 positive integers n and their corresponding term were used as fitness cases (Table 5.5). The mean squared error, evaluated by equation (3.4a), was used as basis for the fitness function, with the fitness being evaluated by equation (3.5) and, therefore, $f_{max} = 1000$. This experiment, with its three different approaches, is summarized in Table 5.6.

Table 5.5
Set of fitness cases for the sequence induction problem.

n	1	2	3	4	5	6	7	8	9	10
a_n	10	98	426	1252	2930	5910	10738	18056	28602	43210

For the function approximation problem, the following "V" shaped function was chosen:

$$y = 4.251a^2 + \ln(a^2) + 7.243e^a \qquad (5.15)$$

where a is the independent variable and e is the irrational number 2.71828183. Evolutionary algorithms have some difficulties in solving exactly (that is,

Table 5.6

Performance and settings used in the sequence induction problem.

	GEA-B	GEP-NC	GEP-RNC
Number of runs	100	100	100
Number of generations	100	100	100
Population size	30	30	30
Number of fitness cases	10 (Table 5.5)	10 (Table 5.5)	10 (Table 5.5)
Function set	+ - * /	+ - * /	+ - * /
Terminal set	n	n 0 1 2 3	n ?
Random constants array length	--	--	10
Random constants type	--	--	Integer
Random constants range	--	--	[0, 3]
Head length	6	6	6
Gene length	13	13	20
Number of genes	5	5	5
Linking function	+	+	+
Chromosome length	65	65	100
Mutation rate	0.044	0.044	0.044
Inversion rate	0.1	0.1	0.1
IS transposition rate	0.1	0.1	0.1
RIS transposition rate	0.1	0.1	0.1
One-point recombination rate	0.3	0.3	0.3
Two-point recombination rate	0.3	0.3	0.3
Gene recombination rate	0.3	0.3	0.3
Gene transposition rate	0.1	0.1	0.1
Dc-specific mutation rate	--	--	0.044
Dc-specific inversion rate	--	--	0.1
Dc-specific transposition rate	--	--	0.1
Random constants mutation rate	--	--	0.01
Fitness function	Eq. (3.5)	Eq. (3.5)	Eq. (3.5)
Average best-of-run fitness	979.472	376.751	766.561
Average best-of-run R-square	0.9999884864	0.999700006	0.999980023
Success rate	97%	35%	75%

with a precision of 0.01, for instance) this kind of problem and, therefore, the performance of all the three approaches will be compared in terms of average best-of-run fitness and average best-of-run R-square. For this problem, a set of 20 random fitness cases chosen from the interval [-1, 1] was used (Table 5.7). Again, the mean squared error, evaluated by equation (3.4a), was used as basis for the fitness function, with the fitness being evaluated by equation (3.5) and, therefore, $f_{max} = 1000$. This experiment, with its three different approaches, is summarized in Table 5.8.

Table 5.7
Set of fitness cases used in the "V" function problem.

a	f(a)
-0.2639725157548	3.19498066265276
0.0578905532656938	1.99052001725998
0.334025290109634	8.39663703997286
-0.236334577564462	3.07088976972825
-0.855744382566804	5.87946763695703
-0.0194437136332785	-0.775326322328458
-0.192134388183304	2.83470225774408
0.529307910124627	12.2154726642137
-0.00788974118728459	-2.49803983418635
0.438969804950631	10.4071734858808
-0.107559292698039	2.09413635645908
-0.274556994377163	3.23927278010839
-0.0595333219604528	1.19701284767347
0.384492993958352	9.35580769189855
-0.874923020736333	6.00642453001302
-0.236546636250546	3.07189729043837
-0.167875941704557	2.67440053130986
0.950682181822091	22.4819639844149
0.946979159577362	22.3750161187355
0.639339910059591	14.5701285332337

For the breast cancer problem, we are going to use the same datasets of section 4.2.1, Diagnosis of Breast Cancer. In this case, however, the fitness function will be based on the number of hits and will be evaluated by equation (3.10). Thus, for this problem with 350 training samples, $f_{max} = 350$. This experiment, with its three different approaches, is summarized in Table 5.9.

For the analog circuit design, we are going to find the transfer function expressing the yield in terms of three parameter tolerances. A training set consisting of $n = 40$ pairs of tolerances and their corresponding yields obtained from n runs of Monte Carlo simulations, kindly provided by Lukasz Zielinski (Zielinski and Rutkowski 2004), will be used. The mean squared error, evaluated by equation (3.4a), will be again used as basis for the fitness function, with the fitness being evaluated by equation (3.5) and, therefore, $f_{max} = 1000$. This experiment, with its three different approaches, is summarized in Table 5.10.

Table 5.8
Performance and settings used in the "V" function problem.

	GEA-B	GEP-NC	GEP-RNC
Number of runs	100	100	100
Number of generations	5000	5000	5000
Population size	30	30	30
Number of fitness cases	10 (Table 5.7)	10 (Table 5.7)	10 (Table 5.7)
Function set	+ - * / Q L E S C	+ - * / Q L E S C	+ - * / Q L E S C
Terminal set	a	a 1 2 3 4 5	a ?
Random constants array length	--	--	10
Random constants type	--	--	Rational
Random constants range	--	--	[-2, 2]
Head length	6	6	6
Gene length	13	13	20
Number of genes	5	5	5
Linking function	+	+	+
Chromosome length	65	65	100
Mutation rate	0.044	0.044	0.044
Inversion rate	0.1	0.1	0.1
IS transposition rate	0.1	0.1	0.1
RIS transposition rate	0.1	0.1	0.1
One-point recombination rate	0.3	0.3	0.3
Two-point recombination rate	0.3	0.3	0.3
Gene recombination rate	0.3	0.3	0.3
Gene transposition rate	0.1	0.1	0.1
Dc-specific mutation rate	--	--	0.044
Dc-specific inversion rate	--	--	0.1
Dc-specific transposition rate	--	--	0.1
Random constants mutation rate	--	--	0.01
Fitness function	Eq. (3.5)	Eq. (3.5)	Eq. (3.5)
Average best-of-run fitness	895.167	714.658	812.840
Average best-of-run R-square	0.9960392190	0.9893241183	0.9936233497

5.6.2 Sequence Induction

For the sequence induction problem, a very simple function set composed of the basic arithmetic operators was chosen for all the three algorithms, that is, $F = \{+, -, *, /\}$. For the GEA-B algorithm, the set of terminals consists obviously of the independent variable n, thus giving $T = \{n\}$. For the GEP-NC algorithm, besides the independent variable, four different integer constants represented by the numerals 0-3 were used, thus giving $T = \{n, 0, 1, 2, 3\}$, where each numeral represents its namesake integer constant. For the GEP-RNC algorithm, the set of terminals consists obviously of the independent variable plus the ephemeral random constant "?", thus giving $T = \{n, ?\}$.

Table 5.9
Performance and settings used in the breast cancer problem.

	GEA-B	GEP-NC	GEP-RNC
Number of runs	100	100	100
Number of generations	1000	1000	1000
Population size	30	30	30
Number of fitness cases	350	350	350
Function set	4(+ - * /)	4(+ - * /)	4(+ - * /)
Terminal set	d0-d8	d0-d8, c0-c4	d0-d8 ?
Random constants array length	--	--	10
Random constants type	--	--	Rational
Random constants range	--	--	[-2, 2]
Head length	7	7	7
Gene length	15	15	23
Number of genes	3	3	3
Linking function	+	+	+
Chromosome length	45	45	69
Mutation rate	0.044	0.044	0.044
Inversion rate	0.1	0.1	0.1
IS transposition rate	0.1	0.1	0.1
RIS transposition rate	0.1	0.1	0.1
One-point recombination rate	0.3	0.3	0.3
Two-point recombination rate	0.3	0.3	0.3
Gene recombination rate	0.3	0.3	0.3
Gene transposition rate	0.1	0.1	0.1
Dc-specific mutation rate	--	--	0.044
Dc-specific inversion rate	--	--	0.1
Dc-specific transposition rate	--	--	0.1
Random constants mutation rate	--	--	0.01
Fitness function	Eq. (3.10)	Eq. (3.10)	Eq. (3.10)
Rounding threshold	0.5	0.5	0.5
Average best-of-run fitness	339.340	339.050	339.130

Furthermore, a set of random numerical constants represented by the numerals 0-9 was used, that is, $R = \{0, 1, 2, 3, 4, 5, 6, 7, 8, 9\}$, and the ephemeral random constant "?" ranged over the integer interval $[0, 3]$ (the complete list of all the parameters used per run is shown in Table 5.6).

As shown in Table 5.6, the probability of success using the GEA-B algorithm is 97%, considerably higher than the 35% obtained using the GEP-NC algorithm or the 75% obtained using the GEP-RNC algorithm, showing that, for this kind of problem, the inclusion of numerical constants in the evolutionary toolkit results in worse performance. Thus, when the required constants to solve a problem are small integer constants, evolutionary algorithms

Table 5.10
Performance and settings used in the analog circuit design problem.

	GEA-B	GEP-NC	GEP-RNC
Number of runs	100	100	100
Number of generations	5000	5000	5000
Population size	30	30	30
Number of fitness cases	40	40	40
Function set	+ - * / P Q E L	+ - * / P Q E L	+ - * / P Q E L
Terminal set	a b c	a b c 1 2 3 4 5	a b c ?
Random constants array length	--	--	10
Random constants type	--	--	Integer
Random constants range	--	--	[0, 100]
Head length	12	12	12
Gene length	25	25	38
Number of genes	6	6	6
Linking function	+	+	+
Chromosome length	150	150	228
Mutation rate	0.044	0.044	0.044
Inversion rate	0.1	0.1	0.1
IS transposition rate	0.1	0.1	0.1
RIS transposition rate	0.1	0.1	0.1
One-point recombination rate	0.3	0.3	0.3
Two-point recombination rate	0.3	0.3	0.3
Gene recombination rate	0.3	0.3	0.3
Gene transposition rate	0.1	0.1	0.1
Dc-specific mutation rate	--	--	0.044
Dc-specific inversion rate	--	--	0.1
Dc-specific transposition rate	--	--	0.1
Random constants mutation rate	--	--	0.01
Fitness function	Eq. (3.5)	Eq. (3.5)	Eq. (3.5)
Average best-of-run fitness	12.9010	36.8972	38.7630
Average best-of-run R-square	0.6871088316	0.9097327677	0.9141464770

find these integer constants more easily if no assumptions whatsoever are a priori made, as small integer constants can be easily created from scratch by performing simple arithmetic operations.

The first perfect solution created by the GEA-B algorithm was found in generation 19 of run 0 (the sub-ETs are linked by addition):

```
0123456789012
*+nn+*nnnnnnn
-//*n-nnnnnnn
*+n-+*nnnnnnn
*+n-*-nnnnnnn
*+***+nnnnnnn
```
(5.16a)

As its expression shows, this program matches exactly the target sequence (5.14) (the contribution of each gene is shown in brackets):

$$a_n = (n^3 + 2n^2) + (n) + (n^3 + n^2) + (n^3 - n^2) + (4n^4) = 4n^4 + 3n^3 + 2n^2 + n \qquad (5.16b)$$

Note how the algorithm creates all the necessary constants from scratch by performing simple arithmetical operations.

The first perfect solution created by the GEP-NC algorithm was discovered in generation 39 of run 1. Its chromosome is shown below (the sub-ETs are linked by addition):

```
0123456789012
*+*++*nn101n1
n1n23nn112311
*+*++*nnn12nn
*+*+-*nnn12nn
0/321+n31331n
```
(5.17a)

As its expression shows, this program matches perfectly the target sequence (5.14) (the contribution of each gene is shown in brackets):

$$a_n = (n^3 + 2n^2) + (n) + (2n^4 + 3n^3) + (2n^4 - n^3) + (0) = 4n^4 + 3n^3 + 2n^2 + n \qquad (5.17b)$$

Note how the algorithm makes use of the ready-made constants, not only by integrating them directly into the sub-ETs but also by combining them to create new ones.

The first perfect solution designed by the GEP-RNC algorithm was discovered in generation 59 of run 0. Its genes and their respective arrays of random numerical constants are shown below (the sub-ETs are linked by addition):

```
Gene 1: *+--*?n??nn??7886912
    C₁: {3, 2, 2, 1, 0, 0, 1, 3, 1, 1}

Gene 2: ?*+++*?n???n?9797539
    C₂: {1, 3, 3, 0, 0, 0, 2, 3, 3, 0}

Gene 3: ----*-?n?nnn?9736635
    C₃: {1, 3, 2, 3, 0, 2, 2, 3, 1, 0}

Gene 4: ****++nnn???n0247720
    C₄: {2, 2, 2, 1, 1, 2, 0, 0, 1, 0}

Gene 5: +/n/??nnn?n??1407127
    C₅: {1, 3, 0, 3, 0, 2, 0, 2, 1, 0}
```
(5.18a)

As its expression shows, the program encoded in this structure matches perfectly the target sequence (5.14) (the contribution of each gene is shown in brackets):

$$a_n = (-n^3 + 3n^2) + (0) + (-n^2) + (4n^4 + 4n^3) + (n) = 4n^4 + 3n^3 + 2n^2 + n \qquad (5.18b)$$

Note how profusely the algorithm uses the random numerical constants at its disposal, either by integrating them directly into the solutions or by combining them to create new ones.

It is worth pointing out that the GEP-RNC algorithm performs considerably better than the GEP-NC algorithm at this task (a success rate of 75% as opposed to 35%), showing that the facility invented for handling random numerical constants in gene expression programming is not only elegant but also extremely efficient. Indeed, thanks to its flexibility and efficiency, this framework can be used to create more complex algorithms that require the swift handling of vast quantities of random numerical constants, such as algorithms for complete neural network induction, algorithms for polynomial induction, parameter optimization, and decision tree induction.

5.6.3 "V" Function

To solve the "V" shaped function problem, a more varied function set consisting of F = {+, -, *, /, Q, L, E, S, C} ("Q" represents the square root function, "L" represents the natural logarithm, "E" represents e^x, "S" represents the sine function, and "C" represents the cosine) was chosen in order to enrich the solution space and, therefore, provide more varied forms of representing numerical constants if need be. As usual, this function set was used by all the three algorithms. For the GEA-B algorithm, the set of terminals consists obviously of the independent variable a, thus giving T = {a}. For the GEP-NC algorithm, besides the independent variable, five different rational constants randomly chosen from the interval [-2, 2] and represented by the numerals 1-5 were used, thus giving T = {a, 1, 2, 3, 4, 5}, where "1" represents the numerical constant -0.977906, "2" represents -0.505524, "3" represents 0.205841, "4" represents 1.6409, and "5" represents 0.371673. For the GEP-RNC algorithm, the set of terminals consists obviously of T = {a, ?} and the set of random numerical constants consisted again of 10 constants represented by the numeral 0-9, thus giving R = {0, 1, 2, 3, 4, 5, 6, 7, 8, 9}. The ephemeral random constant "?" ranged over the rational interval [-2, 2] (see the complete list of all the parameters used per run in Table 5.8).

As shown in Table 5.8, despite requiring rational constants, the "V" function problem was better tackled by the simpler GEA-B approach, as both the average best-of-run fitness (895.167 for the GEA-B algorithm, 714.658 for the GEP-NC algorithm, and 812.840 for the GEP-RNC algorithm) and average best-of-run R-square (0.996039 for the GEA-B algorithm, 0.989324 for the GEP-NC algorithm, and 0.993623 for the GEP-RNC algorithm) indicate. Indeed, with the collection of functions chosen for the function set, most of which extraneous, the algorithm is equipped with different tools for evolving highly accurate models without explicitly using numerical constants. As we will see below, the structures of these extremely accurate models are somewhat unconventional as highly creative ways of representing the numerical constants are discovered by the algorithms. We will also see that, despite performing overall slightly better, the best models designed by the GEA-B algorithm are slightly worse than the best models designed with numerical constants. And not surprisingly, the best models of all were designed by the GEP-RNC algorithm.

The best solution of the GEA-B experiment was found in generation 4698 of run 2 (the sub-ETs are linked by addition):

```
0123456789012
+/L*E*aaaaaaa
ES*+CSaaaaaaa
SSESSEaaaaaaa
EE/*aSaaaaaaa
EE/*aSaaaaaaa
```
$$(5.19a)$$

It has a fitness of 998.730 and an R-square of 0.9999970702 evaluated over the training set of 20 fitness cases and an R-square of 0.999977473 evaluated against a testing set of 100 random points also chosen from the interval [-1, 1] and is therefore a very good solution to the problem at hand. More formally, it corresponds to the following C++ function:

```
double apsModel(double d[])
{
    double dblTemp = 0.0;
    dblTemp  = (((d[0]*d[0])/exp(d[0]))+log((d[0]*d[0])));
    dblTemp += exp(sin(((sin(d[0])+d[0])*cos(d[0])))));
    dblTemp += sin(sin(exp(sin(sin(exp(d[0]))))));
    dblTemp += exp(exp(((sin(d[0])*d[0])/d[0])));
    dblTemp += exp(exp(((sin(d[0])*d[0])/d[0])));
    return dblTemp;
}
```
$$(5.19b)$$

where d_0 corresponds to the independent variable a. It is worth pointing out how unconventional this program is and how the algorithm got around the absence of numerical constants by finding very creative forms of representing them. Notwithstanding, this model is a very good approximation to the target function (5.15) as both the R-square on the testing set and the comparison of the plots for the target function and the model show (Figure 5.13).

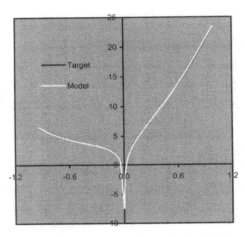

Figure 5.13. Comparison of the target function (5.15) with the model (5.19) evolved by the GEA-B algorithm, that is, without explicitly using numerical constants. The R-square was evaluated over a testing set of 100 random points and is equal to 0.999977473.

The best solution created with the GEP-NC algorithm was found in generation 4220 of run 52. Its genome is shown below (the sub-ETs are linked by addition):

```
0123456789012
*a++*5a5a1314
L*aa1+21141a2
SSSSS5a1a41a2
*a++aaa15544a
/EE+SEa2521a3
```
(5.20a)

It has a fitness of 999.635 and an R-square of 0.999993481 evaluated over the training set of 20 fitness cases and an R-square of 0.999990385 evaluated against the same testing set used in the GEA-B experiment, and therefore generalizes slightly better than the model (5.19) evolved by the GEA-B algorithm. More formally, the model (5.20a) can be expressed by the following C++ function:

```
double apsModel(double d[])
{
    double dblTemp = 0.0;
    dblTemp = (d[0]*((0.371673+d[0])+(0.371673*d[0])));
    dblTemp += log((d[0]*d[0]));
    dblTemp += sin(sin(sin(sin(sin(0.371673)))));
    dblTemp += (d[0]*((d[0]+d[0])+d[0]));
    dblTemp += (exp((exp(0.371673)+d[0]))/exp(sin(-0.977906)));
    return dblTemp;
}
```
                                                                    (5.20b)

where d_0 corresponds to the independent variable a. Note how different this model is from the model (5.19) designed without numerical constants. Here the numerical constants play a prominent role as they were directly integrated into the solution. Note, however, that the designed model is structurally very different from the target function (5.15) despite being an extremely accurate model as both the R-square and the plots of the target function and the model show (Figure 5.14).

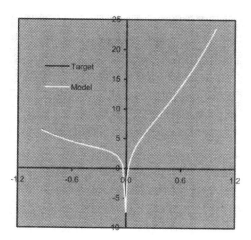

Figure 5.14. Comparison of the target function (5.15) with the model (5.20) evolved by the GEP-NC algorithm, that is, using a fixed set of five numerical constants. The R-square was evaluated over the same testing set of 100 random points and is equal to 0.999990385.

The best solution created with the GEP-RNC algorithm was found in generation 4987 of run 26. Its genes and their corresponding arrays of random numerical constants are shown below (the sub-ETs are linked by addition):

```
Gene 1:  **?***?aa??aa3508773
    C₁:  {-1.68454, 0.798279, 1.39566, 1.081695, -0.497864,
          1.834442, 0.68457, -1.634003, -1.548554, 1.413025}

Gene 2:  *aECC*?a?aaa?7277993
    C₂:  {0.504761, -0.312164, 0.37912, 1.399841, 0.078949,
          1.280243, -0.284759, -0.672516, -0.983948, 0.898652}

Gene 3:  E--SE-??????a3265159
    C₃:  {-1.369293, 0.075317, 0.783661, 0.558258, -0.239593,
          -1.260437, 0.398193, -1.696289, 0.372406, 1.407165}

Gene 4:  -EE-?a?aaaaaa7193625
    C₄:  {-1.263519, -1.874268, 1.334442, 1.417694, 1.470489,
          1.417694, -0.117675, -1.993897, 0.801728, 0.080383}

Gene 5:  -L-*EEaaa?a?a4511699
    C₅:  {0.903077, 1.595642, -0.763794, 1.066101, -0.158966,
          1.764313, -1.357117, 0.955719, -0.027954, -0.122955}           (5.21a)
```

It has a fitness of 999.957 and an R-square of 0.999999672 evaluated over the training set of 20 fitness cases and an R-square of 0.999999204 evaluated against the same testing set used in the previous two approaches, and thus is better than the models (5.19) evolved with the GEA-B algorithm and better than the model (5.20) created with the GEP-NC approach. More formally, the model (5.21a) can be expressed by the following C++ function:

```
double apsModel(double d[])
{
    double dblTemp = 0.0;
    dblTemp = (((((-1.68454*(-1.548544))*1.834442)*
               (d[0]*d[0]))*1.081695);
    dblTemp += (d[0]*exp(cos(cos((-0.672516*d[0]))))));
    dblTemp += exp(((exp(0.783661)-(0.398193-
               (-1.260437)))-sin(0.558258)));
    dblTemp += (exp((d[0]-(-1.874268)))-exp(-1.993897));
    dblTemp += (log((d[0]*d[0]))-(exp(d[0])-exp(-0.158966)));
    return dblTemp;
}
                                                                  (5.21b)
```

where d_0 corresponds to the independent variable a. Once again, the plots of the target function (5.15) and the model (5.21) designed by the GEP-RNC algorithm are compared in Figure 5.15. This model is indeed a very good approximation to the target function as both the R-square on the testing set and the comparison of the plots show.

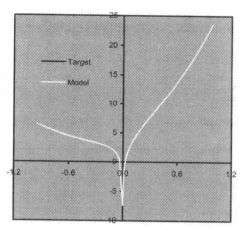

Figure 5.15. Comparison of the target function (5.15) with the model (5.21) evolved by the GEP-RNC algorithm, that is, with the facility for handling random numerical constants. The R-square was evaluated over the same testing set of 100 random points and is equal to 0.999999204.

It is worth noticing how copiously the GEP-RNC algorithm makes use of the random numerical constants to design good solutions. Note, however, that the constants that are integrated in the sub-ETs are, at least in some cases, very different from the expected ones (compare, for instance, the model of the "V" function with the function itself). Indeed, gene expression programming (and I believe, all evolutionary algorithms) can find the expected constants with a precision to the third or fourth decimal place if the target functions are simple polynomial functions with rational coefficients, otherwise a very creative solution will be found. For instance, the simple polynomial function of section 5.1 can be exactly solved by the GEP-RNC algorithm (see the performance and parameters used per run in Table 5.11 and compare these results with the results presented in Table 5.3).

Consider, for instance, the almost perfect solution below discovered in generation 953 of run 31 (the sub-ETs are linked by addition):

```
Gene 1:  *+x*++?x?x?x?1600254
```
C_1:   {1.34436, 1.717926, 1.794709, 0.980042, 1.351318,
          1.551117, 0.453125, 1.718384, 1.597992, 1.111602}

```
Gene 2:  *x?//?xx?x?x?0172474
```
C_2:   {0.378906, 0.76178, 0.437958, 1.168029, 0.307495,
          1.176086, 0.486938, 0.755036, 0.01184, 0.862946}   (5.22a)

It has a fitness of 999.971 an R-square equal to 0.99999999999547 and, therefore, is an almost perfect match to the target function (5.1). Indeed, as

Table 5.11
Settings for the simple polynomial function with rational coefficients.

Number of runs	100
Number of generations	1000
Population size	50
Number of fitness cases	10 (Table 5.2)
Function set	+ - * /
Terminal set	x ?
Random constants array length	10
Random constants type	Rational
Random constants range	[0, 2]
Head length	6
Gene length	20
Number of genes	2
Linking function	+
Chromosome length	40
Mutation rate	0.044
Inversion rate	0.1
IS transposition rate	0.1
RIS transposition rate	0.1
One-point recombination rate	0.3
Two-point recombination rate	0.3
Gene recombination rate	0.3
Gene transposition rate	0.1
Dc-specific mutation rate	0.044
Dc-specific inversion rate	0.1
Dc-specific transposition rate	0.1
Random constants mutation rate	0.01
Fitness function	Equation (3.3b)
Selection range	100%
Precision	0.0%
Average best-of-run fitness	993.270
Average best-of-run R-square	0.9999984024

its mathematical expression shows, the GEP-RNC algorithm discovered the numerical constants of function (5.1) with great precision:

$$y = 2.7179x^2 + 3.1415x \qquad (5.22b)$$

5.6.4 Diagnosis of Breast Cancer

In contrast to the two previous computer-generated problems, the breast cancer problem is a complex real-world problem with nine independent variables, and investigating whether numerical constants are essential to make an accurate diagnosis can give us a more realistic view on the importance of numerical constants to the design of robust mathematical models.

For this classification problem, a very simple function set consisting of the basic arithmetic operators was chosen and used in all the three approaches, that is, F = {+, -, *, /}, where each function was weighted four times. For the GEA-B algorithm, the set of terminals consists obviously of the nine attributes used in this problem and were represented by $T = \{d_0, ..., d_8\}$ which correspond, respectively, to clump thickness, uniformity of cell size, uniformity of cell shape, marginal adhesion, single epithelial cell size, bare nuclei, bland chromatin, normal nucleoli, and mitoses. For the GEP-NC algorithm, besides the nine attributes, five different rational constants randomly chosen from the interval [-2, 2] and represented by c_0-c_4 were used, thus giving $T = \{d_0, d_1, d_2, d_3, d_4, d_5, d_6, d_7, d_8, c_0, c_1, c_2, c_3, c_4\}$, where $c_0 = 0.938233$, $c_1 = 1.859162$, $c_2 = -0.190002$, $c_3 = 1.498413$, $c_4 = -0.242737$. For the GEP-RNC algorithm, the set of terminals consists obviously of the nine attributes plus the ephemeral random constant "?", thus giving $T = \{d_0, ..., d_8, ?\}$. Furthermore, a set of random numerical constants represented by the numerals 0-9 was used, thus giving R = {0, 1, 2, 3, 4, 5, 6, 7, 8, 9}, and the ephemeral random constant "?" ranged over the rational interval [-2, 2] (the complete list of all the parameters used per run is shown in Table 5.9).

And as you can see in Table 5.9, the inclusion of numerical constants in the evolutionary toolkit didn't result in a better performance. Indeed, both the GEP-NC and the GEP-RNC algorithms perform slightly worse than the simpler GEA-B approach, as the average best-of-run fitness indicates (339.340 for the GEA-B algorithm, 339.050 for the GEP-NC algorithm, and 339.130 for the GEP-RNC algorithm). Also interesting is the fact that, even when readily available, numerical constants were seldom integrated in the best-of-run solutions as will next be shown.

One of the best solutions designed with the GEA-B algorithm was found in generation 670 of run 58 (the sub-ETs are linked by addition):

```
-.d3./.d4.+.*.d8.d2.d4.d1.d2.d6.d1.d8.d0
+.+.d7.d5.+.-.d1.d4.d7.d5.d3.d3.d3.d1.d5
+.d0.*.d7.d6.d2.d2.d6.d1.d8.d2.d2.d1.d2.d4
```
(5.23a)

This model classifies correctly 341 out of 350 fitness cases in the training set and 170 out of 174 sample cases in the testing set, which corresponds to a training set classification error of 2.571% and a classification accuracy of 97.429%, and a testing set classification error of 2.299% and a classification accuracy of 97.701%. More formally, the model (5.23) can be translated into the following C++ function:

```
int apsModel(double d[])
{
    const double ROUNDING_THRESHOLD = 0.5;

    double dblTemp = 0.0;
    dblTemp  = (d[3]-(d[4]/((d[2]*d[4])+d[8])));
    dblTemp += ((d[5]+((d[4]-d[7])+d[1]))+d[7]);
    dblTemp += (d[0]+(d[7]*d[6]));

    return (dblTemp >= ROUNDING_THRESHOLD ? 1:0);
}
```
(5.23b)

One of the best solutions created with the GEP-NC algorithm was found in generation 754 of run 47. Its genome is shown below (the sub-ETs are linked by addition):

```
d1.*.+.d4.-.-.+.d1.c2.c1.d2.c1.d4.d0.d0
*.d0.d5.d4.*.*.d6.d8.d3.c0.d5.c1.d8.c4.d5
*.c0.*.+.d8.d3.d6.d1.c1.d6.d5.c1.d4.c4.c0
```
(5.24a)

This model classifies correctly 340 out of 350 fitness cases in the training set and 171 out of 174 fitness cases in the testing set. This corresponds to a training set classification error of 2.857% and a classification accuracy of 97.143%, and a testing set classification error of 1.724% and a classification accuracy of 98.276%. Thus, this model generalizes slightly better than the model (5.23) designed by the GEA-B algorithm. More formally, the model (5.24a) can be expressed by the following C++ function:

```
int apsModel(double d[])
{
    const double ROUNDING_THRESHOLD = 0.5;

    double dblTemp = 0.0;
    dblTemp   = d[1];
    dblTemp += (d[0]*d[5]);
    dblTemp += (0.938233*((d[3]+d[6])*d[8]));

    return (dblTemp >= ROUNDING_THRESHOLD ? 1:0);
}                                                        (5.24b)
```

It is worth pointing out how compact this highly accurate model is: indeed, in terms of nodes, it requires just 11 nodes. Also interesting is that only six of the nine attributes are used in this model. Note also that just one of the five numerical constants available in this experiment was integrated in the fully expressed program.

One of the best solutions created with the GEP-RNC algorithm was discovered in generation 758 of run 93. Its genes and their respective arrays of random numerical constants are shown below (the three sub-ETs are linked by addition):

```
Gene 1: *./.+.d1./.+.d5.d4.d0.d5.d4.d0.d6.d5.?.3.0.2.7.8.9.7.5
    C₁: {-1.08551, 0.060577, -0.614655, -0.041046, 0.717499,
        1.647156, 0.255981, -0.421814, 0.44043, -0.919983}

Gene 2: *.d6.+.*.d5.d8.+.d0.d2.d6.d2.d5.d5.d5.d7.1.4.7.3.6.8.6.8
    C₂: {-1.07431, 1.53714, -0.861328, -0.014801, -1.192108,
        -1.983307, 1.749695, 0.403107, 0.377991, -0.106109}

Gene 3: *.d3.*.d3.d1.d3.*.d7.d4.d7.d5.d8.d0.d5.d4.4.6.1.8.2.5.0.9
    C₃: {-1.08551, 0.060577, 1.801148, 0.737946, -1.972596,
        1.647156, 0.255981, 0.696961, -0.231292, 1.210998}      (5.25a)
```

This model classifies correctly 341 out of 350 fitness cases in the training set and 170 out of 174 fitness cases in the testing set. This corresponds to a training set classification error of 2.571% and a classification accuracy of 97.429%, and a testing set classification error of 2.299% and a classification accuracy of 97.701%. Thus, this model is as good at generalizing as the model (5.23) designed with the GEA-B algorithm. More formally, it corresponds to the following C++ function:

```
int apsModel(double d[])
{
    const double ROUNDING_THRESHOLD = 0.5;

    double dblTemp = 0.0;
    dblTemp   = ((d[1]/(d[4]/d[0]))*((d[5]+d[4])+d[5]));
    dblTemp += (d[6]*((d[8]*(d[0]+d[2]))+d[5]));
    dblTemp += (d[3]*(d[3]*d[1]));

    return (dblTemp >= ROUNDING_THRESHOLD ? 1:0);
}
```
(5.25b)

Note that, of the 30 numerical constants at its immediate disposal, this highly accurate model makes use of none of them as not even one numerical constant is part of the fully expressed program. No wonder the simpler approach without numerical constants produces better results (see Table 5.9), as it seems that, in this case, numerical constants are an unnecessary burden.

5.6.5 Analog Circuit Design

For all the problems analyzed thus far (sequence induction, "V" function, and diagnosis of breast cancer), we have seen that numerical constants were not really crucial for designing good models and, overall, better results were obtained in their absence. This indeed holds true for the vast majority of problems I've dealt with in my professional life, but there are some problems, however, for which numerical constants play a very important role and, in their absence, only mediocre models can be created. The analog circuit we are going to design in this section is particularly interesting because it belongs to this category of problems.

For this problem, a more eclectic function set consisting of the basic arithmetical operators plus the square root function "Q", the exponential function "E", the natural logarithm "L", and the power function "P" was chosen for all the three approaches, that is, F = {+, -, *, /, Q, E, L, P}. For the GEA-B algorithm, the set of terminals consists of the three tolerances L_1, L_2, and C_1 established for this circuit (Zielinski and Rutkowski 2004), which were respectively represented by T = {a, b, c}. For the GEP-NC algorithm, besides the three tolerances, five different integer constants randomly chosen from the interval [0, 100] and represented by the numerals 1-5 were used, thus giving T = {a, b, c, 1, 2, 3, 4, 5}, where "1" corresponds to 40, "2" corresponds to 37,"3" corresponds to 77, "4" corresponds to 4, and "5" corresponds to 12. For the GEP-RNC algorithm, the set of terminals consists obviously of the three tolerances plus the ephemeral random constant "?", thus giving $T = \{a, b, c, ?\}$. Furthermore, a set of random numerical constants represented by the numerals 0-9 was used, thus giving $R = \{0, 1, 2, 3, 4, 5, 6, 7, 8, 9\}$, and the ephemeral random constant "?" ranged over the integer interval $[0, 100]$ (the complete list of all the parameters used per run is shown in Table 5.10).

And as you can see in Table 5.10, for this problem, the inclusion of numerical constants in the evolutionary toolkit is responsible for a considerable increase in performance. Indeed, both the GEP-NC and the GEP-RNC algorithms perform considerably better than the simpler GEA-B approach, as both the average best-of-run fitness (12.9010 for the GEA-B system, 36.8972 for the GEP-NC algorithm, and 38.7630 for the GEP-RNC algorithm) and average best-of-run R-square (0.6871088316 for the GEA-B algorithm, 0.9097327677 for the GEP-NC algorithm, and 0.9141464770 for the GEP-RNC algorithm) indicate. Notice also that, of the two algorithms with numerical constants, the GEP-RNC algorithm performs considerably better than the GEP-NC algorithm, showing again that the facility for handling random numerical constants in gene expression programming is indeed extremely efficient and, therefore, the ideal tool for dealing with great quantities of numerical constants.

The best solution designed with the GEA-B algorithm was found in generation 2696 of run 62 (the sub-ETs are linked by addition):

```
012345678901234567 8901234
EEQQEQQ/EEQ+aabacaacaacaa
EEQQQQQ/EEQ+aacaaaccabbab
EEQQQQQ/EE-+abcaabbaacbbb
EEQQEQ/Q++EEbbcabbaacbbba
EEQQQQQ/*E**aaaaabcbbbbbc
EEQQQQQ/EE-+bacbcccaccbca
```
(5.26a)

It has a fitness of 48.4756 and an R-square of 0.9394514385 evaluated over the training set of 40 fitness cases and is therefore a good solution to the problem at hand. Note that the R-square of this best-of-experiment solution is much higher than the average best-of-run R-square obtained in this experiment (0.6871088316), showing that, although harder to find, good solutions can nevertheless be designed without numerical constants. More formally, the model (5.26) can be expressed by the following C++ function:

```
double apsModel(double d[])
{
    double dblTemp = 0.0;
    dblTemp  = exp(exp(sqrt(sqrt(exp(sqrt(sqrt((exp(sqrt(d[0])))/
               exp((d[0]+d[1])))))))))));
    dblTemp += exp(exp(sqrt(sqrt(sqrt(sqrt(sqrt((exp(sqrt(d[0])))/
               exp((d[0]+d[2])))))))))));
    dblTemp += exp(exp(sqrt(sqrt(sqrt(sqrt((exp((d[0]-d[1])))/
               exp((d[2]+d[0])))))))))));
    dblTemp += exp(exp(sqrt(sqrt(exp(sqrt((sqrt((d[1]+d[1])))/
               (exp(d[2])+exp(d[0])))))))))));
    dblTemp += exp(exp(sqrt(sqrt(sqrt(sqrt(sqrt((((d[0]*d[0])*
               (d[0]*d[0]))/exp(d[0])))))))))));
    dblTemp += exp(exp(sqrt(sqrt(sqrt(sqrt(sqrt((exp((d[1]-d[0])))/
               exp((d[2]+d[1])))))))))));
    return dblTemp;
}
```
$$(5.26b)$$

where d_0-d_2 correspond, respectively, to a-c. It is worth pointing out how different this solution is from the ones created with numerical constants (see models (5.27) and (5.28) below), as the algorithm had to find some creative ways of modeling this function without numerical constants. Notice especially the profuse use of exponential and square root functions in all the sub-ETs and that, even though most function nodes are functions of one argument, this program is considerably larger (with 95 nodes) than the ones created both with the GEP-NC (a total of 72 nodes) and the GEP-RNC (a total of 47 nodes) algorithms.

The best solution created with the GEP-NC algorithm was found in generation 4937 of run 27 (the sub-ETs are linked by addition):

```
012345678901234567 8901234
/E+4*+cc/a3P1cc51aa25125c
-*a/Qc2cc1Q/142b315222254
-3c-L+-1/--23bc35341a22cb
Q+4*/caL+-1/ca4b4bb34a1aa
/E+4P+bb/a++224142a5b3223
/*-cQ--LQb1*232a5c453acb2
```
$$(5.27a)$$

This model has a fitness of 66.7938 and an R-square of 0.9566185279 evaluated against the training set of 40 fitness cases and, therefore, is considerably better than the model (5.26) designed without numerical constants. More formally, the model (5.27a) can be expressed by the following C++ function:

```
double apsModel(double d[])
{
    double dblTemp = 0.0;
    dblTemp  = (exp(4)/((d[2]*d[2])+((77/pow(40,d[2]))+d[0]))) ;
    dblTemp += (((d[2]/37)*sqrt(d[2]))-d[0]) ;
    dblTemp += (77-d[2]) ;
    dblTemp += sqrt((4+((d[0]/log((((d[0]/4)-d[2])+40)))*d[2]))) ;
    dblTemp += (exp(4)/(pow(d[1],d[1])+(((37+37)/(4+40))+d[0]))) ;
    dblTemp += ((d[2]*sqrt(log(37)))/((sqrt(77)-d[1])-
               (40-(37*d[0]))))) ;
    return dblTemp;
}
```
$$(5.27b)$$

It is worth pointing out how profusely the algorithm makes use of the five available numerical constants: in fact, with the exception of one (constant 12), all of them were used to design this accurate model. Note also how structurally different this program is from the model (5.26) designed without numerical constants, especially in terms of the composition of the most abundant function nodes.

The best solution created with the GEP-RNC algorithm was found in generation 4961 of run 30. Its genes and corresponding arrays of random numerical constants are shown below (the sub-ETs are linked by addition):

Gene 1: /?+P?--ab??P??cacaaba??b?4027503030058
   C_1: {11, 70, 24, 15, 83, 29, 82, 21, 16, 49}

Gene 2: QcQE*+??cPE*c?babbabac?b?1370963974448
   C_2: {92, 28, 4, 5, 41, 35, 48, 99, 81, 33}

Gene 3: /?+?Pcbba/-?cccaccab???ca2719004069141
   C_3: {81, 54, 73, 60, 44, 83, 3, 12, 68, 30}

Gene 4: +--/a?c?+-QEccababc??c?ac8512372841014
   C_4: {40, 59, 8, 58, 94, 13, 71, 52, 23, 58}

Gene 5: /P?QQQcQ*a?a?c??ca??acbc?3064768815876
   C_5: {40, 86, 33, 6, 83, 56, 7, 86, 59, 54}

Gene 6: ?--?bL/**a?Qbca?ca?caa?a?4314917061824
   C_6: {92, 74, 34, 96, 56, 49, 34, 33, 58, 4}   (5.28a)

This model has a fitness of 84.4035 and an R-square of 0.9655103219 evaluated against the training set of 40 fitness cases and, therefore, is considerably better than the model (5.26) designed without numerical constants and the model (5.27) designed with a fixed set of numerical constants. More formally, the model (5.28a) can be expressed by the following C++ function:

```
double apsModel(double d[])
{
    double dblTemp = 0.0;
    dblTemp  = (83/(pow((d[0]-d[1]),(24-21))+11));
    dblTemp += sqrt(d[2]);
    dblTemp += (73/(12+pow(d[2],d[1])));
    dblTemp += (((13/((exp(d[0])-d[2])+sqrt(d[2])))-
                d[0])+(23-d[2]));
    dblTemp += (pow(sqrt(sqrt(sqrt((d[0]*40)))),sqrt(d[2]))/6);
    dblTemp += 56;
    return dblTemp;
}
```
(5.28b)

Note how profusely the GEP-RNC algorithm uses the random numerical constants at its disposal. Note also that, structurally, this program resembles the model (5.27) designed with the fixed set of numerical constants, and that both of them are very different from the model (5.26) created without numerical constants.

So, in conclusion, before starting the modeling process, it is advisable to inquire if numerical constants are essential to design good models. When the answer is negative, then the simpler system without numerical constants should be used as it allows a much faster evolution and, consequently, allows the creation of more and better programs in record time. But if the answer is affirmative, then the GEP-RNC algorithm should be used as it provides the ideal tools for designing extremely accurate models with numerical constants.

6 Automatically Defined Functions in Problem Solving

Automatically defined functions were for the first time introduced by Koza as a way of reusing code in genetic programming (Koza 1992). These ADFs obey a rigid syntax in which an S-expression, with a LIST-n function on the root, lists n-1 function-defining branches and one value-returning branch (Figure 6.1). The function-defining branches are used to create ADFs that may or may not be called upon by the value-returning branch. Such rigid structure imposes great constraints on the genetic operators as the different branches of the LIST function are not allowed to exchange genetic material amongst themselves. Furthermore, the ADFs of genetic programming are further constrained by the number of arguments each takes, as the number of arguments must be a priori defined and cannot be changed during evolution.

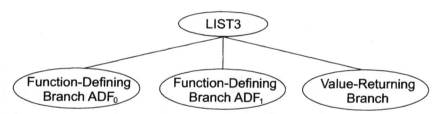

Figure 6.1. The overall structure of an S-expression with two function-defining branches and the value-returning branch used in genetic programming to create automatically defined functions.

We have already seen in section 2.3, Cells and the Evolution of Linking Functions, that in the multigenic system of gene expression programming, the implementation of ADFs can be done elegantly and without any kind of constraints as n different genes are used to encode n different ADFs. The way these ADFs interact with one another and how often they are called upon is encoded in special genes – homeotic genes – that control the overall

Cândida Ferreira: *Gene Expression Programming*, Studies in Computational Intelligence (SCI) **21**, 233–273 (2006)
www.springerlink.com

development of the individual. And the product of expression of such genes consists of the main program or cell. Thus, homeotic genes determine which genes are expressed in which cell and how the different sub-ETs (now working as ADFs) interact with one another. Or stated differently, homeotic genes determine which ADFs are called upon in which main program and how these ADFs are organized in each main program.

The importance of automatically defined functions and the advantages they bring to automatic programming can only be understood if one analyzes how they work and how the algorithm copes with their integration. Is evolution still smooth? Are there gains in performance? Is the system still simple or excessively complicated? How does it compare to simpler systems without ADFs? How does one go about integrating random numerical constants in ADFs? Are these ADFs still manageable or excessively complex? Can the multicellular system be explored to solve problems with multiple outputs? Are there problems that can only be solved with ADFs? These are some of the questions that we will try to address in this chapter by putting ADFs to the test in problem solving. We will start by solving a simple modular function and analyzing a successful run in its entirety in order to understand how such complex systems evolve. Then we will see how the cellular system copes with a more challenging modular problem, the already familiar odd-parity functions. And finally, we will put ADFs to the test by solving four complex real-world problems with them: the first consists of Kepler's Third Law; the second is the challenging analog circuit requiring numerical constants that we studied in section 5.6.5; the third is again the breast cancer classification problem so that we could compare the performance of the cellular system with other simpler systems on a complex real-world problem; and the last one is the interesting three-class prediction iris problem that will be used to show how the multicellular system can be successfully used to solve problems with multiple outputs.

6.1 Solving a Simple Modular Function with ADFs

In this section we are going to analyze a successful run in its entirety in order to understand how populations of such complex entities composed of different cells expressing different ADFs adapt, not only by changing the size and shape of the basic building blocks (the ADFs) but also by creating/choosing the main programs and changing their organization.

The simple problem we are going to solve with the cellular system of gene expression programming is the sextic polynomial $a^6-2a^4+a^2$, the same function chosen by Koza to show that ADFs allow the evolution of modular solutions (Koza 1994). Indeed, its regularity and modularity can be easily guessed by its factorization:

$$a^6-2a^4+a^2 = a^2(a-1)^2(a+1)^2 \tag{6.1}$$

For this problem, a very simple function set composed of the basic arithmetical operators was chosen for both the ADFs and the main programs, that is, $F = F_H = \{+, -, *, /\}$. The set of terminals used in the conventional genes consists obviously of the independent variable a, thus giving $T = \{a\}$; for the homeotic genes, the terminal set consists obviously of the ADFs encoded in the conventional genes (two, in this case) and will be represented by "0" and "1", thus giving $T_H = \{0, 1\}$. For this simple problem of just one variable, a set of 10 random fitness cases chosen from the interval [-1, 1] will be used (Table 6.1). The fitness will be evaluated by equation (3.3b), using a relative error of 100% for the selection range and a precision for the error equal to 0.01%. Thus, for the 10 fitness cases used in this problem, $f_{max} = 1000$. The complete list of the parameters used per run is shown in Table 6.2.

The evolutionary dynamics of the successful run we are going to analyze is shown in Figure 6.2. And as you can see, in this run, a perfect solution to the sextic polynomial was found in generation 8.

Table 6.1
Set of 10 random computer generated fitness cases used to solve the simple modular function using automatically defined functions.

a	f(a)
-0.133270	0.0171355969283171
0.901673	0.0284259762561082
-0.179748	0.0302552836680942
0.221649	0.0444196789826592
-0.920441	0.0197775495439223
-0.633942	0.1437712766244980
0.797516	0.0842569554563209
-0.098480	0.0095111081469302
-0.808197	0.0785663809054614
-0.926209	0.0173313183876865

Table 6.2
Parameters for the sextic polynomial problem.

Number of generations	50
Population size	20
Number of fitness cases	10 (Table 6.1)
Function set of ADFs	+ - * /
Terminal set	a
Number of genes/ADFs	2
Head length	5
Gene length	11
Function set of homeotic genes	+ - * /
Terminal set of homeotic genes	ADFs 0-1
Number of homeotic genes/cells	2
Head length of homeotic genes	4
Length of homeotic genes	9
Chromosome length	40
Mutation rate	0.044
Inversion rate	0.1
RIS transposition rate	0.1
IS transposition rate	0.1
Two-point recombination rate	0.3
One-point recombination rate	0.3
Gene recombination rate	0.3
Gene transposition rate	0.1
Mutation rate in homeotic genes	0.044
Inversion rate in homeotic genes	0.1
RIS transposition rate in homeotic genes	0.1
IS transposition rate in homeotic genes	0.1
Selection range	100%
Precision	0.01%

The initial population of this run and the fitness of each individual in the particular environment of Table 6.1 is shown below (the best cell is indicated after the coma and "3" is used to flag unviable programs that return calculation errors):

```
Generation N: 0
012345678900123456789001234567801234567 8
/*+/aaaaaaa/**/+aaaaaa/--/01100++-/11110,3-[ 0] = 0
*--*+aaaaaa++-a-aaaaaa/1-000101+10110101,3-[ 1] = 0
*-/a*aaaaaa*--aaaaaaaa+/-000001**1*00110,1-[ 2] = 0
/a-a-aaaaaa/-/+*aaaaaa*11/00001+-/001110,1-[ 3] = 0
```

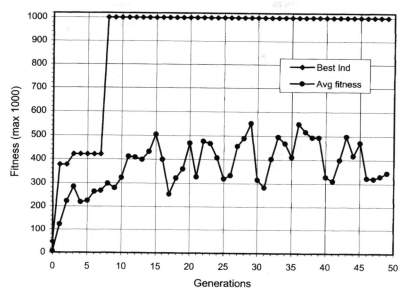

Figure 6.2. Progression of average fitness of the population and the fitness of the best individual for a successful run of the experiment summarized in Table 6.2.

```
///*-aaaaaa*+-**aaaaaa*0+/00111/*0-00011,3-[ 4] = 0
+a*--aaaaaa-/*a-aaaaaa+1*/10101-/*101110,3-[ 5] = 0
+++-/aaaaaa-/*aaaaaaaa/+-011000*+1-01100,1-[ 6] = 0
-*-a-aaaaaa*--++aaaaaa/00/11101/-1100111,3-[ 7] = 0
--a/-aaaaaa-a*/aaaaaaa--1*10101/++/10110,1-[ 8] = 0
//-a+aaaaaa-a/+/aaaaaa-+1100101-+-110101,3-[ 9] = 0
/+*+/aaaaaa*++*-aaaaaa+10+00111-/+-01011,1-[10] = 0
*-/a-aaaaaa+--a+aaaaaa+0-001001/10110101,1-[11] = 0
+*/**aaaaaa*//+*aaaaaa-*1*11100-*/-01000,1-[12] = 0
-a*a-aaaaaa*a-*/aaaaaa-/*100001-/*000000,1-[13] = 0
+a*--aaaaaa/+aa*aaaaaa*+1+01111-**+00100,0-[14] = 24.99622
*aa++aaaaaa-**/+aaaaaa+0*-11010**-011010,1-[15] = 46.24757
--//+aaaaaa/-a-aaaaaaa+-+*00011+*1*01000,1-[16] = 0
+++/aaaaaaa--+/+aaaaaa*11/01000**/010010,0-[17] = 28.95218
*a++aaaaaaaa+***aaaaaaa-00-11101*-/000110,1-[18] = 0
*/**-aaaaaa*+/-*aaaaaa+1+001010/0**10100,3-[19] = 0
```

It is worth noticing that 17 out of 20 individuals of the initial population are unviable, either because they return calculation errors such as division by zero (those flagged by "3") or because they didn't solve a single fitness case within the chosen selection range of 100% (a total of 10 individuals in the population above).

The expression of the best individual of this population (chromosome 15) and the corresponding mathematical expression are shown in Figure 6.3. Note that both ADFs are used twice in the active cell, resulting in a program considerably more complex than the basic building blocks that it uses, although, as we will see next, not necessarily fitter.

This model has a fitness of 46.24757 and an R-square of 0.02292 evaluated against the training set and, as you can see by plotting both the target and predicted values (Figure 6.4), this model is indeed a very rough approximation to the target function (6.1).

a.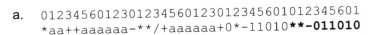
```
0123456012301234560123012345601012345601
*aa++aaaaaa-**/+aaaaaa+0*-11010**-011010
```

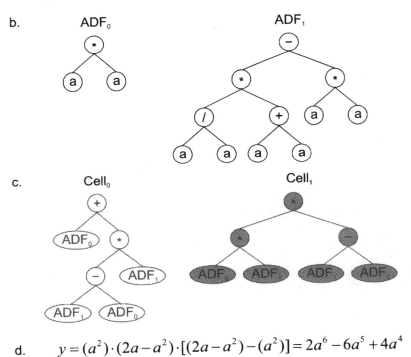

b.   ADF$_0$ ADF$_1$

c.   Cell$_0$ Cell$_1$

d. $y = (a^2) \cdot (2a - a^2) \cdot [(2a - a^2) - (a^2)] = 2a^6 - 6a^5 + 4a^4$

Figure 6.3. Best individual of generation 0 (chromosome 15). It has a fitness of 46.24757 evaluated against the set of fitness cases presented in Table 6.1. **a)** The chromosome of the individual with the best cell shown in bold. **b)** The ADFs codified by each gene. **c)** The cells or main programs encoded in the homeotic genes (the best cell is highlighted). **d)** The mathematical expression encoded in the best cell (the contribution of each ADF is shown in parentheses).

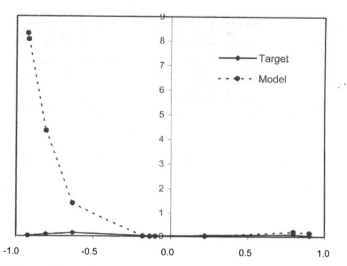

Figure 6.4. Comparison of the target function (6.1) with the best model of generation 0 (chromosome 15; see its expression in Figure 6.3). It has a fitness of 46.24757 and an R-square of 0.02292.

In the next generation two new individuals were created, chromosomes 13 and 19, that are considerably better than the best individual of the previous generation. Note that they code for exactly the same ADFs and the sequence of the best cell, although different, results in identical main programs. You can see the complete expression of chromosome 19 in Figure 6.5. The complete population is shown below (the best of generation are shown in bold):

```
Generation N: 1
012345678900123456789001234567 8012345678
*aa++aaaaaa-**/+aaaaaa+0*-11010**-011010,1-[ 0] = 46.24757
+aa++aaaaaa-a*/+aaaaaa+0*-11010*--011010,1-[ 1] = 0
*aa++aaaaaa-**/+aaaaaa+0*-11010**-010010,1-[ 2] = 0
+a*--aaaaaa*/+aaaaaaaa*+1+11010***-11010,1-[ 3] = 46.85482
*a*--aaaaaa/aa*+aaaaaa*+1+01111-**+00100,1-[ 4] = 0
*aa+/aaaaaa-**/-aaaaaa+0*-11010**-011010,0-[ 5] = 353.5659
+++/aaaaaaa--+/+aaaaaa*11/01000**/010010,0-[ 6] = 28.95218
+a*--aaaaaa/+aa*aaaaaa*1++01011-**+00100,1-[ 7] = 0
+*-/+aaaaaa/+aa*aaaaaa*+1+01111-**+00100,1-[ 8] = 0
++a*aaaaaaa--+/+aaaaaa/*1101000**/010010,1-[ 9] = 74.28295
*aa++aaaaaa+-**-aaaaaa+0*-11010**-011010,0-[10] = 353.5659
+++/aaaaaaa--+/+aaaaaa*11/01001**/010010,0-[11] = 28.95218
*aa++aaaaaa-**/+aaaaaa+0*-11010**-011010,1-[12] = 46.24757
```

```
*aa++aaaaaa-**/+aaaaaa+0*-01111-**+00100,0-[13]  = 376.9933
*aa++aaaaaa-**/+aaaaaa+0*-01010**/010010,0-[14]  = 264.9838
*aa++aaaaaa-**/+aaaaaa+0*-11010**-011010,1-[15]  = 46.24757
+++aaaaaaaa--+/+aaaaaa*11101000**-011110,0-[16]  = 28.95218
*aa++aaaaaa-aa+*aaaaaa+0*-11010**-011010,0-[17]  = 376.9933
--+/+aaaaaa+++/aaaaaaa*11/01000**/010010,1-[18]  = 0
*aa++aaaaaa-**/+aaaaaa+0*-11110**-011010,0-[19]  = 376.9933
```

a.   0123456012301234560123012345601012345601
     *aa++aaaaaa-**/+aaaaaa+0*-11110**-011010

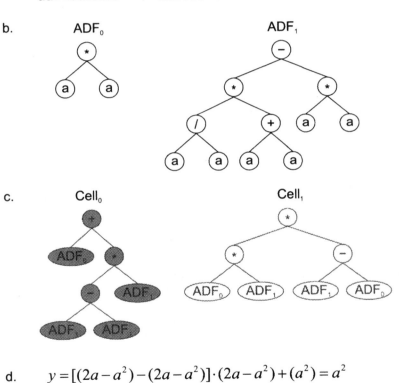

b.   ADF₀   ADF₁

c.   Cell₀   Cell₁

d. $y = [(2a - a^2) - (2a - a^2)] \cdot (2a - a^2) + (a^2) = a^2$

Figure 6.5. One of the best individuals of generation 1 (chromosome 19). It has a fitness of 376.9933 evaluated against the set of fitness cases presented in Table 6.1. **a)** The chromosome of the individual with the best cell shown in bold. **b)** The ADFs codified by each gene. **c)** The cells or main programs encoded in the homeotic genes (the best cell is shown in gray). **d)** The mathematical expression encoded in the best cell (the contribution of each ADF is shown in parentheses). Note that the calls to ADF₁ in the active cell have no impact in the overall solution as they are part of a neutral block.

The best-of-generation models (chromosomes 13 and 19) have an R-square of 0.01346 and a fitness of 376.9933 and, therefore, are better (in terms of fitness alone) than the best program of the previous generation. The plots for the target and predicted values are shown in Figure 6.6.

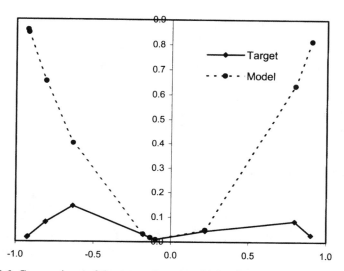

Figure 6.6. Comparison of the target function (6.1) with one of the best models of generation 1 (chromosome 19; see its expression in Figure 6.5). It has a fitness of 376.9933 and an R-square of 0.01346.

It's worth noticing that these best-of-generation individuals are direct descendants of the best individual of the previous generation. In fact, chromosome 19 codes not only for the same ADFs but also for a cell that is exactly the same – $Cell_1$ – which, interestingly, was the best cell in the parental chromosome. Thus, thanks to a small point mutation in the homeotic gene 0, the focus shifted from $Cell_1$ to $Cell_0$, which now codes for the best or active cell.

In the next generation there was no improvement in best fitness and the best ten individuals of this generation are all descendants of either chromosome 13 or 19 of the previous generation:

```
Generation N: 2
012345678900123456789001234567801234567 8
*aa++aaaaaa-**/+aaaaaa+0*-11110**-011010,0-[ 0] = 376.9933
-**-+aaaaaa--*/*aaaaaa+0*-01010**/010010,1-[ 1] = 0
*aa++aaaaaa-**/+aaaaaa0+*-11010**-011010,0-[ 2] = 376.9933
```

```
*aa++aaaaaa-**/+aaaaaa10*-01010**/010110,1-[ 3] = 99.37589
*aaa+aaaaaa*aa++aaaaaa+0*-01111-**+00100,0-[ 4] = 376.9933
++a*aaaaaaa--+/+aaaaaa/*1101000**/000010,1-[ 5] = 0
*aa++aaaaaa-aa++aaaaaa+0*-11010**-001010,0-[ 6] = 376.9933
*aa/+aaaaaa-aa+*aaaaaa+0*-11010**-011010,0-[ 7] = 376.9933
*aa++aaaaaa-**/+aaaaaa+0*-01010**/010010,0-[ 8] = 264.9838
++a-aaaaaaa--+/+aaaaaa/*1101000*//011010,1-[ 9] = 0
+++/aaaaaaa--+/+aaaaaa011/01001**0/10110,1-[10] = 0
*aa++aaaaaa-**/+aaaaaa+0*-01010**/010010,0-[11] = 264.9838
+aa*aaaaaaa+a*--aaaaaa+0*-11010**-*10010,1-[12] = 0
*aa++aaaaaa-**/+aaaaaa*-+011110**-011010,1-[13] = 46.24757
*aa++aaaaaa+-*--aaaaaa+0*-11010**-011010,0-[14] = 376.9933
*aa++aaaaaa*aa++aaaaaa+0*-11010***-10010,0-[15] = 376.9933
*aa++aaaaaa-**/+aaaaaa+0*-11110**-011010,0-[16] = 376.9933
-***aaaaaaa-**/+aaaaaa+01-11110**-011010,1-[17] = 0
*aa++aaaaaa-**/+aaaaaa+0*-01111-**+00101,0-[18] = 376.9933
*aa++aaaaaa-**/+aaaaaa+0*-01111-**+00100,0-[19] = 376.9933
```

It is worth noticing how neutral mutations accumulate quickly in the neutral regions of these best-of-generation genomes (the entire gene 1 and the entire homeotic gene 1), allowing the quick divergence of these genes. Note also that gene 0, the one encoding the most important ADF, shows almost no variation and, in all cases, encodes exactly the same structure.

In the next generation a new individual was created, chromosome 3, considerably better than the best individuals of the previous generations. Its expression is shown in Figure 6.7. The complete population is shown below (the best of generation is shown in bold):

```
Generation N: 3
0123456789001234567890012345678012345678
*aa++aaaaaa-**/+aaaaaa+0*-01111-**+00100,0-[ 0] = 376.9933
*aa++aaaaaa*aa++aaaaaa+0*-11010***-10110,0-[ 1] = 376.9933
aa*+aaaaaaa-**/+aaaaaa+0*-11111*+**11010,1-[ 2] = 94.71163
*aa++aaaaaa-aa++aaaaaa+0*-01010****01010,0-[ 3] = 421.5158
*aa++aaaaaa+-*--aaaaaa+0*-11010**-011010,0-[ 4] = 376.9933
+*-aaaaaaaa*aaa/aaaaaa+0*-11010***-10010,0-[ 5] = 376.9933
*aa++aaaaaa-**/+aaaaaa+0*-01010**/010010,0-[ 6] = 264.9838
*aa++aaaaaa+-a*-aaaaaa+0*-11010**-011010,1-[ 7] = 148.9205
*aa++aaaaaa-aa++aaaaaa+0*-110101*-001010,0-[ 8] = 376.9933
*aaa+aaaaaa*+aa+aaaaaa+0*-111100*-011010,1-[ 9] = 376.9933
*aaa+aaaaaa*aa++aaaaaa+0*-01111-**+00100,0-[10] = 376.9933
*aa+*aaaaaa-**/+aaaaaa+0*-11110**-011010,0-[11] = 376.9933
-**/-aaaaaa*aa++aaaaaa+0*-11110-***10010,1-[12] = 0
*aa++aaaaaa-**/+aaaaaa+0*-01011-**+10100,0-[13] = 264.9838
```

```
*aa++aaaaaa*+a++aaaaaa+0*-11010**-*11010,0-[14]  = 236.4288
*aa/+aaaaaa-aa+*aaaaaa0+*-11010*/-011010,3-[15]  = 0
*aa++aaaaaa*aa++aaaaaa+0*-01111-+**00100,1-[16]  = 376.9933
*aaa+aaaaaa*aa++aaaaaa+**-11111-*-+00100,0-[17]  = 99.37589
*aaa+aaaaaa++*aaaaaaaa+0*-110100**-10010,1-[18]  = 376.9933
*aa++aaaaaa*aa++aaaaaa+0*-01111-*-+00100,0-[19]  = 376.9933
```

As you can see in Figure 6.7, ADF_0 continues to be the ADF that counts, since ADF_1, which encodes zero, is used just as a neutral block in $Cell_0$. Note also that this new individual invokes ADF_0 three different times from the best main program, creating a modular solution that fits the target function (6.1) much more accurately than its predecessors (Figure 6.8).

a.   012345601230123456012301234560101234560101234560
     *aa++aaaaaa-aa++aaaaaa**+0**-01010****01010

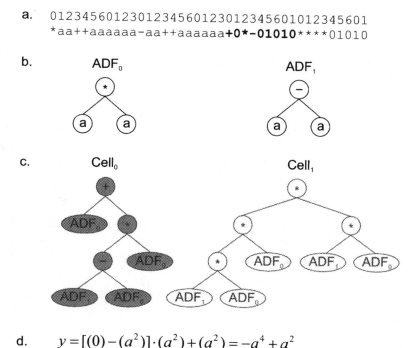

d. $y = [(0) - (a^2)] \cdot (a^2) + (a^2) = -a^4 + a^2$

Figure 6.7. Best individual of generation 3 (chromosome 3). It has a fitness of 421.5158 evaluated against the set of fitness cases presented in Table 6.1. **a)** The chromosome of the individual with the best cell shown in bold. **b)** The ADFs codified by each gene. **c)** The cells or main programs encoded in the homeotic genes (the best cell is shown in gray). **d)** The mathematical expression encoded in the best cell (the contribution of each ADF is shown in parentheses). Note that ADF_1 is doing nothing in the best cell.

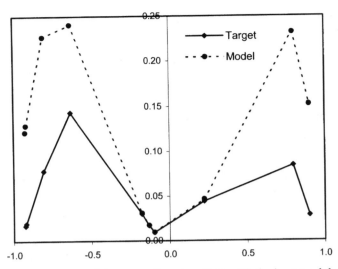

Figure 6.8. Comparison of the target function (6.1) with the best model of generation 3 (chromosome 3; see its expression in Figure 6.7). It has a fitness of 421.5158 and an R-square of 0.59947.

In the next four generations the best individuals are either clones of the best of the previous generations or variants (some minor, other not so minor) of the best (these chromosomes are highlighted in all the populations):

```
Generation N: 4
0123456789001234567890012345678012345678
*aa++aaaaaa-aa++aaaaaa+0*-01010****01010,0-[ 0] = 421.5158
*a*a+aaaaaa*aa++aaaaaa*0*-01111-**+00100,1-[ 1] = 3.155922
*aa++aaaaaa*+aa+aaaaaa+0*-111100*-011010,1-[ 2] = 376.9933
*aa++aaaaaa*aa++aaaaaa++-001111-+**00100,1-[ 3] = 376.9933
*+aa+aaaaaa+-*--aaaaaa+0*-11010**-011010,1-[ 4] = 0
*aa+*aaaaaa-**/+aaaaaa+0*-11110**-011010,0-[ 5] = 376.9933
*aa++aaaaaa*aa++aaaaaa00*-01010**1*01010,0-[ 6] = 376.9933
*aa++aaaaaa-**/+aaaaaa-+0*01110-/*+00100,1-[ 7] = 0
*aa++aaaaaa-**/+aaaaaa+0*-01011-*-+00100,0-[ 8] = 264.9838
-aa++aaaaaa+aa**aaaaaa+0*-11010**-011010,1-[ 9] = 0
*aaa+aaaaaa-aa++aaaaaa+0-*11010***-10110,0-[10] = 376.9933
*aa++aaaaaa*aa++aaaaaa+0*-010101*-001010,1-[11] = 376.9933
*aa++aaaaaa*aa++aaaaaa+0--11111-*-+00100,1-[12] = 23.42741
-aaaaaaaaaa+*aa+aaaaaa+0*-11110**-011010,1-[13] = 0
*aa++aaaaaa-aa++aaaaaa-*0+110101**001010,1-[14] = 0
+aa+*aaaaaa-**/+aaaaaa-+0*11110**-011010,1-[15] = 0
*aa++aaaaaa+**--aaaaaa+0*-01010****01010,0-[16] = 376.9933
```

```
*aaaaaaaaaa*aa++aaaaaa+0*-01110**/010010,0-[17]  = 376.9933
*aa++aaaaaa-aa++aaaaaa+0*+01111+*-+00100,0-[18]  = 376.9933
*aa+*aaaaaa-**/+aaaaaa+0*-01010****01010,0-[19]  = 264.9838
```

```
Generation N: 5
012345678900123456789001234567801234567 8
*aa++aaaaaa-aa++aaaaaa+0*-01010****01010,0-[ 0]  = 421.5158
/aa++aaaaaa*aaa-aaaaaa*0+-01010****01010,0-[ 1]  = 376.9933
*+a++aaaaaa*aa++aaaaaa++-011111-+**00100,1-[ 2]  = 0
aaa*+aaaaaa-aa++aaaaaa+0-*11010***-10110,1-[ 3]  = 0
*aa++aaaaaa*aa++aaaaaa+0*-01010****01010,0-[ 4]  = 376.9933
*aa*aaaaaa-aa++aaaaaa+0*-00010****01010,0-[ 5]  = 376.9933
*+*+*aaaaaa-**/+aaaaaa-*0+11110**-011010,1-[ 6]  = 7.94248
/aa+*aaaaaa-**/+aaaaaa+0*-01010****01010,1-[ 7]  = 0
*aa+*aaaaaa-**/+aaaaaa+0*-01010***001010,0-[ 8]  = 264.9838
/aa++aaaaaa+*aa+aaaaaa++-001111-+**00100,1-[ 9]  = 0
*aa++aaaaaa*a+a+aaaaaa+0*-010101*-001010,0-[10]  = 353.5659
*aa++aaaaaa*aa++aaaaaa+0*-010101*-001010,1-[11]  = 376.9933
*aa+*aaaaaa**-**aaaaaa+0*-01010**-+00100,0-[12]  = 421.5158
*aa++aaaaaa*aa+aaaaaaa+0*-111100*-*10010,1-[13]  = 376.9933
*aa++aaaaaa-aa++aaaaaa+0*+01111+**-01110,0-[14]  = 376.9933
*aa++aaaaaa-aa++aaaaaa*-+00101010-000010,1-[15]  = 0
*a+++aaaaaa-aa++aaaaaa+0*+01111+*-+00100,1-[16]  = 0
*a+aaaaaaaaa+aa+aaaaaa+0*-111100*-011010,1-[17]  = 0
*aa-+aaaaaa*aa++aaaaaa00+-01010**1*01010,0-[18]  = 376.9933
*aa++aaaaaa-aa++aaaaaa+0*-01110***011010,0-[19]  = 376.9933
```

```
Generation N: 6
012345678900123456789001234567801234567 8
*aa+*aaaaaa**-**aaaaaa+0*-01010**-+00100,0-[ 0]  = 421.5158
*+aa+aaaaaa*aa++aaaaaa+0--01110*--011010,1-[ 1]  = 0
/aa++aaaaaa*aaa-aaaaaa*0+-01010****01010,0-[ 2]  = 376.9933
*aa++aaaaaa*aa++aaaaaa+0*-01010****01010,0-[ 3]  = 376.9933
*aaa-aaaaaa-aa+*aaaaaa+0*+01111+**-01110,0-[ 4]  = 376.9933
*aa+*aaaaaa*aa++aaaaaa+0+-01010**1*01010,1-[ 5]  = 23.93407
*aa+*aaaaaa-**/+aaaaaa+0+-01011****01010,1-[ 6]  = 74.18751
++aa/aaaaaa-*aaaaaaaaa*0*-01010**0*01010,0-[ 7]  = 65.07316
*aa-+aaaaaa-**/+aaaaaa+0*-01010***001010,0-[ 8]  = 264.9838
*aa+*aaaaaa*aa++aaaaaa00+-01010***011010,0-[ 9]  = 376.9933
*aa++aaaaaa*aaaaaaaaaa+0*-111100*--10010,1-[10]  = 376.9933
*aa++aaaaaa*aa+aaaaaaa+0*-111100*-*11000,1-[11]  = 376.9933
*aa++aaaaaa*aa+aaaaaaa0+*-111100*-*10010,1-[12]  = 376.9933
*aa-+aaaaaa*aa++aaaaaa00*-01010****01011,0-[13]  = 376.9933
*aa++aaaaaa**-**aaaaaa*-+001010**-+00100,1-[14]  = 91.13461
*aa+/aaaaaa-aa++aaaaaa+0*-010101*-001010,0-[15]  = 421.5158
*aa++aaaaaa*aa++aaaaaa+0*-01010****01010,0-[16]  = 376.9933
```

```
*a*aaaaaaaa/aa++aaaaaa*0+-01010****01010,1-[17]  = 45.10381
*aa++aaaaaa/aaa+aaaaaa*0+-01010*0**01010,0-[18]  = 376.9933
*aa+*aaaaaa**-**aaaaaa+/*-01010**-+00100,1-[19]  = 91.13461
```

```
Generation N: 7
012345678900123456789001234567801234 5678
*aa+/aaaaaa-aa++aaaaaa+0*-010101*-001010,0-[ 0]  = 421.5158
*a-++aaaaaa*aa++aaaaaa+0*-111100*--10010,1-[ 1]  = 0
*aa+aaaaaa*aa++aaaaaa+0*-011100*-*11000,1-[ 2]  = 376.9933
*aa++aaaaaa*aa++aaaaaa00*-01011****01010,0-[ 3]  = 376.9933
/aa++aaaaaa*aaa-aaaaaa/0+-01010****01010,1-[ 4]  = 99.37589
*aa++aaaaaa*aaaaaaaaaa+0*-00010**-+00100,0-[ 5]  = 376.9933
*aa+*aaaaaa*aa++aaaaaa00+-01010***011010,0-[ 6]  = 376.9933
*a*+/aaaaaa*aa*aaaaaaa-0+*111100*-*10010,0-[ 7]  = 421.5158
+aa*+aaaaaa-a*++aaaaaa+0*-010101*-001011,1-[ 8]  = 56.58727
*+a+*aaaaaa*aa+aaaaaaa+0*+01011+**-01110,1-[ 9]  = 56.16257
*aa-+aaaaaa*aa+aaaaaaa+-*-01010-***01011,1-[10]  = 0
+/aa+aaaaaa*a*aaaaaaaa**0+01010****01010,0-[11]  = 93.40756
/aa++aaaaaa*aa++aaaaaa*0+-01010****01010,0-[12]  = 376.9933
*aa+/aaaaaa*aa++aaaaaa+0*-01010****01010,0-[13]  = 376.9933
*aa++aaaaaa*aaa-aaaaaa1+0*01010****01010,0-[14]  = 376.9933
*aa++aaaaaa*aa+aaaaaaa+0*-111100*-*11000,1-[15]  = 376.9933
*aa+*aaaaaa**-*/aaaaaa+0*-01010****01010,0-[16]  = 421.5158
*aa-+aaaaaa-**/+aaaaaa*+0*01010***001010,0-[17]  = 11.77589
*aa+*aaaaaa*aa++aaaaaa00+-01010**+*11010,0-[18]  = 376.9933
*aaa-aaaaaa-a+a+aaaaaa00+-01110***001010,0-[19]  = 376.9933
```

Note, again, how divergence takes over in these best-of-generation programs: although encoding the same solution, their genomes started to differ, generating genetic diversity through the accumulation of neutral mutations that might be used in future generations to create better adapted programs.

Finally, by generation 8 an individual with maximum fitness (chromosome 11), matching exactly the target function (6.1), was created. Its expression is shown in Figure 6.9. The complete population is shown below (the best of generation is shown in bold):

```
Generation N: 8
012345678900123456789001234567801234 5678
*aa+*aaaaaa**-*/aaaaaa+0*-01010****01010,0-[ 0]  = 421.5158
*aa+*aaaaaa-+a+aaaaaaa+0*-01010***011010,0-[ 1]  = 265.8303
*+a+*aaaaaa*a-*/aaaaaa+0*-01010****01110,1-[ 2]  = 51.42732
*aa++aaaaaa-+a++aaaaaa1+0*01010****01010,1-[ 3]  = 36.86388
*aa+/aaaaaa*aaa-aaaaaa+0*-0101010-*01010,1-[ 4]  = 376.9933
/*aaaaaaaaa*aa++aaaaaa1+0*01010****01010,0-[ 5]  = 376.9933
*aa++aaaaaa*aaaaaaaaaa1+0*01010****01010,0-[ 6]  = 376.9933
```

```
*aa+*aaaaaa**-*/aaaaaa+0*-11010***01010,0-[ 7] = 376.9933
*aa++aaaaaa*aaa-aaaaaa-0+*111100*-*10011,1-[ 8] = 376.9933
*aa+*aaaaaa*+aaaaaaaaa00+-01110***01010,0-[ 9] = 376.9933
+-*aaaaaaaa*aa+aaaaaaa+01-111100*-*11000,1-[10] = 376.9933
*aa+*aaaaaa*-**/aaaaaa+0*-01010***01011,0-[11] = 1000
**-*/aaaaaa*aa+*aaaaaa+0*+01011+**-01010,1-[12] = 0
*aa++aaaaaa*aaa-aaaaaa+0*-00010*+*-00100,0-[13] = 376.9933
*aa++aaaaaa*aa++aaaaaa00*-01011***01010,0-[14] = 376.9933
*aa++aaaaaa*aa/aaaaaa1+0*01010*-**01010,0-[15] = 376.9933
*/a*aaaaaaa*a*+/aaaaaa-0+-01010*+**01010,1-[16] = 38.04328
```

a.   012345601230123456012301234560101 2345601
      *aa+*aaaaaa*-**/aaaaaa**+0*-01010******01011

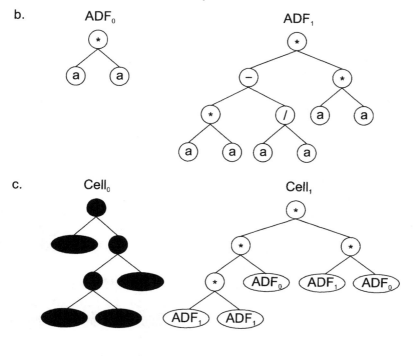

b.   ADF₀   ADF₁

c.   Cell₀   Cell₁

d. $y = [(a^4 - a^2) - (a^2)] \cdot (a^2) + (a^2) = a^6 - 2a^4 + a^2$

Figure 6.9. Perfect solution to the sextic polynomial function (6.1). This model, with maximum fitness and an R-square of 1.0, was created in generation 8. **a)** The chromosome of the individual with the best cell shown in bold. **b)** The ADFs codified by each gene. **c)** The cells or main programs encoded in the homeotic genes (the best cell is shown in gray). **d)** The mathematical expression encoded in the best cell (the contribution of each ADF is shown in parentheses).

```
*a++/aaaaaa*aa++aaaaaa+0*-01010*+**01010,0-[17]  = 54.46888
*aa+/aaaaaa*aa++aaaaaa*-+001010**--01010,1-[18]  = 0
/aa++aaaaaa-*aaaaaaaaa/0+-01010****01010,1-[19]  = 320.8196
```

As you can see in Figure 6.9, both the active cell (Cell$_0$) and ADF$_0$ remained unchanged not only in this perfect solution but also in most of the best models of the previous generations (compare Figures 6.7 and 6.9). Indeed, what was at the heart of this improvement in fitness was a modification in gene 1 (probably a small inversion) during the reproduction of chromosome 16 of generation 7, resulting in the perfect solution of generation 8:

```
0123456789001234567890012345678012345678
*aa+*aaaaaa**-*/aaaaaa+0*-01010****01010,0-[7,16]  = 421.516
*aa+*aaaaaa*-**/aaaaaa+0*-01010****01011,0-[8,11]  = 1000
```

As you can see, in both models, the active main programs have exactly the same structure and both invoke ADF$_0$ three times, whereas ADF$_1$ is used just once. But thanks to the small inversion in gene 1, a completely new program matching perfectly the target function (6.1) is expressed in the active cell.

6.2 Odd-parity Functions

The motivation behind the implementation of automatically defined functions in genetic programming, was the belief that ADFs allow the evolution of modular solutions and, consequently, improve the performance of the GP technique (Koza 1992, 1994; Koza et al. 1999). Koza proved this by solving the sextic polynomial problem we analyzed in the previous section and the even-parity functions (Koza 1994).

Here, we are going to solve the already familiar odd-parity functions in order to compare this new form of dealing with modular solutions with other simpler forms, namely, the simpler acellular systems with and without UDFs we studied in chapter 4 (section Odd-parity Functions).

So, we have seen already in chapter 4 that finding a solution to higher order odd-parity functions is extremely difficult using just ANDs, ORs, and NOTs in the function set (see Table 4.14). But with the help of UDFs we were able to discover the pattern underlying the construction of these functions and, consequently, create a scheme to design any higher order odd-parity function (see Table 4.15 and Figure 4.9). Automatically defined functions can also be used to discover patterns that might be useful to design good solutions. We will see, however, that these patterns are not as clear cut as the patterns discovered with the help of user defined functions as the system is highly inventive and extremely flexible. Indeed, the system can discover a large variety of building blocks that it then uses to build extremely interesting modular solutions.

So, in this study, we will continue to work with the Boolean system of ANDs, ORs, and NOTs and, therefore, we are going to use these functions both in the ADFs and in the main programs, that is, $F = F_H = \{A, O, N\}$. The set of terminals used in the genes encoding the ADFs consists obviously of the independent variables (two for the odd-2-parity function; three for the odd-3-parity; and four for the odd-4-parity function); for the homeotic genes, the terminal set consists obviously of the ADFs encoded in the conventional genes (just two, in all the experiments) and will be represented by "0" and "1", thus giving $T_H = \{0, 1\}$. The fitness function is based on the number of hits and will be evaluated by equation (3.8). The complete list of the parameters used per run, together with the performance obtained in each experiment, is shown in Table 6.3.

And as you can see in Table 6.3, the use of ADFs results in a considerable increase in performance for the odd-3 and odd-4-parity functions (compare Tables 4.14 and 6.3). This is the telltale mark that there is indeed some regularity in the organization of these functions that the system is successfully exploring to build modular solutions. Let's then analyze some of the perfect solutions (three, in total) designed in these experiments to see what kind of modules are discovered and reused by the main programs.

The structures of three perfect solutions to the odd-3-parity function are shown in Figure 6.10 and, as you can see, the system has no problems in creating a wide variety of building blocks (ADFs) and then combine them in different ways, calling each of them as many times as necessary in the main program. For instance, in (a), the system created an extremely simple building block (ADF_0) encoding a one-argument function that returns the value of the argument c and a more elaborated one (ADF_1) involving the arguments a and b. Then, in the main program, these building blocks are each called twice and combined in different ways through the use of different linkers, forming a perfect solution. It's worth noticing that the pattern discovered here matches exactly the one discovered with UDFs (see Figure 4.9 in chapter 4). In (b), a small module (ADF_0) involving arguments b and c, and a larger one (ADF_1) involving the three arguments to the odd-3-parity function, are combined in the main program through different linkers to design a perfect solution. Note,

Table 6.3

Performance and parameters for the odd-n-parity problem with ADFs.

	Odd-2	Odd-3	Odd-4
Number of runs	100	100	100
Number of generations	50	10000	100000
Population size	30	30	30
Number of fitness cases	4	8	16
Function set of ADFs	A O N	A O N	A O N
Terminal set	a b	a b c	a b c d
Head length	7	7	10
Gene length	15	15	21
Number of genes/ADFs	2	2	2
Function set of homeotic genes	A O N	A O N	A O N
Terminal set of homeotic genes	ADFs 0-1	ADFs 0-1	ADFs 0-1
Number of homeotic genes/cells	3	3	3
Head length of homeotic genes	7	7	7
Length of homeotic genes	15	15	15
Chromosome length	75	75	87
Mutation rate	0.044	0.044	0.044
Inversion rate	0.1	0.1	0.1
IS transposition rate	0.1	0.1	0.1
RIS transposition rate	0.1	0.1	0.1
One-point recombination rate	0.3	0.3	0.3
Two-point recombination rate	0.3	0.3	0.3
Gene recombination rate	0.3	0.3	0.3
Gene transposition rate	0.1	0.1	0.1
Mutation rate in homeotic genes	0.044	0.044	0.044
Inversion rate in homeotic genes	0.1	0.1	0.1
RIS transposition rate in homeotic genes	0.1	0.1	0.1
IS transposition rate in homeotic genes	0.1	0.1	0.1
Fitness function	Eq. (3.8)	Eq. (3.8)	Eq. (3.8)
Success rate	100%	97%	44%

again, that each ADF is called twice from the main program. In (c), the ADFs designed by the system are both of them functions of three arguments. And both these functions are called twice from different places in the main program. These are just some examples of the variety of building blocks the system is able to invent and then immediately use to build modular solutions to the problem at hand.

In Figure 6.11 is shown a small sample of modular solutions to the more complex odd-4-parity function designed with the cellular system. For

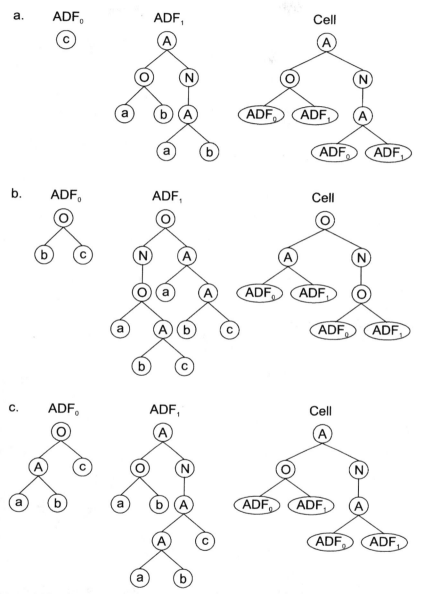

Figure 6.10. Three perfect solutions to the odd-3-parity function designed with ADFs. Note that different building blocks are discovered and immediately used to build the main program or cell. Note also that the number of arguments of each ADF is not fixed and the system is free to create and then try any kind of building block. The programs presented here have been simplified to better perceive the underlying pattern and only the active cell is shown. Note that the pattern in **(a)** is the already familiar pattern discovered in section 4.3.3 with the help of UDFs.

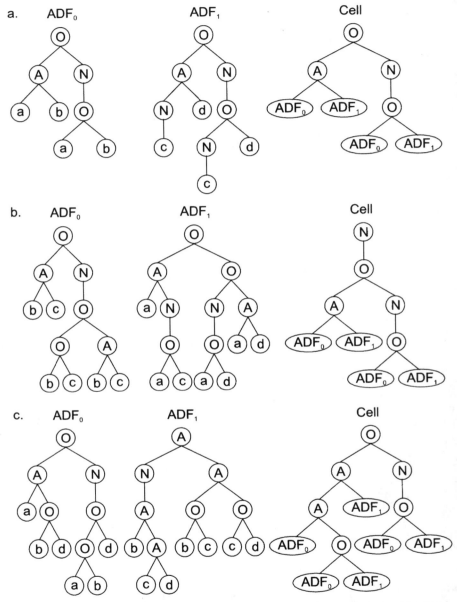

Figure 6.11. Three perfect solutions to the odd-4-parity function designed with ADFs. Note that different building blocks are discovered and immediately used to build the main program or cell. Note also that the number of arguments of each ADF is not fixed and the system is free to create and then try any kind of building block. The programs presented here have been simplified to better perceive the underlying pattern and only the active cell is shown.

instance, in (a), a small building block (ADF$_0$) involving arguments a and b, and a larger one (ADF$_1$) involving arguments c and d, were created and then combined in different ways in the main program in order to design a perfect solution to the odd-4-parity function. And as you can see, they are both called twice from the main program. In (b), two relatively large building blocks, each involving a different set of variables (ADF$_0$ is a function of b and c, whereas ADF$_1$ is a function of a, c, and d) were discovered. Then these ADFs are linked together in the main program to form a perfect solution. Note, again, that each building block is expressed twice in the cell. And in (c), a more verbose solution was created, involving two relatively large building blocks, each using a different set of arguments (ADF$_0$ is a function of a, b, and d, whereas ADF$_1$ is a function of b, c, and d). Then these ADFs are each invoked from three different places in the main program and a perfect solution to the odd-4-parity function is also created with them.

6.3 Kepler's Third Law

Kepler's Third Law is an interesting problem to solve with the cellular system of gene expression programming as it can be used to illustrate the advantages of the flexible linking that takes place in the cells. Moreover, this scientific law, considered by Langley at al. (1987) one of the most challenging to have been rediscovered by a family of heuristic techniques for inducing scientific laws from empirical data, was also rediscovered by genetic programming using populations of 500 individuals for 51 generations (Koza 1992). We will see here that gene expression programming can discover Kepler's Third Law using population sizes of just 10 individuals evolving for 50 generations.

Kepler's Third Law states that $P^2 = cD^3$, where P is the orbital period of a planet and D is the average distance from the Sun. When P is expressed in units of Earth's years and D is expressed in units of Earth's average distance from the Sun, $c = 1$.

The fitness cases used to rediscover Kepler's Third Law are the astronomical data for the six planets presented in Table 6.4, where D is given in units of Earth's semimajor axis of orbit and P is given in units of Earth's years (Urey 1952). So, the goal here consists of finding a relationship describing the orbital period of a planet as a function of its average distance from the Sun. For that we could define the fitness to be the number of outputs that are within 20% of the correct value and, therefore, evaluate the fitness by equation (3.1), thus giving $f_{max} = 6$.

Table 6.4
Planetary data used to rediscover Kepler's Third Law.

Planet	Distance	Orbital Period
Venus	0.72	0.61
Earth	1.00	1.00
Mars	1.52	1.84
Jupiter	5.20	11.9
Saturn	9.53	29.4
Uranus	19.1	83.5

In this study, we are going to use three different algorithms to solve the orbital period problem. The first one consists of the basic gene expression algorithm with just one gene (UGS); the second consists again of the basic GEA, but a multigenic system composed of two genes with the sub-ETs linked by multiplication will be used (MGS); and the last one consists of the cellular system with two ADFs (MCS). For all the experiments, in the function set we will use the basic arithmetical operators plus the square root function "Q", thus giving $F = \{+, -, *, /, Q\}$ (and, in the cellular system, the same function set will be used in the homeotic genes, that is, $F_H = \{+, -, *, /, Q\}$); the terminal set will consist obviously of the independent variable D, which will be represented by "a", thus giving $T = \{a\}$. This study, with its three different experiments, is summarized in Table 6.5.

And as you can see in Table 6.5, in all the experiments, the discovery of Kepler's Third Law was an easy task for gene expression programming. Indeed, using populations of just 10 individuals evolving for just 50 generations, the unigenic system discovered Kepler's Third Law in 76 out of 100 runs; the multigenic system in 96% of the runs; and the multicellular system in 99% of the runs. Let's now analyze some of the perfect solutions discovered in the three experiments.

The first 10 perfect solutions discovered in the UGS experiment are shown below (the numbers in square brackets indicate, respectively, the run and the generation by which they were created; and the number in parentheses after the fitness indicates the program size):

```
012345678901234
Q*a*/a*aaaaaaaa-[01,01] = 6 (10)
/*/aaaQaaaaaaaa-[04,25] = 6 (8)
*Q**Q/aaaaaaaaa-[05,15] = 6 (11)
```

```
/Q*a///aaaaaaaa-[07,16] = 6 (12)
*Qaa/a/aaaaaaaa-[08,01] = 6 (4)
+-Qaa**aaaaaaaa-[09,01] = 6 (10)
*QaaaQ+aaaaaaaa-[10,49] = 6 (4)
///++Q*aaaaaaaa-[11,05] = 6 (14)
*aQaaa*aaaaaaaa-[13,39] = 6 (4)
*Qaaaaaaaaaaaaa-[17,08] = 6 (4)
```

As you can see by their expression, all of them are exact formulations of Kepler's Third Law. Take, for instance, the last of the solutions presented

Table 6.5
Performance and parameters for the orbital period problem using a unigenic system (**UGS**), a multigenic system (**MGS**), and a cellular system (**MCS**).

	UGS	MGS	MCS
Number of runs	100	100	100
Number of generations	50	50	50
Population size	10	10	10
Number of fitness cases	6 (Table 6.4)	6 (Table 6.4)	6 (Table 6.4)
Function set of normal genes/ADFs	+ - * / Q	+ - * / Q	+ - * / Q
Terminal set	a	a	a
Head length	7	7	7
Gene length	15	15	15
Number of genes/ADFs	1	2	2
Linking function	--	*	--
Function set of homeotic genes	--	--	+ - * / Q
Terminal set of homeotic genes	--	--	ADFs 0-1
Number of homeotic genes/cells	--	--	3
Head length of homeotic genes	--	--	4
Length of homeotic genes	--	--	9
Chromosome length	15	30	49
Mutation rate	0.044	0.044	0.044
Inversion rate	0.1	0.1	0.1
IS transposition rate	0.1	0.1	0.1
RIS transposition rate	0.1	0.1	0.1
One-point recombination rate	0.3	0.3	0.3
Two-point recombination rate	0.3	0.3	0.3
Gene recombination rate	--	0.3	0.3
Gene transposition rate	--	0.1	0.1
Mutation rate in homeotic genes	--	--	0.044
Inversion rate in homeotic genes	--	--	0.1
RIS transposition rate in homeotic genes	--	--	0.1
IS transposition rate in homeotic genes	--	--	0.1
Fitness function	Eq. (3.1)	Eq. (3.1)	Eq. (3.1)
Precision	20%	20%	20%
Success rate	76%	96%	99%

above, one of the most compact programs discovered with the unigenic system. More formally, its expression results in the following equation:

$$P = a \cdot \sqrt{a} = \sqrt{a^3}$$ (6.2)

which is a perfect match for Kepler's Third Law.

The first 10 perfect solutions discovered in the MGS experiment are shown below (the sub-ETs are linked by multiplication):

```
01234567890123401234 5678901234
/*aQ*/aaaaaaaaaQaaaaQaaaaaaaaa-[0,12] = 6 (12)
QaaQ-+aaaaaaaaa*a+/-++aaaaaaaa-[1,00] = 6 (15)
Qa/aa/QaaaaaaaaaQ++QQ+aaaaaaaa-[2,11] = 6 (3)
QQ/Q***aaaaaaaa/***aQQaaaaaaaa-[3,14] = 6 (23)
Qa*-+a/aaaaaaaaQ***a//aaaaaaaa-[4,08] = 6 (14)
-a-aaQ/aaaaaaaaQQ*aa*Qaaaaaaaa-[5,10] = 6 (10)
*/*Q/Q/aaaaaaaaQaQQQa*aaaaaaaa-[6,30] = 6 (15)
*a-+++/aaaaaaaa-a+a--Qaaaaaaaa-[7,06] = 6 (23)
/*QaQQQaaaaaaaa-+aa+-Qaaaaaaaa-[8,13] = 6 (19)
Q//////aaaaaaaa+aQ**Q-aaaaaaaa-[9,03] = 6 (25)
```

As their expression shows, all of them are exact formulations of Kepler's Third Law. Consider, again, the most parsimonious solution presented above, the one discovered in generation 11 of run 2. More formally, it can be represented by the following expression (the contribution of each gene is shown in brackets):

$$P = (\sqrt{a}) \cdot (a) = \sqrt{a^3}$$ (6.3)

which matches perfectly Kepler's Third Law.

And finally, the first 10 perfect solutions discovered by the cellular system using two ADFs and three cells are shown below (the best cell is indicated after the coma):

```
01234567890012345678900123456780123456780 12345678
+/-a/aaaaaa/*a/Qaaaaaa/*0*11100*--110110/Q*011110,2-[0,00] = 6
**a*/aaaaaaQ/aa+aaaaaa+0+Q11100Q+Q*01011Q/*Q01101,2-[1,12] = 6
+/a+/aaaaaa*Qaa+aaaaaaQ*1111010*-1/11110+*Q001011,0-[2,05] = 6
*a/**aaaaaa*aaaaaaaaaa/Q/001011Q11+11110//0Q00010,0-[3,03] = 6
*a*+/aaaaaa*aQaaaaaaaa10/+100010/QQ00011+QQ/01100,0-[4,27] = 6
*--/-aaaaaa/*a*QaaaaaaQ0+Q01001Q1Q001000*+/101110,2-[5,08] = 6
+QQ+aaaaaaaa*+/*aaaaaa/0Q100010*1Q101000/1QQ10000,1-[6,07] = 6
Qa--Qaaaaaa///*aaaaaaa*1+-11000*0/*01000+11/10001,1-[7,04] = 6
```

```
//***aaaaaa**aaaaaaaaa+*+100100Q1+110110+00Q11100,1-[8,09] = 6
*Q*/*aaaaaaQaa+-aaaaaaQ+-101110+-Q/10001//QQ00010,0-[9,00] = 6
```

As their expression shows, they also are exact formulations of Kepler's Third Law. Consider, for instance, the perfect solution discovered in generation 3 of run 3. It encodes the following expression (the contribution of each ADF is shown in brackets):

$$P = \frac{\sqrt{(a)}}{\dfrac{(a)}{(a^2)}} = \sqrt{a^3} \qquad (6.4)$$

and, therefore, is also a perfect formulation of Kepler's Third Law.

6.4 RNCs and the Cellular System

The complexity of both the uni- and multicellular systems is considerably higher than that of the acellular systems we have dealt with so far. And since this complexity was not something created naturally by the system itself, it is important to know how plastic these systems are, or in other words, how efficient is evolution in these systems.

We have seen already in the previous section that, at least for simple problems, the cellular system is extremely flexible and allows an efficient evolution. This gives hope that perhaps it is possible to increase the complexity of the cellular system even more in order to introduce new features, such as a facility for handling random numerical constants.

The analog circuit we are going to design in this section with the cellular system is the same of section 5.6.5 and, from the results obtained in that section, we know that random numerical constants are absolutely necessary to design this particular circuit. Thus, in this section, we are going to see first how random numerical constants are implemented in the cellular system and then proceed with the testing of this new algorithm on this challenging problem of analog circuit design.

6.4.1 Incorporating RNCs in ADFs

The incorporation of random numerical constants in automatically defined functions is not difficult. As you probably guessed, the gene structure used to accomplish this includes the special domain Dc for encoding the random numerical constants, which, for the sake of simplicity and efficiency, is only implemented in the genes encoding the ADFs (one can obviously extend this organization to the homeotic genes, but I have the feeling that nothing is gained from that except a considerable increase in computational effort). The structure of the homeotic genes remains exactly the same and they continue to control how often each ADF is called upon and how these ADFs interact with one another.

Consider, for instance, the chromosome below with two homeotic genes and two conventional genes encoding ADFs with random numerical constants (the Dc's are shown in bold):

```
0123456789001234567890012345678012345678
**?b?aa4701+/Q?ba?8536*0Q-10010/Q-+01111          (6.5)
```

and the respective arrays of random numerical constants:

C_0 = {0.664, 1.703, 1.958, 1.178, 1.258, 2.903, 2.677, 1.761, 0.923, 0.796}
C_1 = {0.588, 2.114, 0.510, 2.359, 1.355, 0.186, 0.070, 2.620, 2.374, 1.710}

The genes encoding the ADFs are expressed exactly as normal genes with a Dc domain (see section 5.2, Genes with Multiple Domains to Encode RNCs) and, therefore, the ADFs will, most probably, include random numerical constants (Figure 6.12). Then these ADFs with RNCs are called upon as many times as necessary from any of the main programs encoded in the homeotic genes. As you can see in Figure 6.12, in this case, ADF_0 is invoked twice in $Cell_0$ and once in $Cell_1$, whereas ADF_1 is used just once in $Cell_0$ and called three different times in $Cell_1$. Let's now see how the system copes with the evolution of these complex entities by solving a difficult problem requiring random numerical constants.

6.4.2 Designing Analog Circuits with the ADF-RNC Algorithm

The analog circuit we are going to design in this section is the same of section 5.6.5 and, therefore, whenever possible, we will keep the same settings, namely, the number of runs, population size, number of generations, function set, training set, fitness function, number of genes and head sizes and, of course, identical rates of modification in order to facilitate the comparisons between the different systems. Although we know that random numerical constants play a crucial

a.   0123456789001234567890012345678012345678
      ?b?aa4701**+/Q?ba?**8536***0Q-10010/Q-+01111

      C_0 = {0.664, 1.703, 1.958, 1.178, 1.258, 2.903, 2.677, 1.761, 0.923, 0.796}
      C_1 = {0.588, 2.114, 0.510, 2.359, 1.355, 0.186, 0.070, 2.620, 2.374, 1.710}

b.

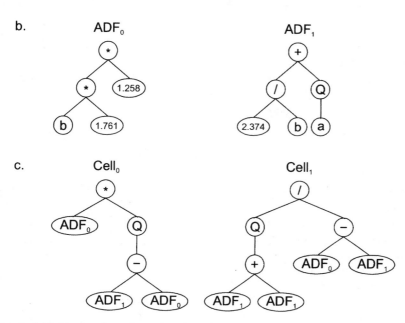

Figure 6.12. Expression of a multicellular program with ADFs containing random numerical constants. **a)** The genome of the individual with two homeotic genes and two normal genes plus their arrays of random numerical constants (the Dc's are shown in bold). **b)** The ADFs encoded by each normal gene. **c)** Two different programs expressed in two different cells.

role in the design of this particular circuit, we are also going to analyze the performance of the simpler cellular system without RNCs on this problem in order to compare it with the performance of the cellular system with RNCs (see Table 6.6 for the complete list of the parameters used per run in both experiments).

And as expected, the cellular system with RNCs performs considerably better than the simpler cellular system without RNCs at this task (average best-of-run fitness of 31.4709 against 13.6953). And taking into account the considerable increase in complexity, it is not surprising that the cellular

Table 6.6

Performance and parameters used to solve the analog circuit problem using a unicellular system without random numerical constants (**UCS**) and a unicellular system with ADFs containing RNCs (**UCS-RNC**).

	UCS	UCS-RNC
Number of runs	100	100
Number of generations	5000	5000
Population size	30	30
Number of fitness cases	40	40
Function set of ADFs	+ - * / P Q E L	+ - * / P Q E L
Terminal set	a b c	a b c ?
Random constants array length	--	10
Random constants type	--	Integer
Random constants range	--	[0, 100]
Head length	12	12
Gene length	25	38
Number of genes/ADFs	6	6
Function set of homeotic genes	+ - * /	+ - * /
Terminal set of homeotic genes	ADFs 0-5	ADFs 0-5
Number of homeotic genes/cells	1	1
Head length of homeotic genes	10	10
Length of homeotic genes	21	21
Chromosome length	171	249
Mutation rate	0.044	0.044
Inversion rate	0.1	0.1
IS transposition rate	0.1	0.1
RIS transposition rate	0.1	0.1
One-point recombination rate	0.3	0.3
Two-point recombination rate	0.3	0.3
Gene recombination rate	0.3	0.3
Gene transposition rate	0.1	0.1
Dc-specific mutation rate	--	0.044
Dc-specific inversion rate	--	0.1
Dc-specific transposition rate	--	0.1
Random constants mutation rate	--	0.01
Mutation rate in homeotic genes	0.044	0.044
Inversion rate in homeotic genes	0.1	0.1
RIS transposition rate in homeotic genes	0.1	0.1
IS transposition rate in homeotic genes	0.1	0.1
Fitness function	Eq. (3.5)	Eq. (3.5)
Average best-of-run fitness	13.6953	31.4709
Average best-of-run R-square	0.6708783722	0.8965381674

systems perform slightly worse than their acellular counterparts (see Table 5.10 for a comparison); these systems obviously require more time or larger populations to fine-tune their complex structures. Let's now see what kind of circuits can be designed with these systems.

The best solution designed with the cellular system without RNCs was found in generation 4571 of run 59:

```
012345678901234567 8901234
QQEa-QLP*+-+cbaacbbbacaab
EQ+QEQEQP+E-aabbbcaccbccc
L/QQQQ-QE-Eaabaacababacbb
QQ+b-+aL*Pcabbacccaacaabb
PPQabcaPb+PEabacbabccabcb
QQEcaab*abaPacbaacbbaacab
++-//1*+++53211505510
```
                                                                    (6.6a)

It has a fitness of 46.4137 and an R-square of 0.9358798486 evaluated over the training set of 40 fitness cases and is therefore a good solution to the problem at hand. Note that the R-square of this best-of-experiment solution is much higher than the average best-of-run R-square obtained in this experiment (0.6708783722), showing that, although harder to find, good solutions can nevertheless be designed with the cellular system without random numerical constants. Note also how flexible the linking is with this system, with not only addition being used to link the ADFs but also subtraction, multiplication, and division. And as we know, this kind of dynamic linking also allows code reuse. For instance, in the program above, ADF_1 is called from three different places in the main program. More formally, the model (6.6a) can be expressed by the following C++ program (ADF_4 is not used in the main program and, therefore, is not listed):

```
double ADF0(double d[])
{
    double dblTemp = 0.0;
    dblTemp = sqrt(sqrt(exp(d[0])));
    return dblTemp;
}

double ADF1(double d[])
{
    double dblTemp = 0.0;
    dblTemp = exp(sqrt((sqrt(sqrt(sqrt((d[0]+d[0])))))+
              exp(exp(pow(exp(d[1]),(d[1]-d[1]))))))));
    return dblTemp;
}
```

```
double ADF2(double d[])
{
    double dblTemp = 0.0;
    dblTemp = log((sqrt(sqrt((exp(d[0])-(d[0]-d[1])))))/
              sqrt(sqrt(sqrt(exp(d[0])))))));
    return dblTemp;
}

double ADF3(double d[])
{
    double dblTemp = 0.0;
    dblTemp = sqrt(sqrt((d[1]+((log(pow(d[1],d[1])))+
              (d[2]*d[0]))-d[0]))));
    return dblTemp;
}

double ADF5(double d[])
{
    double dblTemp = 0.0;
    dblTemp = sqrt(sqrt(exp(d[2])));
    return dblTemp;
}

double apsModel(double d[])
{
    double dblTemp = 0.0;
    dblTemp = ((((ADF1(d)+ADF1(d))/(ADF5(d)+ADF0(d)))+
              ((ADF5(d)+ADF5(d))/ADF5(d)))+
              (ADF1(d)-(ADF3(d)*ADF2(d)))));
    return dblTemp;
}
```
                                                                    (6.6b)

where d_0-d_2 correspond, respectively, to *a-c*. It is worth pointing out how different this solution is from the ones created with numerical constants (see model (6.7) below), as the algorithm had to find some creative ways of modeling this function without numerical constants at its disposal.

The best solution created by the cellular system with RNCs was found in generation 4799 of run 46. Its genes and corresponding arrays of random numerical constants are shown below:

```
Gene 0: P?c?*E?/-+**abbbbccac?cba7381293446232
     C0: {76, 4, 59, 83, 37, 52, 1, 45, 66, 42}

Gene 1: a/?/?c-+Lba/cc?cabaabcaa?9486967847808
     C1: {88, 4, 59, 48, 2, 66, 83, 79, 66, 42}
```

Gene 2: *a?E-+EQQEbcaacbab??baaab1126452672558
   C_2: {66, 6, 39, 22, 7, 27, 12, 98, 85, 71}

Gene 3: +?—bL/Qb-Q*??cbbccc?bb?c7785717851803
   C_3: {79, 92, 92, 20, 74, 34, 40, 5, 21, 67}

Gene 4: -+-*?c?E*-a-??c?cc?b?bbab6113140654932
   C_4: {80, 14, 56, 18, 97, 92, 75, 33, 34, 61}

Gene 5: Q?PEaEL*Ea*L?c????bacbc?c1072195508656
   C_5: {14, 88, 68, 39, 1, 11, 80, 85, 33, 35}

Homeotic Gene: -+14/4+3/*31220320254   (6.7a)

This model has a fitness of 66.1181 and an R-square of 0.9591176321 evaluated against the training set of 40 fitness cases and, therefore, is considerably better than the model (6.6) designed without random numerical constants. More formally, the model (6.7a) can be expressed by the following C++ program:

```cpp
double ADF0(double d[])
{
    double dblTemp = 0.0;
    dblTemp = pow(45,d[2]);
    return dblTemp;
}

double ADF1(double d[])
{
    double dblTemp = 0.0;
    dblTemp = d[0];
    return dblTemp;
}

double ADF2(double d[])
{
    double dblTemp = 0.0;
    dblTemp = (d[0]*6);
    return dblTemp;
}

double ADF3(double d[])
{
    double dblTemp = 0.0;
    dblTemp = (5+((log(sqrt(sqrt(21)))-(d[1]/
              ((d[2]*d[1])-5)))-d[1]));
    return dblTemp;
}
```

```
double ADF4(double d[])
{
    double dblTemp = 0.0;
    dblTemp = (((exp((14-18))*(d[0]*(d[2]-14)))+75)-
               (d[2]-14));
    return dblTemp;
}
double ADF5(double d[])
{
    double dblTemp = 0.0;
    dblTemp = sqrt(88);
    return dblTemp;
}
double apsModel(double d[])
{
    double dblTemp = 0.0;
    dblTemp = ((ADF4(d)+(ADF4(d)/(ADF3(d)+
               ((ADF1(d)*ADF2(d))/ADF3(d))))))-ADF1(d));
    return dblTemp;
}
```
                                                              (6.7b)

Note how profusely the cellular system with the facility for handling random numerical constants uses the RNCs at its disposal. Note also that structurally this program is very different from the model (6.6) created by the cellular system without random numerical constants. Indeed, as observed for the acellular systems, the models built with RNCs are usually much more compact and are also usually less varied in terms of the function nodes used in their construction.

6.5 Diagnosis of Breast Cancer

In this section we are going to solve a complex, real-world classification problem with the cellular system, both with and without random numerical constants, in order to evaluate the evolvability of these complex learning systems. Again, we are going to use settings very similar to the ones used in section 5.6.4 in order to compare the complex cellular systems with the much simpler acellular systems. Both the performance and parameters used per run are shown in Table 6.7.

And as you can see, despite their higher complexity, the cellular systems perform quite well at this difficult task and, as expected, the cellular system

Table 6.7
Performance and parameters used to diagnose breast cancer using a
multicellular system without random numerical constants (**MCS**) and a
multicellular system with ADFs containing RNCs (**MCS-RNC**).

	MCS	MCS-RNC
Number of runs	100	100
Number of generations	1000	1000
Population size	30	30
Number of fitness cases	350	350
Function set of ADFs	4(+ - * /)	4(+ - * /)
Terminal set	d0-d8	d0-d8 ?
Random constants array length	--	10
Random constants type	--	Rational
Random constants range	--	[-2, 2]
Head length	7	7
Gene length	15	23
Number of genes/ADFs	3	3
Function set of homeotic genes	+ - * /	+ - * /
Terminal set of homeotic genes	ADFs 0-2	ADFs 0-2
Number of homeotic genes/cells	3	3
Head length of homeotic genes	7	7
Length of homeotic genes	15	15
Chromosome length	90	114
Mutation rate	0.044	0.044
Inversion rate	0.1	0.1
IS transposition rate	0.1	0.1
RIS transposition rate	0.1	0.1
One-point recombination rate	0.3	0.3
Two-point recombination rate	0.3	0.3
Gene recombination rate	0.3	0.3
Gene transposition rate	0.1	0.1
Dc-specific mutation rate	--	0.044
Dc-specific inversion rate	--	0.1
Dc-specific transposition rate	--	0.1
Random constants mutation rate	--	0.01
Mutation rate in homeotic genes	0.044	0.044
Inversion rate in homeotic genes	0.1	0.1
RIS transposition rate in homeotic genes	0.1	0.1
IS transposition rate in homeotic genes	0.1	0.1
Fitness function	Eq. (3.10)	Eq. (3.10)
Rounding threshold	0.5	0.5
Average best-of-run fitness	339.19	339.06

without RNCs performs slightly better than the ADF-RNC algorithm (average best-of-run fitnesses of 339.19 in the absence of RNCs and 339.06 in the presence of RNCs), corroborating the results obtained in section 5.6.4 that RNCs are not important to make an accurate diagnosis of breast cancer.

The comparison of the performances of the acellular systems with the cellular shows that adaptation in the cellular systems occurs more slowly than in the simpler systems without ADFs (compare Tables 5.9 and 6.7). But, obviously, with more time or larger populations, good solutions can also be found with the cellular systems as adaptation in them still occurs smoothly.

Consider, for instance, one of the best solutions designed with the simpler cellular system without RNCs. It was found in generation 890 of run 75 (the active cell is shown in bold):

```
Normal Genes:
+.*.d3.d8.+.-.+.d0.d3.d6.d0.d2.d3.d7.d6
+.d5.d6.+.d6.-.+.d0.d3.d1.d0.d2.d1.d7.d2
/.-.*.-.+.*.d0.d0.d4.d1.d0.d4.d3.d3.d6

Homeotic Genes:
*.-.+.+.2.-.+.0.2.2.2.1.1.1.1
+.0./.+.+.+.+.1.1.0.1.0.2.0.2
1.1.+.*./.*.2.0.0.2.2.2.2.1.0
```
(6.8a)

This model classifies correctly 341 out of 350 fitness cases in the training set and 172 out of 174 sample cases in the testing set, which corresponds to a training set classification error of 2.571% and a classification accuracy of 97.429%, and a testing set classification error of 1.149% and a classification accuracy of 98.851%. More formally, the model (6.8a) can be translated into the following C++ program:

```
double ADF0(double d[])
{
    double dblTemp = 0.0;
    dblTemp = ((d[8]*((d[0]-d[3])+(d[6]+d[0])))+d[3]);
    return dblTemp;
}

double ADF1(double d[])
{
    double dblTemp = 0.0;
    dblTemp = (d[5]+d[6]);
    return dblTemp;
}
```

```
double  ADF2(double  d[])
{
    double dblTemp = 0.0;
    dblTemp = (((d[0]-d[4])-(d[1]+d[0]))/((d[4]*d[3])*d[0]));
    return dblTemp;
}

double apsModel(double d[])
{
    double ROUNDING_THRESHOLD = 0.5;
    double dblTemp = 0.0;
    dblTemp = (((ADF0(d)+ADF2(d))-ADF2(d))*
              ((ADF2(d)-ADF2(d))+(ADF1(d)+ADF1(d))));
    return (dblTemp >= ROUNDING_THRESHOLD ? 1:0);
}
```
$$\text{(6.8b)}$$

Note that, despite being called from four different places in the main program, the calls to ADF_2 cancel themselves out and, therefore, only ADF_0 and ADF_1 have an impact on the decision. As you can see, ADF_0 is called once from the main program, whereas ADF_1 is called twice. It is also interesting to see that this extremely accurate model uses only five of the nine available analyses to make the diagnosis, namely, clump thickness, marginal adhesion, bare nuclei, bland chromatin, and mitoses.

One of the best solutions created with the ADF-RNC algorithm was discovered in generation 759 of run 4. Its homeotic gene plus all the normal genes and their respective arrays of random numerical constants are shown below (the neutral cells are not shown):

```
Gene 0: -.*.?./.+.d8.+.d1.d5.d6.d0.d7.d1.d2.d6.3.9.5.5.2.6.3.6
     C₀: {1.777649, -1.266601, -0.487305, -0.586578, -0.147583,
          0.402771, -1.077301, -0.016326, 1.421356, -1.279846}

Gene 1: *.+.d8.*.d2./.d3.d6.d4.?.d7.d8.d0.d3.?.5.2.3.7.1.0.6.1
     C₁: {-0.384216, 0.10672, 0.475586, -0.800049, 0.122528,
          -1.120483, 1.075531, 1.535095, 0.461121, 0.364654}

Gene 2: /.*.?.+.+.d2.+.d1.d5.d6.d0.d7.d1.d2.d0.5.2.7.6.0.8.1.0
     C₂: {1.777649, -1.266601, -0.487305, -0.586578, -0.147583,
          0.316253, -1.077301, -0.016326, 1.421356, -1.279846}

Homeotic Gene: *.+.*.2.*./.+.2.2.1.2.2.0.1.0
```
$$\text{(6.9a)}$$

This model classifies correctly 340 out of 350 fitness cases in the training set and 171 out of 174 fitness cases in the testing set. This corresponds to a training set classification error of 2.857% and a classification accuracy of 97.143%, and a testing set classification error of 1.724% and a classification accuracy of 98.276%. Thus, this model is a slightly worse predictor than the model (6.8) created without random numerical constants. More formally, the model (6.9a) can be automatically translated into the following C++ model:

```cpp
double ADF0(double d[])
{
    double dblTemp = 0.0;
    dblTemp = (((d[8]/(d[6]+d[0]))*(d[1]+d[5]))-(-0.586578));
    return dblTemp;
}

double ADF1(double d[])
{
    double dblTemp = 0.0;
    dblTemp = ((((d[6]/d[4])*d[3])+d[2])*d[8]);
    return dblTemp;
}

double ADF2(double d[])
{
    double dblTemp = 0.0;
    dblTemp = (((d[2]+(d[6]+d[0]))*(d[1]+d[5]))/0.316253);
    return dblTemp;
}

double apsModel(double d[])
{
    double ROUNDING_THRESHOLD = 0.5;
    double dblTemp = 0.0;
    dblTemp = ((ADF2(d)+(ADF2(d)*ADF2(d)))*
               ((ADF1(d)/ADF2(d))*(ADF2(d)+ADF0(d))));
    return (dblTemp >= ROUNDING_THRESHOLD ? 1:0);
}
```

(6.9b)

Note that different linking functions are used to connect the ADFs in the main program. Note also that, although ADF_2 is called from five different places in the main program, this program could be simplified so that this ADF would be called just twice. Nonetheless, this model is much more verbose than the model (6.8) designed without RNCs as all the ADFs are invoked in the main program (ADF_0 and ADF_1 are both used just once). And as you can see, in this case, with the exception of one (normal nucleoli), all the analyses are used to make the diagnosis of breast cancer.

6.6 Multiple Cells for Multiple Outputs: The Iris Problem

We have seen already in this chapter some of the advantages of the cellular systems, especially the discovery of modular solutions, code reuse, and totally unsupervised linking. Now we are going to see how the multicellular system can be successfully used to solve problems with multiple outputs.

In all the problems we have solved so far with the multicellular system, the different cells were engaged in finding the same kind of solution and only the performance of the best cell mattered to evaluate the fitness. But, here, n different cells will be used to classify n different classes and, therefore, the fitness of the individual will depend on the accuracy of the different sub-models expressed in the different cells.

This is indeed very similar to what we did in chapter 4 (section Multiple Genes for Multiple Outputs), where the multigenic system of gene expression programming was used to solve problems of multiple outputs. But there is a very important difference between these systems: whereas in the multigenic system we could only have n genes for classifying n classes, in the multicellular system we can have as many normal genes as necessary encoding different automatically defined functions. These ADFs are then invoked as many times as necessary from the n different cells. This means that, for instance, if there is a sub-task that is very easy, the cell in charge of solving it might only need to use one or two ADFs; but if one of the sub-tasks is very complex, the cell in charge might need to combine more ADFs, say four or five, in order to build a good model. We will see below that the multicellular system with multiple outputs works exactly like this, creating extremely compact and intricately connected models.

As a comparison, in the multigenic system with multiple outputs, we don't have this kind of flexibility: for all the sub-tasks a single gene encoding a single sub-ET is all the system has to work with, irrespective of the complexity of the sub-task.

So, in this section, firstly, we are going to analyze the performance of the multicellular system with multiple outputs by solving the already familiar iris problem of section 4.2.3. Here, we are going to use pretty much the same settings (number of runs, number of generations, population size, function set, training set, and fitness function) so that we can compare both algorithms (see Table 6.8 and compare with Table 4.7). And secondly, we are going to study the structure of these intricately connected models by analyzing the structure of one of the best-of-experiment solutions.

Table 6.8
Performance and settings used in the iris problem with multiple outputs.

Number of runs	100
Number of generations	20,000
Population size	30
Number of fitness cases	150
Function set of ADFs	+ - * /
Terminal set	d0 - d3
Head length	7
Gene length	15
Number of genes/ADFs	5
Function set of homeotic genes	+ - * /
Terminal set of homeotic genes	ADFs 0-4
Number of homeotic genes/cells	3
Head length of homeotic genes	7
Length of homeotic genes	15
Chromosome length	120
Mutation rate	0.044
Inversion rate	0.1
IS transposition rate	0.1
RIS transposition rate	0.1
One-point recombination rate	0.3
Two-point recombination rate	0.3
Gene recombination rate	0.3
Gene transposition rate	0.1
Mutation rate in homeotic genes	0.044
Inversion rate in homeotic genes	0.1
RIS transposition rate in homeotic genes	0.1
IS transposition rate in homeotic genes	0.1
Fitness function	Eq. (4.15)
Rounding threshold	0.5
Average best-of-run fitness	147.20

And as you can see in Tables 4.7 and 6.8, the multicellular system with multiple outputs (MCS-MO, for short) is extremely efficient, significantly surpassing the multigenic system with multiple outputs (average best-of-run number of hits of 147.20 compared to 146.48). Furthermore, the best-of-experiment model (4.20) designed with the GEP-MO algorithm, classified correctly only 148 out of 150 fitness cases, whereas the best-of-experiment models designed with the MCS-MO algorithm classify correctly 149 sample cases and, therefore, these models are as good as the model (4.14) designed using the three-step approach of section 4.2.3. Let's now analyze the structure of these intricately connected models.

One of the best solutions discovered in this experiment has a fitness of 149.037 and was created in generation 13819 of run 12 (for compactness, ADFs 0-4 are, respectively, represented by the numerals 0-4):

```
Normal Genes/ADFs:
-.d1.*./.d2.d0.d0.d1.d0.d3.d2.d2.d0.d3.d1
d2.*.d1.d1.d3.d3.d0.d2.d3.d3.d2.d1.d1.d3.d1
-.*.*.d2.d2.d0./.d0.d3.d0.d0.d3.d1.d1.d3
+.*.d1.-.+.*.d2.d0.d3.d3.d1.d0.d0.d2.d0
-.d1.d2.d3./.d3.+.d3.d2.d2.d0.d3.d2.d3.d1

Homeotic Genes/Cells:
4.1.1.-.4.1.-.1.1.0.3.2.0.0.0
-.*.+.4.2.-./.4.4.3.0.3.3.4.3
-.*.+.0.*.0.3.0.2.4.4.4.3.0.3
```
$$\text{(6.10)}$$

In terms of number of hits, this model classifies correctly 149 out of 150 fitness cases. This corresponds to a classification error of 0.667% and a classification accuracy of 99.33% and, therefore, is an almost perfect solution to the iris problem.

As you can see by its expression in Figure 6.13, $Cell_0$ (the cell that distinguishes the *setosa* variety from the other two) invokes just one ADF (ADF_4) and again we can see that just the difference between sepal width and petal length is enough to distinguish *Iris setosa* from *Iris versicolor* and *Iris virginica*. Also worth pointing out is that ADF_4 is also invoked from $Cell_1$ (three times, although two of the calls cancel themselves out); ADF_0 is used to design not only the sub-model that is responsible for distinguishing the *versicolor* variety (ADF_0 is used once in $Cell_1$) but also the *virginica* variety ($Cell_2$ calls this function from three different places); ADF_2 and ADF_3 are both of them part of the sub-models that classify *Iris versicolor* and *Iris virginica*, each being invoked just once from each of these cells.

So, apart from some neutral motifs that can be easily removed by hand, it is worth noticing how compact these models are thanks to code reuse. Indeed, small building blocks are discovered in the ADFs and then these building blocks are used as many times as necessary by the different main programs, creating an extremely compact and intricately connected computer program.

a.   Normal Genes/ADFs:
```
-.d1.*./.d2.d0.d0.d1.d0.d3.d2.d2.d0.d3.d1
d2.*.d1.d1.d3.d3.d0.d2.d3.d3.d2.d1.d1.d3.d1
-.*.*.d2.d2.d0./.d0.d3.d0.d0.d3.d1.d1.d3
+.*.d1.-.+.*.d2.d0.d3.d3.d1.d0.d0.d2.d0
-.d1.d2.d3./.d3.+.d3.d2.d2.d0.d3.d2.d3.d1
```

Homeotic Genes/Cells:
```
4.1.1.-.4.1.-.1.1.0.3.2.0.0.0
-.*.+.4.2.-./.4.4.3.0.3.3.4.3
-.*.+.0.*.0.3.0.2.4.4.4.3.0.3
```

b.

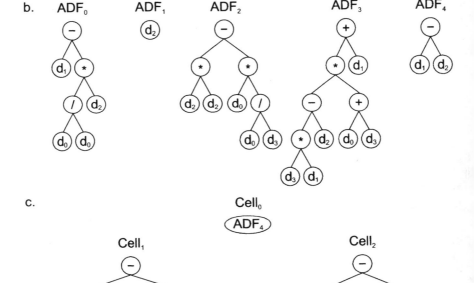

c.

Figure 6.13. Model evolved by the multicellular system with multiple outputs to classify three different kinds of irises. **a)** The genome of the individual with five normal genes encoding five ADFs and three homeotic genes encoding three different sub-models. **b)** The ADFs codified by each normal gene. **c)** The cells or sub-models codified by each homeotic gene. Note that the expression of this individual is only complete after applying the rounding threshold of 0.5 to all the three sub-models. This intricately connected model classifies correctly 149 out of 150 irises.

In the next chapter we will see how to induce Kolmogorov-Gabor polyno-mials in gene expression programming, a complex implementation that re-quires even more random numerical constants than the ADF-RNC or the GEP-RNC algorithms.

7 Polynomial Induction and Time Series Prediction

In this chapter we are going to explore further the idea of domains of random numerical constants in order to evolve high-order multivariate polynomials. Then we are going to discuss the importance of these so called Kolmogorov-Gabor polynomials in evolutionary modeling by comparing the performance of this new algorithm, GEP-KGP (GEP for inducing Kolmogorov-Gabor polynomials), with much simpler and much more intelligible GEP systems. The chapter finishes with the detailed description of all the steps in time series prediction, from preparing the data and building the model through to making predictions about future events.

7.1 Evolution of Kolmogorov-Gabor Polynomials

Kolmogorov-Gabor polynomials have been widely used to evolve general nonlinear models (Iba and Sato 1992, Iba et al. 1994, Ivakhnenko 1971, Kargupta and Smith 1991, Nikolaev and Iba 2001). Evolving such polynomials with gene expression programming is very simple and only requires the implementation of special functions of two arguments. For instance, the expression tree represented below:

corresponds to the following mathematical expression:

$$y = a_0 + a_1 x_1 + a_2 x_2 + a_3 x_1 x_2 + a_4 x_1^2 + a_5 x_2^2 \qquad (7.1)$$

for the complete second-order bivariate basis polynomial function (the complete series is represented in Table 7.1). The six coefficients a_0- a_5 can be

Cândida Ferreira: *Gene Expression Programming*, Studies in Computational Intelligence (SCI) **21**, 275–295 (2006)
www.springerlink.com

easily evolved using a special domain Dc for handling random numerical constants. As for the GEP-RNC algorithm, the Dc comes after the tail but, in this case, has a length equal to $6h$ in order to cover for all the cases, including when all the nodes in the head are function nodes requiring the maximum value of six coefficients.

Consider, for instance, the chromosome below with a head length of two (the Dc is shown in bold):

```
01234567890123456
FFabc252175089728
```
(7.2)

where the numerals 0-9 represent the polynomial coefficients. Then, for the set of random numerical constants:

$$C = \{-0.606, -0.398, -0.653, -0.818, -0.047, 0.036, 0.889, 0.148, -0.377, -0.841\}$$

its expression gives:

$$y = -0.606 - 0.377y_1 - 0.841a + 0.148y_1a - 0.653y_1^2 - 0.377a^2$$
(7.3)

where $y_1 = -0.653 + 0.036b - 0.653c - 0.398bc + 0.148b^2 + 0.036c^2$.

Table 7.1
Bivariate basis polynomials.

$$F_1 = a_0 + a_1x_1 + a_2x_2 + a_3x_1x_2.$$
$$F_2 = a_0 + a_1x_1 + a_2x_2$$
$$F_3 = a_0 + a_1x_1 + a_2x_2 + a_3x_1^2 + a_4x_2^2$$
$$F_4 = a_0 + a_1x_1 + a_2x_1x_2 + a_3x_1^2$$
$$F_5 = a_0 + a_1x_1 + a_2x_2^2$$
$$F_6 = a_0 + a_1x_1 + a_2x_2 + a_3x_1^2$$
$$F_7 = a_0 + a_1x_1 + a_2x_1^2 + a_3x_2^2$$
$$F_8 = a_0 + a_1x_1^2 + a_2x_2^2$$
$$F_9 = a_0 + a_1x_1 + a_2x_2 + a_3x_1x_2 + a_4x_1^2 + a_5x_2^2$$
$$F_{10} = a_0 + a_1x_1 + a_2x_2 + a_3x_1x_2 + a_4x_1^2$$
$$F_{11} = a_0 + a_1x_1 + a_2x_1x_2 + a_3x_1^2 + a_4x_2^2$$
$$F_{12} = a_0 + a_1x_1x_2 + a_2x_1^2 + a_3x_2^2$$
$$F_{13} = a_0 + a_1x_1 + a_2x_1x_2 + a_3x_2^2$$
$$F_{14} = a_0 + a_1x_1 + a_2x_1x_2$$
$$F_{15} = a_0 + a_1x_1x_2$$
$$F_{16} = a_0 + a_1x_1x_2 + a_2x_1^2$$

It is worth pointing out how complex these polynomials might become: as you can see, a simple organization like the one above with just two functions in the head already gives rise to a very complex expression.

The two-parameter function "F" described above is the function used in the STROGANOFF system that composes complete bivariate basis polynomial functions (Ivakhnenko 1971). In Table 7.1, this function is represented by F_9. The implementation of similar functions in gene expression programming, such as the incomplete bivariate basis polynomial functions used in the enhanced STROGANOFF (Nikolaev and Iba 2001), is also very simple and is done exactly as illustrated above for function F_9. The definition of all these functions is given in Table 7.1.

Another function easily implemented is the following:

$$y = x_1 + x_2 + x_1 x_2 + x_1^2 + x_2^2 \tag{7.4}$$

which, except for the coefficients, is identical to the function (7.1) or function F_9 in Table 7.1. This and similar functions based on the 16 bivariate functions of Table 7.1 are going to be included in the function kit of GEP in order to analyze the importance of random numerical constants in the evolution of polynomials. Their description is given in Table 7.2.

Table 7.2
Bivariate basis polynomials with unit coefficients.

$F_1 = x_1 + x_2 + x_1 x_2$
$F_2 = x_1 + x_2$
$F_3 = x_1 + x_2 + x_1^2 + x_2^2$
$F_4 = x_1 + x_1 x_2 + x_1^2$
$F_5 = x_1 + x_2^2$
$F_6 = x_1 + x_2 + x_1^2$
$F_7 = x_1 + x_1^2 + x_2^2$
$F_8 = x_1^2 + x_2^2$
$F_9 = x_1 + x_2 + x_1 x_2 + x_1^2 + x_2^2$
$F_{10} = x_1 + x_2 + x_1 x_2 + x_1^2$
$F_{11} = x_1 + x_1 x_2 + x_1^2 + x_2^2$
$F_{12} = x_1 x_2 + x_1^2 + x_2^2$
$F_{13} = x_1 + x_1 x_2 + x_2^2$
$F_{14} = x_1 + x_1 x_2$
$F_{15} = x_1 x_2$
$F_{16} = x_1 x_2 + x_1^2$

7.2 Simulating STROGANOFF in GEP

So, by choosing a function set consisting exclusively of function F_9 of Table 7.1, we are exactly simulating the original STROGANOFF system in gene expression programming. And by choosing a function set containing functions F_1 - F_{16} of Table 7.1, we are exactly simulating the enhanced STROGANOFF system in GEP. The performance of both these systems, the GEP-OS and the GEP-ES, will be analyzed in the next section by comparing them with other conventional GEP systems.

As stated previously, the implementation of both STROGANOFF systems (the original and the enhanced) in gene expression programming requires a Dc six times larger than the head, thus, much larger than the Dc domain of the GEP-RNC algorithm, which, as you recall, equals the tail length. So, given the much larger number of random numerical constants required for polynomial induction, it is convenient to use a slightly different notation to represent in the genome the RNCs that correspond to the polynomial coefficients. For instance, for the gene below with an $h = 3$:

```
F9.d6.F7.d8.d2.d0.d5
```
(7.5)

a Dc with length 18 is already necessary in order to express correctly this gene. If we choose a set of random numerical constants with 20 members and represent them by R = $\{r_0, ..., r_{19}\}$, then the following structure will be a valid Dc domain for chromosome (7.5) above:

```
r19.r12.r15.r9.r2.r19.r13.r6.r5.r19.r12.r7.r17.r10.r7.r15.r7.r11
```

As before, the values of all the random numerical constants are kept in an array and are retrieved as needed. For each function node evaluated in the expression tree, a block of six elements in Dc is processed, independently of the number of coefficients required for that particular function. So, in the example above, function F_7 will use the first block of six random constants:

```
r19.r12.r15.r9.r2.r19
```

even though it requires just the first four coefficients (see its definition in Table 7.1), and function F_9 will use the next block of six coefficients:

```
r13.r6.r5.r19.r12.r7
```

As you can see, the remaining constants are simply not needed to express that particular gene and are, therefore, part of a neutral region in the gene.

Due to the huge dimensions of the random constants' domains, it is important to be able to control the degree of genetic variation in them. Consequently, special mutation operators were implemented that allow the autonomous control of the mutation rate both on the head/tail domain and on the Dc domain. The inversion and IS transposition operators are also controlled separately and, therefore, a Dc-specific inversion and a Dc-specific transposition were also implemented in the GEP-KGP algorithm. The other operators (one-point and two-point recombination, gene recombination, and gene transposition) remain unchanged and work exactly as described in chapter 5.

7.3 Evaluating the Performance of STROGANOFF

In this section, we are going to evaluate the performance of three STROGANOFF systems (both the original and the enhanced STROGANOFF, and a multigenic implementation of the enhanced STROGANOFF) by comparing them with four simpler GEP systems (the basic GEA both with and without RNCs and the cellular system both with and without RNCs) on the sunspots problem. The performance of all the different systems will be compared in terms of average best-of-run fitness and average best-of-run R-square.

For this prediction task, 100 observations of the Wolfer sunspots series were used (Table 7.3) with an embedding dimension of 10 and a delay time of one (see section 7.4 below to learn how to prepare time series data for symbolic regression). Thus, for the sunspots series presented in Table 7.3,

Table 7.3
Wolfer sunspots series (read by rows).

101	82	66	35	31	7	20	92
154	125	85	68	38	23	10	24
83	132	131	118	90	67	60	47
41	21	16	6	4	7	14	34
45	43	48	42	28	10	8	2
0	1	5	12	14	35	46	41
30	24	16	7	4	2	8	17
36	50	62	67	71	48	28	8
13	57	122	138	103	86	63	37
24	11	15	40	62	98	124	96
66	64	54	39	21	7	4	23
55	94	96	77	59	44	47	30
16	7	37	74				

the chosen embedding dimension and the chosen delay time result in a total of 90 fitness cases. This dataset will be used for training in all the experiments of this section. As basis for the fitness function, we are going to use the mean squared error evaluated by equation (3.4a), being the fitness evaluated by equation (3.5) and, therefore, $f_{max} = 1000$.

7.3.1 Original and Enhanced STROGANOFF

For the original STROGANOFF, the function set consists obviously of just one kind of function, namely, function F_9 of Table 7.1, but this function will be weighted 16 times in order to facilitate the comparison with the enhanced implementation. For the enhanced STROGANOFF, all the 16 functions of Table 7.1 will be used in the function set, thus giving $F = \{F_1, ..., F_{16}\}$. The set of terminals $T = \{a, b, c, d, e, f, g, h, i, j\}$, which correspond, respectively, to t-10, t-9, ..., t-1. In all the experiments, a total of 120 random numerical constants, ranging over the rational interval [-1, 1], will be used per chromosome (in the multigenic system 40 RNCs will be used per gene, giving also 120 RNCs per chromosome). The performance and the parameters used in all the three experiments are shown in Table 7.4.

And as expected, the enhanced implementation is considerably better than the original STROGANOFF. Note, however, that, in the GEP-ESM experiment, a three-genic system was used and, therefore, the system we are simulating is not a straightforward port of the enhanced STROGANOFF as described by Nikolaev and Iba (2001) but is rather a much more efficient algorithm, as it benefits from the multigenic nature of gene expression programming. Indeed, the multigenic system works considerably better than the unigenic one (average best-of-run R-square of 0.8472991731 for the unigenic implementation and 0.8566823219 for the multigenic). It is worth emphasizing that the implementation of multiple parse trees in genetic programming is unfeasible and so is a system similar to the one used in the GEP-ESM experiment. Furthermore, the facility for the manipulation of random numerical constants in GP is not appropriate for handling the huge amount of random numerical constants that are necessary for polynomial induction. In fact, genetic programming usually uses a neural network to discover a posteriori the coefficients of the Kolmogorov-Gabor polynomials. This obviously raises the question of what is exactly the GP doing in this case for, without the coefficients, the evolution of just the polynomial skeleton is not particularly useful (see Table 7.5 below).

Table 7.4

Settings used in the GEP simulation of the original STROGANOFF (**GEP-OS**) and
the GEP simulation of the enhanced STROGANOFF using a unigenic system (**GEP-
ESU**) and a multigenic system (**GEP-ESM**).

	GEP-OS	GEP-ESU	GEP-ESM
Number of runs	100	100	100
Number of generations	5000	5000	5000
Population size	100	100	100
Number of fitness cases	90 (Table 7.3)	90 (Table 7.3)	90 (Table 7.3)
Function set	16(F9)	F1 - F16	F1 - F16
Terminal set	a-j	a-j	a-j
Random constants array length	120	120	40
Random constants type	Rational	Rational	Rational
Random constants range	[-1,1]	[-1,1]	[-1,1]
Head length	21	21	7
Gene length	169	169	57
Number of genes	1	1	3
Linking function	--	--	+
Chromosome length	169	169	171
Mutation rate	0.044	0.044	0.044
Inversion rate	0.1	0.1	0.1
IS transposition rate	0.1	0.1	0.1
RIS transposition rate	0.1	0.1	0.1
One-point recombination rate	0.3	0.3	0.3
Two-point recombination rate	0.3	0.3	0.3
Gene recombination rate	--	--	0.3
Gene transposition rate	--	--	0.1
Dc-specific mutation rate	0.06	0.06	0.06
Dc-specific inversion rate	0.1	0.1	0.1
Dc-specific transposition rate	0.1	0.1	0.1
Random constants mutation rate	0.25	0.25	0.25
Fitness function	Eq. (3.5)	Eq. (3.5)	Eq. (3.5)
Average best-of-run fitness	0.95220	5.30666	5.81873
Average best-of-run R-square	0.2530739782	0.8472991731	0.8566823219

Indeed, continuing our discussion about the importance of random numerical constants in evolutionary modeling, there is a simple experiment we could
do. We could implement the bivariate polynomials with unit coefficients
presented in Table 7.2 and try to evolve complex polynomial solutions with
them (Table 7.5). And one thing that immediately strikes us is that both the
fitness and the R-square of every single best-of-run solution have exactly the
same value that was obtained for average best-of-run fitness and average
best-of-run R-square. This obviously means that the system is trapped in

Table 7.5
The role of coefficients in polynomial evolution.

Number of runs	100
Number of generations	5000
Population size	100
Number of fitness cases	90 (Table 7.3)
Function set	F1 - F16 (Table 7.2)
Terminal set	a-j
Head length	21
Gene length	43
Number of genes	1
Chromosome length	43
Mutation rate	0.044
Inversion rate	0.1
IS transposition rate	0.1
RIS transposition rate	0.1
One-point recombination rate	0.3
Two-point recombination rate	0.3
Fitness function	Eq. (3.5)
Average best-of-run fitness	0.02355
Average best-of-run R-square	0.3500310276

some local optimum from which it is incapable of escaping. Indeed, such local optima are pervasive trappings of most time series prediction tasks, of which the most common consists of using the present state for predicting the next. Indeed, in that case, the simplistic models created have usually excellent statistics but nil predictive power. For instance, in the particular problem of this section, the simplistic solution $y = j$ has already a fitness of 2.38177 and an R-square of 0.6961811429. And as you can deduce by comparing these values with the averaged values obtained with the GEP-OS system and in the experiment summarized in Table 7.5, something had to have been done to prevent the rediscovery of this solution in those systems, otherwise much higher values for both the average best-of-run fitness and average best-of-run R-square would have been obtained. Indeed, all the STROGANOFF experiments of this chapter use a terminal control that checks the total number of different terminals in each expression tree and selects only solutions that use at least five different terminals, making all the solutions with less than five arguments unviable. It is worth pointing out, though, that this kind of measure was not implemented in the simpler GEP systems, as they are extremely flexible and can easily find ways out of all these local optima.

As a matter of fact, the greatest challenge in time series prediction consists of avoiding these strong attractors in the solution landscape. And this is no doubt one of the reasons why Kolmogorov-Gabor polynomials are particularly suited for this task, for it is almost impossible to stumble upon these points in the solution space with them. Notwithstanding, as we will see in the next section, Kolmogorov-Gabor polynomials are much less efficient than the much simpler GEP systems. Furthermore, we will also see that these polynomials have another great disadvantage: they are immensely complex, making it almost impossible to extract knowledge from them.

Just for the sake of curiosity, let's take a look at the structure of the best-of-experiment solutions designed with Kolmogorov-Gabor polynomials. The best solution of the GEP-OS experiment was discovered in generation 4927 of run 45. Its structure is shown below:

```
F9.F9.F9.j.F9.e.d.b.i.a.j.F9.a.F9.F9.g.g.F9.F9.F9.F9.g.d.e.j.d.g.c.b.
b.e.d.j.e.b.h.e.c.g.b.b.h.e.r112.r58.r105.r18.r18.r18.r6.r66.r13.r18.
r41.r41.r13.r12.r109.r64.r63.r63.r80.r21.r85.r87.r87.r87.r103.r18.
r10.r100.r83.r89.r51.r42.r14.r106.r58.r68.r92.r3.r67.r51.r87.r103.
r20.r105.r8.r9.r98.r72.r17.r72.r99.r84.r29.r55.r40.r47.r56.r1.r116.
r80.r80.r45.r79.r88.r24.r98.r45.r51.r21.r38.r32.r44.r40.r24.r43.
r99.r12.r93.r11.r118.r33.r14.r57.r107.r36.r5.r62.r12.r55.r117.r50.
r87.r116.r76.r100.r72.r117.r29.r52.r61.r13.r108.r8.r42.r103.r21.
r32.r64.r62.r103.r109.r91.r3.r16.r30.r7.r92.r61.r90.r35.r99.r61.
r12.r78.r0.r110
```

C = {-0.771332, 0.042266, -0.443359, -0.478027, -0.365906,
    0.437317, -0.749542, -0.30246, -0.751587, 0.050018,
    -0.300842, 0.646515, -0.932831, -0.369415, -0.692658,
    0.234894, -0.192352, -0.707092, -0.003479, -0.703156,
    -0.570465, -0.995209, -0.845459, -0.756562, -0.072815,
    -0.67981, -0.274902, -0.528076, 0.051422, -0.221344,
    -0.573303, -0.833222, 0.280884, -0.180175, -0.343201,
    0.426483, 0.355102, -0.797303, 0.382172, -0.835694,
    0.043457, -0.002349, -0.961548, 0.287567, 0.840149,
    -0.263611, 0.44754, -0.083435, 0.153839, -0.817536,
    0.270538, -0.256012, -0.587372, 0.469818, 0.378082,
    0.09375, 0.302521, -0.56604, 0.847077, -0.699921,
    -0.080535, 0.599335, -0.084106, 0.04776, -0.093383,
    -0.561371, -0.995942, -0.322906, -0.039794, 0.830567,
    -0.999817, -0.671631, -0.217529, 0.15271, 0.84372,

-0.852814, -0.605652, 0.142334, 0.045288, 0.886292,
0.616364, 0.888611, -0.669922, 0.815094, -0.629089,
0.038024, 0.504761, 0.000396, -0.632202, 0.651642,
0.086883, 0.015319, -0.266907, 0.219177, -0.606659,
0.393951, 0.097168, 0.835389, 0.680176, -0.068634,
0.742584, -0.862275, 0.749787, 0.705231, 0.277801,
-0.739319, -0.835602, 0.541199, -0.216095, -0.115631,
0.664246, 0.094177, 0.933045, -0.018493, 0.094513,
0.337219, -0.804444, -0.009307, 0.646362, -0.584412}   (7.6)

It has a fitness of 6.58292 and an R-square of 0.8806859744 evaluated against the training set of 90 fitness cases and, therefore, is not a bad solution to the problem at hand. Note that, despite the complexity of this structure, the K-expression of this individual has just nine elements and, therefore, codes for a much simpler expression than the allowable maximum complexity of 43 tree nodes.

The best-of-experiment solution designed with the GEP-ESU system is slightly better in terms of fitness but not less complicated than the one above:

```
F2.j.F2.j.F2.i.F2.F3.F1.j.F2.F14.f.c.h.h.j.g.F2.F14.g.e.h.a.c.j.d.
g.c.h.j.f.a.j.c.h.c.i.j.b.i.h.d.r113.r105.r24.r94.r114.r31.r7.r111.
r112.r20.r57.r58.r85.r7.r100.r15.r28.r102.r18.r34.r115.r73.r19.
r65.r102.r117.r96.r77.r30.r34.r79.r28.r89.r102.r94.r17.r50.r55.
r11.r66.r88.r106.r0.r7.r49.r84.r90.r50.r75.r27.r16.r115.r13.r98.
r91.r47.r80.r83.r65.r105.r21.r100.r55.r9.r35.r22.r108.r61.r93.
r103.r2.r45.r70.r25.r106.r20.r113.r116.r109.r51.r33.r20.r2.r103.
r111.r41.r5.r92.r17.r104.r116.r18.r38.r68.r2.r119.r39.r87.r81.
r119.r49.r68.r18.r17.r53.r97.r87.r23.r58.r106.r91.r82.r106.r60.
r83.r24.r11.r42.r86.r70.r21.r106.r9.r83.r75.r11
```

C = {0.881379, 0.108337, -0.365173, 0.098724, 0.598511,
      0.744446, -0.10202, 0.719605, 0.357757, -0.354553,
      -0.140228, 0.875122, -0.535553, -0.086608, 0.916382,
      0.304809, 0.793152, -0.092559, -0.787476, -0.958893,
      0.314483, -0.326111, -0.485535, -0.165649, 0.768799,
      -0.402161, -0.225952, 0.476105, -0.657257, -0.565643,
      0.932129, -0.03598, 0.635437, 0.493622, -0.125366,
      0.562286, 0.025451, -0.423736, -0.645783, 0.97464,
      0.60791, -0.18518, -0.176452, -0.306152, 0.444031,
      0.719361, 0.161895, 0.117309, -0.10086, 0.998414,
      0.635651, -0.423431, 0.171112, 0.463592, 0.845185,

0.853363, -0.373535, -0.043579, -0.160919, -0.88617,
-0.102905, -0.243988, -0.5607, 0.786652, -0.727112,
-0.014068, 0.878937, -0.849091, 0.577972, 0.915894,
-0.993561, 0.89328, -0.843506, -0.001434, 0.732331,
0.39801, 0.539948, -0.49115, -0.212158, -0.697907,
-0.186248, -0.288727, -0.669739, -0.560303, 0.413574,
0.770386, 0.751618, -0.091766, 0.737488, 0.108642,
0.818726, 0.946137, -0.009643, -0.698914, -0.831513,
0.058075, -0.355438, -0.958924, 0.054229, -0.243988,
-0.30249, -0.135406, 0.251983, -0.520142, -0.54834,
-0.92212, -0.619782, -0.920594, 0.612183, 0.01358,
-0.719635, -0.789856, -0.187286, 0.674286, -0.561249,
0.083618, -0.784943, -0.045196, 0.755402, -0.476898}   (7.7)

It was discovered in generation 4948 of run 44 and has a fitness of 7.42215 and an R-square of 0.8915609535. Note that it codes for an expression tree with 17 nodes and that most of the function nodes are "F2", one of the simplest bivariate polynomials. Indeed, the analysis of the structures of the best-of-run solutions shows that less verbose functions work much better than the more complicated ones, as they are preferentially selected as building blocks.

The best-of-experiment solution designed with the GEP-ESM was created in generation 4972 of run 19. Its three genes and respective arrays of random constants are shown below:

```
F2.i.b.F12.F8.c.F7.d.a.a.g.h.d.h.b.r22.r2.r17.r14.r30.r9.r0.r8.r15.
r13.r38.r3.r37.r39.r7.r23.r13.r1.r32.r17.r9.r14.r21.r21.r26.r16.r3.
r1.r26.r24.r28.r34.r3.r6.r39.r24.r25.r38.r5.r24.r0.r13
```

```
F1.F2.F2.F2.F2.F2.F2.h.j.i.b.i.j.c.e.r2.r10.r36.r9.r4.r26.r30.r38.r18.
r36.r10.r7.r24.r23.r24.r0.r34.r26.r4.r35.r0.r14.r5.r13.r36.r16.r3.r3.
r1.r17.r21.r36.r8.r34.r19.r35.r2.r1.r29.r8.r12.r18
```

```
j.F1.F2.F1.F10.e.F16.i.e.a.c.c.f.j.h.r4.r20.r7.r9.r28.r27.r7.r6.r19.r39.
r39.r20.r36.r31.r18.r38.r1.r30.r30.r30.r14.r24.r19.r9.r24.r3.r19.r26.
r20.r21.r14.r24.r12.r14.r25.r30.r14.r0.r4.r11.r30.r33
```

$C_0 = \{$-0.275146, 0.891419, -0.18515, 0.41275, -0.388275,
-0.935364, -0.450287, 0.510437, -0.168457, 0.007629,
0.839447, -0.250213, -0.673462, 0.65799, 0.882569,
0.081878, -0.575134, 0.168701, -0.054168, 0.459106,
-0.281494, -0.610748, 0.450317, 0.827942, 0.398163,

-0.486023, 0.834504, 0.890992, 0.025726, -0.428284,
-0.598419, -0.559448, -0.713593, -0.917023, -0.481476,
-0.299591, -0.868195, -0.244323, -0.798615, 0.481018}

C_1 = {0.542053, 0.420563, 0.891358, -0.707062, 0.006195,
-0.2052, -0.200714, -0.415863, -0.040832, 0.296112,
0.188751, 0.789734, -0.577301, 0.438232, 0.382141,
0.264282, 0.209289, 0.245178, 0.106659, 0.116241,
0.501892, 0.830902, -0.762116, 0.716706, -0.846863,
0.945435, -0.559113, 0.446686, 0.367798, -0.921967,
-0.904847, 0.162902, 0.845185, -0.816712, -0.583527,
0.652588, 0.823944, -0.451233, 0.640656, -0.627686}

C_2 = {0.702881, 0.375091, 0.978302, 0.820649, -0.314178,
0.529114, 0.813904, 0.364746, -0.060516, -0.734009,
-0.006195, 0.427917, -0.658539, 0.103332, 0.256897,
-0.392517, 0.379822, -0.767029, -0.777375, -0.856202,
0.491455, -0.804261, -0.632691, -0.414032, -0.888764,
0.27832, -0.55777, 0.572449, 0.418579, -0.390381,
-0.568359, 0.407379, -0.532379, 0.245025, -0.648712,
-0.686615, 0.506744, 0.078186, -0.013549, 0.147033}   (7.8)

It has a fitness of 8.24180 and an R-square of 0.9025703053 and, therefore, is better than the models (7.6) and (7.7) created with the unigenic STROGA-NOFF systems. Note that gene 0 encodes a very simple expression with just three nodes and that gene 2 codes for a tree with just one node (argument j); gene 1, though, codes for a complex solution with a total of 15 nodes. Note also that, again, seven out of eight function nodes consist of function F_2.

Let's now see what kind of solutions can be designed by the much simpler GEP systems.

7.3.2 Simpler GEP Systems

For all the GEP systems, a very simple function set composed of the basic arithmetical operators was chosen, that is, F = {+, -, *, /} with each function weighted four times, thus also making for a total of 16 functions in the function set. For both the GEA-B and multicellular systems, the set of terminals T = {a, b, c, d, e, f, g, h, i, j}, which correspond, respectively, to t-10, t-9, ..., t-1. For both the GEP-RNC and MCS-RNC experiments, the set of terminals consists obviously of the independent variables plus the ephemeral random constant "?", thus giving T = {a, b, c, d, e, f, g, h, i, j, ?}. The set of random numerical constants R = {0, 1, 2, 3, 4, 5, 6, 7, 8, 9}, and "?" ranged over the rational interval [-1, 1]. The performance and the parameters used in the acellular experiments are shown in Table 7.6 and in the multicellular systems in Table 7.7.

And as you can see, all the four systems perform quite well at this difficult task and considerably better than all the STROGANOFF systems studied in

Table 7.6

General settings for the sunspots prediction task using the basic gene expression algorithm (**GEA-B**) and the GEP-RNC algorithm (**GEP-RNC**).

	GEA-B	GEP-RNC
Number of runs	100	100
Number of generations	5,000	5,000
Population size	100	100
Number of fitness cases	90 (Table 7.3)	90 (Table 7.3)
Function set	4(+ - * /)	4(+ - * /)
Terminal set	a-j	a-j ?
Random constants array length	--	10
Random constants type	--	Rational
Random constants range	--	[-1, 1]
Head length	7	7
Gene length	15	23
Number of genes	3	3
Linking function	+	+
Chromosome length	45	69
Mutation rate	0.044	0.044
Inversion rate	0.1	0.1
IS transposition rate	0.1	0.1
RIS transposition rate	0.1	0.1
One-point recombination rate	0.3	0.3
Two-point recombination rate	0.3	0.3
Gene recombination rate	0.3	0.3
Gene transposition rate	0.1	0.1
Dc-specific mutation rate	--	0.044
Dc-specific inversion rate	--	0.1
Dc-specific transposition rate	--	0.1
Random constants mutation rate	--	0.01
Fitness function	Eq. (3.5)	Eq. (3.5)
Average best-of-run fitness	5.64991	6.70823
Average best-of-run R-square	0.8798383868	0.8803029253

Table 7.7
General settings for the sunspots prediction problem using a multicellular system with three automatically defined functions (**MCS**) and a multicellular system encoding ADFs with random numerical constants (**MCS-RNC**).

	MCS	MCS-RNC
Number of runs	100	100
Number of generations	5,000	5,000
Population size	100	100
Number of fitness cases	90 (Table 7.3)	90 (Table 7.3)
Function set of ADFs	4(+ - * /)	4(+ - * /)
Terminal set	a-j	a-j ?
Random constants array length	--	10
Random constants type	--	Rational
Random constants range	--	[-1, 1]
Head length	7	7
Gene length	15	23
Number of genes/ADFs	3	3
Function set of homeotic genes	+ - * /	+ - * /
Terminal set of homeotic genes	ADFs 0-2	ADFs 0-2
Number of homeotic genes/cells	3	3
Head length of homeotic genes	7	7
Length of homeotic genes	15	15
Chromosome length	90	114
Mutation rate	0.044	0.044
Inversion rate	0.1	0.1
IS transposition rate	0.1	0.1
RIS transposition rate	0.1	0.1
One-point recombination rate	0.3	0.3
Two-point recombination rate	0.3	0.3
Gene recombination rate	0.3	0.3
Gene transposition rate	0.1	0.1
Dc-specific mutation rate	--	0.044
Dc-specific inversion rate	--	0.1
Dc-specific transposition rate	--	0.1
Random constants mutation rate	--	0.01
Mutation rate in homeotic genes	0.044	0.044
Inversion rate in homeotic genes	0.1	0.1
RIS transposition rate in homeotic genes	0.1	0.1
IS transposition rate in homeotic genes	0.1	0.1
Fitness function	Eq. (3.5)	Eq. (3.5)
Average best-of-run fitness	4.84324	6.01525
Average best-of-run R-square	0.8670323293	0.8701697733

the previous section (compare Tables 7.6 and 7.7 with Table 7.4). Note also that, for this problem, the presence of random numerical constants results in a slight increase in performance both in the acellular (average best-of-run fitness of 5.64991 in the GEA-B experiment against 6.70823 in the GEP-RNC) and cellular systems (average best-of-run fitness of 4.84324 in the multicellular system without RNCs compared to 6.01525 in the MCS-RNC experiment).

Let's now analyze the structures of the best-of-experiment solutions designed in all the four GEP systems in order to compare their simplicity with the overwhelming complexity of the Kolmogorov-Gabor polynomials created in the previous section.

In the GEA-B experiment, the best solution was found in generation 4528 of run 5 (the sub-ETs are linked by addition):

```
012345678901234
/*+-ch+ghgadbgd
/*+-+h+jigcbieh
-j/-+++deghbbbf
```
(7.9a)

It has a fitness of 8.29243 and an R-square of 0.9159703184 evaluated over the training set of 90 fitness cases and, thus, is considerably better than all the models designed with the STROGANOFF systems. More formally, the model (7.9a) above can be expressed by the following function:

$$y = \frac{(g-h)\cdot c}{h+g+a} + \frac{(j-i)\cdot(g+c)}{h+b+i} + j - \frac{g+h-2b}{d+e}$$
(7.9b)

In the GEP-RNC system, the best solution was designed in run 62 after 4862 generations. Its genes and respective arrays of random numerical constants are shown below (the sub-ETs are linked by addition):

```
Gene 0:  /*+*-++?cgiiiia68585856
```
C_0: {-0.955719, 0.919098, -0.821198, 0.897492, 0.102966, -0.827301, 0.504425, -0.117828, -0.410248, -0.597534}

```
Gene 1:  /*++-i+acjibiab56995766
```
C_1: {-0.955719, 0.771973, 0.456207, 0.557037, -0.687439, 0.62616, -0.989319, -0.484497, 0.861939, 0.636414}

```
Gene 2:  *j+j-?jidc?ahbj85025038
```
C_2: {-0.288116, -0.868805, -0.85849, -0.445801, 0.172882, 0.623901, -0.907898, -0.41098, 0.849945, 0.384094}   (7.10a)

It has a fitness of 10.7883 and an R-square of 0.9259986046 evaluated over the training set of 90 fitness cases and, thus, is slightly better than the model (7.9) created without random numerical constants and considerably better than all the models designed with Kolmogorov-Gabor polynomials. More formally, the model (7.10a) can be expressed by the following function:

$$y = \frac{0.504425c \cdot (g-i)}{3i + a} + \frac{(a+c) \cdot (j-i)}{2i + b} + 0.849945j \qquad (7.10b)$$

In the multicellular system with ADFs without random numerical constants, the best solution was found in generation 1996 of run 2 (the neutral cells are not shown):

```
012345678901234
i*f++//ajghdbje
+/jb-bfdbhehfei
+*j/a++hcjibhcc
/*++1+022021010
```
(7.11a)

It has a fitness of 8.63869 and an R-square of 0.9100736178 evaluated over the training set of 90 fitness cases and, therefore, is also considerably better than all the models designed with the STROGANOFF systems. More formally, the model (7.11a) above can be expressed by the following function (the contribution of each ADF is shown in brackets):

$$y = \frac{2 \cdot \left(\frac{h+c}{j+i} \cdot a + j \right) \cdot \left(\frac{b}{b-f} + j \right)}{2 \cdot (i) + \left(\frac{h+c}{j+i} \cdot a + j \right)} \qquad (7.11b)$$

In the multicellular system with RNCs, the best solution was designed in run 71 after 4561 generations (the neutral cells are not shown):

```
Gene 0:  +j/++-+egajhhj?21492592
     C₀:  {0.631439, -0.25534, -0.833344, 0.831757, -0.645355,
            -0.542816, -0.72107, 0.228332, -0.170013, -0.259155}

Gene 1:  i/bd/-*ghfgjjde86346814
     C₁:  {0.369568, 0.184173, -0.173675, 0.604859, -0.843994,
            -0.537872, -0.44339, 0.228332, -0.951661, -0.916932}

Gene 2:  *+?i**??didg?aj77428535
     C₂:  {-0.306518, -0.864869, -0.668701, -0.181671, 0.701142,
            0.77948, -0.183654, -0.518463, 0.567505, -0.48941}

Homeotic gene:  /*+0+-100022110
```
(7.12a)

It has a fitness of 9.00806 and an R-square of 0.9110742575 evaluated against the training set of 90 fitness cases and, therefore, is slightly better than the model (7.11) designed with the cellular system without random numerical constants and also considerably better than all the models designed with Kolmogorov-Gabor polynomials. More formally, the model (7.12a) can be expressed by the following function (the contribution of each ADF is shown in brackets):

$$y = \frac{2 \cdot \left(j + \dfrac{a - j + 2h}{e + g} \right)^2}{\left(j + \dfrac{a - j + 2h}{e + g} \right) - (0.0018847d - 0.518463i) + (i)} \tag{7.12b}$$

Several conclusions can be drawn from the experiments presented here. First, a STROGANOFF-like system exploring second-order bivariate basis polynomials, although mathematically appealing, is extremely inefficient in evolutionary terms. For one thing, its performance is significantly worse than all the much simpler GEP systems. For another, the structural complexity of the solutions it designs is overwhelmingly complicated, making it almost impossible to extract knowledge from them. And second, any conventional GEP system with a simple set of arithmetical functions is not only much more efficient but also considerably simpler than the extravagantly complicated and computationally expensive STROGANOFF systems.

In the next section we are going to use the settings of the simplest of the GEP systems (the basic GEA with the basic arithmetical operators) to design a model and then make predictions about sunspots with it.

7.4 Predicting Sunspots with GEP

In this section we are going to explore all the fundamental steps in time series prediction. As an example, we are going to learn how to make predictions about sunspots but, indeed, we could be predicting anything from financial markets to the price of peanuts for all such tasks are solved using exactly the same time series prediction framework.

Time series analysis is a special case of symbolic regression and, therefore, can be easily done using the familiar framework of gene expression programming. Indeed, one has only to prepare the data to fit a conventional symbolic regression task, which means that we must have a dependent variable and a set of independent variables. For this problem, we are going to use

again the Wolfer sunspots series of Table 7.3. And a different way of representing this time series is shown in Figure 7.1. Let's see then how the data
should be prepared for time series prediction.

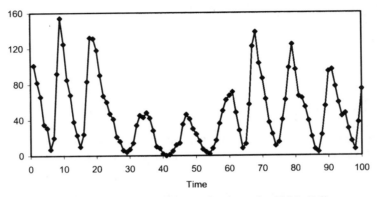

Figure 7.1. Wolfer sunspots series (see also Table 7.3).

In time series analysis, the data represent a series of observations taken at
certain intervals, for instance, a year or a day. The idea behind time series
prediction is that past observations determine future ones. This means that,
in practical terms, one is trying to find a prediction model that is a function
of a certain number of past observations. This certain number of past observations is what is called the embedding dimension d in time series analysis.
For the sunspots prediction task of this section, we are going to use $d = 10$.
There is also another important parameter in time series analysis – the delay
time τ – that determines how data are processed. A delay time of one means
that the data are processed continuously, whereas higher values of τ indicate
that some observations are skipped. For instance, using $d = 10$ and $\tau = 1$
(exactly the same settings used in all the experiments of this chapter), the
sunspots series of Table 7.3 gives:

	t-10	t-9	t-8	t-7	t-6	t-5	t-4	t-3	t-2	t-1	t
1.	101	82	66	35	31	7	20	92	154	125	85
2.	82	66	35	31	7	20	92	154	125	85	68
3.	66	35	31	7	20	92	154	125	85	68	38
...					...						
89.	55	94	96	77	59	44	47	30	16	7	37
90.	94	96	77	59	44	47	30	16	7	37	74

As you can see, the time series data is now ready to be used in a normal symbolic regression task, where (*t*-10) through (*t*-1) are the independent variables and *t* is the dependent variable.

In real-world time series prediction the goal is to find a model and then use that model to make predictions about future events. And the model is usually put to the test the next day or hour, depending on the frequency of the observations. But here, as an illustration, we can simulate a real situation using only the first 80 observations of the Wolfer sunspots to evolve the model (that is, for training) and the last 10 for prediction (for testing, in this case). This way we will be able to evaluate the accuracy of the predictions that can be made by the models created by the algorithm.

So, let's try to design a model to explain and predict sunspots with gene expression programming. A good starting point would be to choose the kind of parameters used in the sunspots experiment summarized in the first column of Table 7.6. Then, with Gepsoft APS, through a series of optimization runs, one can exploit the current chromosomal structure to the fullest, that is, let the system evolve for a couple of thousands of generations and go through three of four mass extinction cycles until it stops improving. After that, a neutral gene is added to the system and again the system is exploited to the fullest, and so forth until the introduction of another neutral gene is no longer accompanied by a burst in genetic innovation. Theoretically, this kind of procedure allows one to approximate any continuous function to an arbitrary precision if there is a sufficient number of terms. Here, the added neutral genes are potential new terms and Gepsoft APS allows their fruitful integration in the equation. For instance, the C++ model below was created using a total of five such cycles (one for each added neutral gene):

```
double apsModel(double d[])
{
    double dblTemp = 0;
    dblTemp  = (d[9]+((d[8]/(d[5]+d[4]))+((d[9]-d[9])*d[0])));
    dblTemp += (d[9]/((((d[3]+d[3])+d[2])-d[9])+d[8]));
    dblTemp += (d[9]/(d[4]+d[6]));
    dblTemp += (d[9]/(d[4]+d[2]));
    dblTemp += (d[5]/(d[2]-d[6]));
    dblTemp += (d[3]/(d[1]-d[5]));
    dblTemp += (d[1]/(d[7]-d[0]));
    dblTemp += ((((d[2]-d[8])*d[9])+(d[0]+d[0]))/
                (d[7]+(d[8]+d[2])));
    return dblTemp;
}
```
(7.13)

where d_0 - d_9 represent, respectively, $(t$-10$)$ - $(t$-1$)$. This model has an R-square equal to 0.94974095 in the training set of 80 data points. And as you can see in Figures 7.2 and 7.3, the model evolved by GEP is an extremely good predictor, with an R-square of 0.8785305643 on the testing set. Note, in Figure 7.3, how the most accurate predictions are the most immediate: the more one ventures into the future the less accurate they become.

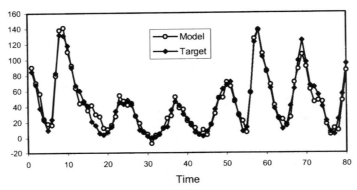

Figure 7.2. Comparing the model (7.13) designed by gene expression programming with the target sunspots series on the training data.

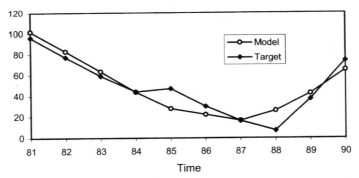

Figure 7.3. Comparing the mode (7.13) designed by gene expression programming with the target sunspots series on the testing data.

The remarkable thing about these time series prediction models (the model (7.13) above and the models (7.9) - (7.12) of the previous section) is that they are all composed of simple terms involving a quotient. This is something the algorithm discovered on its own without any kind of prompting on my part. Furthermore, this method of approaching a complex function by introducing one term at a time seems to me a very creative and useful framework for designing accurate predictive models.

In the next chapter we will study yet another interesting application of gene expression programming in which the facility for handling random numerical constants is instrumental in finding the parameters that optimize a function.

8 Parameter Optimization

Genetic algorithms have been widely used for parameter optimization (see, e.g., Goldberg 1989, Michalewicz 1996, and Haupt and Haupt 1998). In that case, the simple chromosomes of the GA encode the values of the different parameters. Whether in binary or floating-point format, the different parameters of a function are directly encoded in the simple GA chromosomes.

Gene expression programming can also be used to do parameter optimization very efficiently and, in this chapter, we will introduce two different algorithms for parameter optimization: the HZero algorithm and the GEP-PO algorithm.

The HZero algorithm is very similar to the GA in the sense that the different candidate parameters are not transformed by mathematical operations, being directly used instead. Consequently, although more flexible than the GA and easier to implement, this algorithm behaves in much the same fashion as the GA, that is, it is quick and goes straight to the point. The only problem is that that point is not always the best and there is not much these algorithms can do to take themselves out of there.

The GEP-PO algorithm is not as afflicted by this problem as the HZero system as it designs the parameter values by performing mathematical operations on a great variety of random numerical constants. This means that the GEP-PO algorithm has at its disposal much more precise tools for fine-tuning the parameter values, which, consequently, also means that it also has the right tools to search all the peaks and valleys of the solution space and, therefore, increase the odds of finding the global optimum.

Both these algorithms explore the facility for handling random numerical constants described in chapter 5. And they both explore the multigenic nature of gene expression programming to find the parameters that optimize a multidimensional function. As you probably guessed, the values of the N parameters of a function are encoded in N different genes.

Cândida Ferreira: *Gene Expression Programming*, Studies in Computational Intelligence (SCI) **21**, 297–336 (2006)
www.springerlink.com

Let's first study the architectures of both these algorithms and how they learn and then proceed with the analysis of their performance by optimizing two different functions: a well-studied two-parameter function and a more complex five-parameter function.

8.1 The HZero Algorithm

The HZero algorithm is an implementation of the GEP-RNC algorithm at its utmost simplicity, that is, when the head size h is equal to zero (see chapter 5 for a description of the GEP-RNC algorithm). Indeed, when $h = 0$, the length of the tail evaluated by equation (2.4) gives $t = 1$. And since the Dc length is equal to the tail length, this means that the Dc's in the HZero algorithm have just one element. As we will see next, this algorithm is even easier to implement than the simple GA, and has also the additional advantage of being more flexible than the GA in evolutionary terms.

8.1.1 The Architecture

Thus, when the head is equal to zero, a normal gene with a Dc domain has the following structure (the Dc is shown in bold):

$$01$$
$$?\mathbf{3} \tag{8.1}$$

where "?" represents the ephemeral random constants and "3" is a numeral representing a specific random numerical constant, which, for convenience, also indicates the order that the corresponding numerical constant occupies in the array of constants. For instance, for the five-element array of random numerical constants presented below:

$C = \{-1.144714, -1.80484, 0.936646, 1.509033, -1.157348\}$

the chromosome (8.1) above encodes the parameter value $p = 1.509033$.

It is worth pointing out that the number of RNCs associated with each gene is arbitrary and the HZero algorithm is in fact very flexible regarding the amount of constants it can handle. From configurations with just one constant per gene to configurations with a few hundreds, the algorithm handles them very fluidly. However, better and faster results are obtained using simpler configurations of just 1-10 constants per gene.

For a multidimensional optimization task, multigenic chromosomes are obviously used. Consider, for instance, the chromosome below composed of three genes (the Dc's are shown in bold):

```
010101
?2?4?1
```
(8.2)

and respective arrays of random numerical constants:

C_1 = {-1.777252, -0.281341, -1.666779, -1.060455, 1.65213},
C_2 = {1.340088, -0.242614, -0.592712, 0.729187, -1.63266},
C_3 = {-0.113372, -0.763123, 0.138122, 1.322388, -1.060913}.

Its expression results in three different parameter values:

p_1 = -1.666779,
p_2 = -1.63266,
p_3 = -0.763123.

These parameters are then used to find the maximum (or minimum) of a function. For instance, for the function below:

$$f(p_1, p_2, p_3) = 2.5 + \sin\sqrt{p_1^2 + p_2^2} + 1.02 \cdot \sin(4p_3)$$
(8.3)

they give:

$$f(-1.666779, -1.63266, -0.763123) = 3.1324168397$$

which is the output or value returned by function (8.3) for that particular set of parameter values.

So, each chromosome in a parameter optimization task encodes a particular set of parameter values. And for each set of parameter values, the evaluation of the function at hand returns a corresponding output. And this output is what will be used as basis for evaluating the fitness of the individual. But since the value returned by a particular set of parameters can be anywhere in the real line, something must be done about negative or zero fitnesses. To solve this problem, for each generation, the worse-of-generation fitness f_{min} is evaluated and, if f_{min} is zero or negative, the absolute value of f_{min} plus 1 is added to the fitness of all individuals. This way one guarantees that all the individuals will have positive fitness, with the less fit having fitness equal to one. This transformed fitness can now be used to select individuals to reproduce with modification and will be used in all the problems of this chapter.

8.1.2 Optimization of a Simple Function

From the presentation above and from what you know about the GEP-RNC algorithm and the GA, you can see why the HZero algorithm is more flexible than the GA. Basically, when the arrays of the RNCs contain just one random constant, we get the same functionality of a GA. But by increasing the number of elements in the arrays of RNCs, we are improving the odds of finding the right combination of parameters that optimize our function. For most problems, I use 10 random constants per gene, not only because it is very convenient in terms of representation but also because it produces good results. Let's see then how the HZero algorithm discovers the parameters that optimize a function.

In parameter optimization the goal is to find a set of parameter values that maximize (or minimize) a complex multidimensional function. Sometimes minimizing a function makes more sense, and when that is the case, one just has to multiply the function by minus one and then maximize. Obviously, in gene expression programming, it is more convenient to maximize a function as, in that case, better solutions have corresponding higher fitnesses, making the selection process much easier.

The simple function we are going to maximize in this section, is the same chosen by Michalewicz to illustrate the workings of the GA (Michalewicz 1996) and, therefore, serves also to show the differences between the two algorithms. This simple, one variable, function is defined as:

$$f(x) = x \cdot \sin(10\pi \cdot x) + 1.0 \tag{8.4a}$$

and the goal consists of finding, in the domain [-1, 2], the value x_0 that maximizes the function, that is:

$$f(x_0) \geq f(x), \text{ for all } x \in [-1, 2] \tag{8.4b}$$

The global maximum of this function in the given domain is obviously known and, for simplicity, will be considered 2.85.

For this problem, a set of just five random numerical constants will be used so that we can fit it all together (chromosomes and random constants) in a row. As usual, the RNCs will be represented by the numerals 0-4, thus giving R = {0, 1, 2, 3, 4}. The ephemeral random constant "?" will range obviously over the rational interval [-1, 2] as the random constants are used in their natural state by the algorithm, that is, no new constants can be created through simple mathematical operations such as the addition or multi-

plication of two constants. We will also use small populations of just 20 individuals so that we can analyze the evolutionary process in its entirety and still
keep a readable presentation. The complete list of parameters used per run is
given in Table 8.1. And as you can see, for this simple chromosomal structure, we can only benefit from mutation, both of the gene sequence and the
random constants themselves, and from one-point recombination (it still works
because the RNCs stay in the same place). Note, however, that for multidimensional optimization problems, we can also benefit from two-point recombination, gene recombination, and gene transposition for fine-tuning the parameters. Let's see then how these structures evolve, finding good solutions
to the problem at hand.

Table 8.1
Settings for the simple optimization problem using the HZero algorithm.

Number of generations	50
Population size	20
Terminal set	?
Random constants array length	5
Random constants type	Rational
Random constants range	[-1, 2]
Head length	0
Gene length	2
Number of genes	1
Chromosome length	2
Random constants mutation rate	0.55
Dc-specific mutation rate	0.2
One-point recombination rate	0.3

The evolutionary dynamics of the successful run we are going to analyze is
shown in Figure 8.1. And as you can see, in this run, the global maximum of
function (8.4) was found in generation 8.

Both the structures and the outputs of all the individuals of the initial
population are shown below (the best of generation is shown in bold):

```
GENERATION N: 0
Structures:
?1-[ 0]-{0.317169, -0.356384, 0.231598, -0.356384, -0.916901}
?1-[ 1]-{1.95847, 0.743866, 1.47241, 0.06842, -0.480194}
?4-[ 2]-{-0.72638, -0.827637, 0.578095, 1.4209, 1.32874}
?3-[ 3]-{0.769379, 0.231598, 1.28174, -0.642487, -0.808625}
```

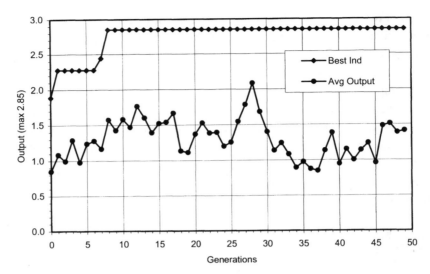

Figure 8.1. Progression of average output of the population and the output of the best individual for a successful run of the experiment summarized in Table 8.1.

```
?4-[ 4]-{-0.96225, -0.098144, 1.0191, 1.95749, 1.3316}
?1-[ 5]-{-0.438599, 1.90933, 1.32874, 0.415588, 0.28598}
?0-[ 6]-{-0.89972, 0.644684, 1.42151, 1.82889, -0.065002}
?0-[ 7]-{0.584931, -0.177093, 1.33887, 0.54953, 1.03934}
?1-[ 8]-{1.53787, 0.701203, 1.15833, -0.539063, 0.425629}
?4-[ 9]-{1.57269, 0.160522, 1.41119, -0.014373, -0.850556}
?0-[10]-{1.95749, -0.580597, 0.743866, -0.438599, 0.54953}
?0-[11]-{-0.642761, 0.743866, -0.945466, -0.72638, -0.515564}
?4-[12]-{0.425629, 0.061004, 1.95303, -0.580597, -0.2294}
?4-[13]-{1.4209, 1.42151, -0.438599, -0.648102, 1.15833}
?3-[14]-{-0.850556, 1.1875, 1.98966, 0.701203, 0.061004}
?2-[15]-{-0.643494, 1.68057, 1.41119, 1.41119, -0.945466}
?2-[16]-{1.47241, 0.908783, -0.065002, 1.53787, 0.160522}
?3-[17]-{0.644684, 0.05429, 1.95749, 1.42151, 0.701203}
?3-[18]-{0.644684, -0.065002, 1.57269, -0.916901, -0.69754}
?1-[19]-{-0.827637, 1.61771, -0.79599, -0.342102, -0.79599}

Outputs:
[ 0] = f(-0.356384) = 0.65076
[ 1] = f(0.743866) = 0.269903
[ 2] = f(1.32874) = -0.0430903
[ 3] = f(-0.642487) = 1.62467
[ 4] = f(1.3316) = -0.115347
[ 5] = f(1.90933) = 0.448161
[ 6] = f(-0.89972) = 1.00791
```

```
[ 7] = f(0.584931) = 0.733318
[ 8] = f(0.701203) = 0.973505
[ 9] = f(-0.850556) = 1.85043
[10] = f(1.95749) = -0.903561
[11] = f(-0.642761) = 1.62621
[12] = f(-0.2294) = 1.18301
[13] = f(1.15833) = -0.118934
[14] = f(0.701203) = 0.973505
[15] = f(1.41119) = 1.48611
[16] = f(-0.065002) = 1.05792
[17] = f(1.42151) = 1.8891
[18] = f(-0.916901) = 0.535716
[19] = f(1.61771) = 1.85416
```

The best individual of this generation, chromosome 17, encodes the parameter value $x_0 = 1.42151$, which corresponds to $f(x_0) = 1.8891$.

In the next generation a better solution was found (chromosome 7):

```
GENERATION N: 1
Structures:
?3-[ 0]-{0.644684, 0.05429, 1.95749, 1.42151, 0.701203}
?4-[ 1]-{-0.438599, 1.41833, 1.32874, 0.415588, 0.13858}
?4-[ 2]-{0.13858, 0.160522, 1.41119, 0.011413, 1.76102}
?1-[ 3]-{0.425629, 0.061004, -0.572632, -0.580597, -0.2294}
?4-[ 4]-{1.4209, -0.531799, -0.438599, -0.90857, 1.15833}
?3-[ 5]-{0.584931, -0.177093, 1.33887, 0.689667, 1.03934}
?2-[ 6]-{1.47241, 0.908783, 1.71039, 1.53787, 0.160522}
?2-[ 7]-{-0.684418, -0.401642, 1.46631, 1.53787, 0.160522}
?4-[ 8]-{0.425629, 1.81119, 1.91766, 1.982, -0.2294}
?3-[ 9]-{-0.850556, 1.27115, 1.98966, 0.701203, 1.91766}
?4-[10]-{-0.850556, 1.1875, 1.98966, 0.701203, 0.061004}
?0-[11]-{0.317169, -0.049316, 0.664765, 0.819458, -0.916901}
?3-[12]-{0.9581, 0.061004, -0.049316, 1.04767, -0.2294}
?3-[13]-{0.644684, -0.065002, 1.81119, 0.425842, 0.110962}
?4-[14]-{1.57269, 1.74732, 1.41119, -0.014373, 0.843506}
?0-[15]-{-0.049316, 1.91766, 1.98966, 0.701203, 0.061004}
?4-[16]-{0.689667, 0.061004, 1.55521, -0.719116, -0.2294}
?3-[17]-{0.769379, 0.231598, 1.28174, -0.642487, 0.620331}
?3-[18]-{0.644684, -0.531799, -0.709107, -0.916901, 0.13858}
?2-[19]-{1.47241, -0.584045, -0.065002, 1.74732, 0.160522}

Outputs:
[ 0] = f(1.42151) = 1.8891
[ 1] = f(0.13858) = 0.870243
[ 2] = f(1.76102) = -0.656588
[ 3] = f(0.061004) = 1.05739
```

```
[ 4]  =  f(1.15833)    =  -0.118934
[ 5]  =  f(0.689667)   =  1.21997
[ 6]  =  f(1.71039)    =  0.451673
[ 7]  =  f(1.46631)    =  2.27802
[ 8]  =  f(-0.2294)    =  1.18301
[ 9]  =  f(0.701203)   =  0.973505
[10]  =  f(0.061004)   =  1.05739
[11]  =  f(0.317169)   =  0.837101
[12]  =  f(1.04767)    =  2.04486
[13]  =  f(0.425842)   =  1.30897
[14]  =  f(0.843506)   =  1.82601
[15]  =  f(-0.049316)  =  1.0493
[16]  =  f(-0.2294)    =  1.18301
[17]  =  f(-0.642487)  =  1.62467
[18]  =  f(-0.916901)  =  0.535716
[19]  =  f(-0.065002)  =  1.05792
```

As you can see, the best of this generation is considerably better than the best of the initial population. In this case, the parameter value $x_0 = 1.46631$ was discovered, resulting in $f(x_0) = 2.27802$.

It is worth pointing out that, even though this is an evolutionary process where most of the times some kind of learning takes place, for this simple problem of just one parameter, the system benefits little from past experience and each new best-of-generation individual is just a lucky winner that suffered a very beneficial mutation. Note, however, that for multidimensional optimization problems, past experience counts, as some genes might just be too precious to lose and are passed on from generation to generation.

Then for the next five generations there was no improvement in best output and all the best individuals of these populations are either clones of the best created by elitism or variants of the best created mostly by replacing old random constants by new ones (all these individuals are shown in bold):

```
GENERATION N: 2
Structures:
?2-[ 0]-{-0.684418, -0.401642, 1.46631, 1.53787, 0.160522}
?3-[ 1]-{1.05344, 0.05429, 1.17279, 0.677033, 0.701203}
?1-[ 2]-{1.05344, -0.275757, 1.41119, -0.014373, 0.843506}
?4-[ 3]-{-0.850556, -0.275757, 1.98966, 1.17444, 1.72028}
?4-[ 4]-{-0.850556, 1.1875, 1.74033, -0.825592, 0.061004}
?3-[ 5]-{0.644684, -0.065002, 1.2999, 1.17444, -0.805512}
?0-[ 6]-{0.769379, 0.231598, 1.28174, -0.642487, 0.620331}
?3-[ 7]-{0.769379, 0.231598, -0.444427, 1.40485, -0.006225}
?4-[ 8]-{1.47241, 1.2999, -0.065002, 1.74732, 0.160522}
?4-[ 9]-{1.31094, -0.531799, -0.709107, -0.916901, -0.101165}
```

```
?4-[10]-{-0.850556, 1.7684, 1.98966, 0.701203, 0.061004}
?4-[11]-{0.732422, 0.790375, 0.664765, 0.819458, -0.916901}
?0-[12]-{0.317169, -0.049316, 1.17279, -0.233673, -0.444427}
?4-[13]-{-0.850556, 1.05344, 1.98966, 0.701203, 0.061004}
?0-[14]-{0.790375, -0.401642, 1.46631, 1.53787, 1.17279}
?3-[15]-{0.644684, 1.17279, 1.81119, 1.77588, 0.110962}
?3-[16]-{-0.850556, 1.27115, 1.98966, 0.701203, -0.641846}
?3-[17]-{0.644684, -0.065002, 1.31094, 1.05344, 0.110962}
?3-[18]-{-0.850556, -0.413696, 1.98966, 1.05344, 1.91766}
?3-[19]-{1.95157, 1.2999, 1.95749, 0.420013, 0.701203}
```

Outputs:
```
[ 0] = f(1.46631) = 2.27802
[ 1] = f(0.677033) = 1.4472
[ 2] = f(-0.275757) = 1.1903
[ 3] = f(1.72028) = -0.0231821
[ 4] = f(0.061004) = 1.05739
[ 5] = f(1.17444) = 0.155015
[ 6] = f(0.769379) = 0.368856
[ 7] = f(1.40485) = 1.21305
[ 8] = f(0.160522) = 0.848168
[ 9] = f(-0.101165) = 0.996298
[10] = f(0.061004) = 1.05739
[11] = f(-0.916901) = 0.535716
[12] = f(0.317169) = 0.837101
[13] = f(0.061004) = 1.05739
[14] = f(0.790375) = 0.764633
[15] = f(1.77588) = -0.220585
[16] = f(0.701203) = 0.973505
[17] = f(1.05344) = 2.0473
[18] = f(1.05344) = 2.0473
[19] = f(0.420013) = 1.24702
```

GENERATION N: 3
Structures:
```
?2-[ 0]-{-0.684418, -0.401642, 1.46631, 1.53787, 0.160522}
?1-[ 1]-{0.904419, 1.26227, 1.41119, -0.014373, 0.269256}
?4-[ 2]-{0.436615, -0.065002, -0.455322, 1.05344, 0.110962}
?3-[ 3]-{-0.850556, 1.83643, 0.269256, 0.701203, 0.921143}
?3-[ 4]-{-0.850556, 1.7684, 1.98966, 0.701203, 0.061004}
?3-[ 5]-{1.05344, 1.14948, 1.17279, 0.677033, 0.701203}
?4-[ 6]-{0.904419, -0.275757, 1.98966, 1.17444, 0.606445}
?3-[ 7]-{-0.850556, 1.27115, 0.120758, 0.701203, -0.641846}
?2-[ 8]-{-0.699127, -0.401642, 1.46631, 0.635193, 0.160522}
?3-[ 9]-{1.95157, 1.2999, 1.79532, 0.420013, 0.809845}
?3-[10]-{-0.850556, 1.05344, 1.98966, -0.253265, -0.091796}
?3-[11]-{-0.455322, 1.1478, -0.582733, 1.05344, 1.91766}
```

?4-[12]-{0.809845, -0.948304, 1.2999, 1.17444, -0.805512}
?4-[13]-{1.79532, -0.253265, 1.02157, 1.05344, 1.61856}
?2-[14]-{0.236633, -0.275757, 1.61856, -0.014373, 0.436615}
?4-[15]-{-0.850556, 1.7684, 1.98966, 0.701203, 0.061004}
?3-[16]-{0.644684, 1.14948, 1.2999, 1.17444, 0.759918}
?1-[17]-{1.69095, -0.036529, 1.17279, 1.26227, -0.424988}
?4-[18]-{-0.850556, 1.05344, -0.033874, -0.113586, 1.14948}
?3-[19]-{-0.92215, 1.07703, 1.98966, 0.091522, 0.309265}

Outputs:
[0] = f(1.46631) = 2.27802
[1] = f(1.26227) = 2.16967
[2] = f(0.110962) = 0.962538
[3] = f(0.701203) = 0.973505
[4] = f(0.701203) = 0.973505
[5] = f(0.677033) = 1.4472
[6] = f(0.606445) = 1.12195
[7] = f(0.701203) = 0.973505
[8] = f(1.46631) = 2.27802
[9] = f(0.420013) = 1.24702
[10] = f(-0.253265) = 1.25193
[11] = f(1.05344) = 2.0473
[12] = f(-0.805512) = 1.13879
[13] = f(1.61856) = 1.89122
[14] = f(1.61856) = 1.89122
[15] = f(0.061004) = 1.05739
[16] = f(1.17444) = 0.155015
[17] = f(-0.036529) = 1.03331
[18] = f(1.14948) = -0.149319
[19] = f(0.091522) = 1.02409

GENERATION N: 4
Structures:
?2-[0]-{-0.699127, -0.401642, 1.46631, 0.635193, 0.160522}
?4-[1]-{-0.339233, 1.26227, 0.537354, -0.014373, 0.269256}
?3-[2]-{-0.388489, 1.4407, 1.98966, 1.29065, -0.091796}
?0-[3]-{0.436615, -0.065002, -0.455322, -0.963288, 1.33404}
?1-[4]-{1.69095, -0.036529, 1.10434, 1.26227, -0.424988}
?0-[5]-{1.29065, 0.431702, 1.79532, -0.508575, 1.17502}
?1-[6]-{0.610077, -0.036529, -0.482391, 1.26227, -0.424988}
?1-[7]-{0.904419, 1.51877, 0.587067, 1.17444, 0.009002}
?3-[8]-{-0.684418, -0.401642, 1.17502, 1.53787, 1.10434}
?3-[9]-{1.33404, 1.07703, 1.98966, 1.53879, 0.309265}
?2-[10]-{0.598267, -0.401642, 1.46631, 0.635193, 0.160522}
?4-[11]-{-0.958619, -0.253265, 1.02157, 1.20904, -0.022888}
?3-[12]-{0.956604, -0.482391, 1.79532, 0.420013, 0.809845}
?3-[13]-{-0.92215, 1.07703, 1.98966, 0.955597, -0.388489}

?3-[14]-{-0.850556, 1.83643, 0.269256, 0.431702, 0.921143}
?2-[15]-{0.598267, -0.699249, 1.29065, 1.53787, 0.160522}
?3-[16]-{0.644684, 1.14948, -0.388489, 1.17444, 1.67844}
?1-[17]-{1.17502, -0.388489, 1.17279, 1.26227, -0.424988}
?3-[18]-{1.29065, 0.955597, 1.53879, -0.022888, 1.91766}
?1-[19]-{0.904419, -0.275757, 1.98966, 1.17444, 0.606445}

Outputs:
[0] = f(1.46631) = 2.27802
[1] = f(0.269256) = 1.22147
[2] = f(1.29065) = 1.37372
[3] = f(0.436615) = 1.39858
[4] = f(-0.036529) = 1.03331
[5] = f(1.29065) = 1.37372
[6] = f(-0.036529) = 1.03331
[7] = f(1.51877) = 0.155504
[8] = f(1.53787) = -0.427589
[9] = f(1.53879) = -0.444312
[10] = f(1.46631) = 2.27802
[11] = f(-0.022888) = 1.01508
[12] = f(0.420013) = 1.24702
[13] = f(0.955597) = 0.0591375
[14] = f(0.431702) = 1.36232
[15] = f(1.29065) = 1.37372
[16] = f(1.17444) = 0.155015
[17] = f(-0.388489) = 0.862553
[18] = f(-0.022888) = 1.01508
[19] = f(-0.275757) = 1.1903

GENERATION N: 5
Structures:
?2-[0]-{0.598267, -0.401642, 1.46631, 0.635193, 0.160522}
?3-[1]-{-0.850556, 1.83643, 1.9292, 0.431702, 0.921143}
?1-[2]-{0.697174, -0.611939, 1.98966, 1.29065, -0.091796}
?2-[3]-{-0.699127, -0.401642, 1.46631, 0.635193, 0.160522}
?3-[4]-{-0.958619, 0.316925, -0.82428, -0.416962, -0.022888}
?3-[5]-{0.644684, 1.14948, -0.388489, 1.17444, 0.351868}
?3-[6]-{0.904419, -0.275757, 1.26807, 1.17444, -0.611939}
?3-[7]-{-0.75528, -0.75528, 0.268921, 1.48795, 0.268921}
?0-[8]-{0.436615, -0.402405, -0.455322, -0.963288, 1.33404}
?3-[9]-{0.956604, -0.733795, 1.48795, 1.11856, -0.733795}
?3-[10]-{1.33404, 1.07703, -0.611939, 1.53879, -0.402405}
?4-[11]-{0.351868, -0.75528, 0.707672, 1.88525, 0.269256}
?3-[12]-{1.29065, 0.955597, -0.416962, -0.022888, -0.348816}
?0-[13]-{1.29065, 0.431702, -0.348816, -0.508575, 1.68961}
?3-[14]-{-0.92215, 1.07703, 1.4397, 0.955597, 1.33374}
?2-[15]-{-0.329284, 1.72037, 1.46631, 0.635193, 0.884491}

?4-[16]-{-0.388489, -0.554718, -0.611939, 1.29065, 0.884491}
?2-[17]-{-0.827027, -0.401642, 1.46631, -0.348816, 0.160522}
?1-[18]-{0.598267, 1.24966, 1.46631, 0.635193, 0.160522}
?2-[19]-{0.598267, -0.699249, 1.29065, 1.56586, 0.160522}

Outputs:
[0] = f(1.46631) = 2.27802
[1] = f(0.431702) = 1.36232
[2] = f(-0.611939) = 1.22418
[3] = f(1.46631) = 2.27802
[4] = f(-0.416962) = 1.21182
[5] = f(1.17444) = 0.155015
[6] = f(1.17444) = 0.155015
[7] = f(1.48795) = 1.5501
[8] = f(0.436615) = 1.39858
[9] = f(1.11856) = 0.384124
[10] = f(1.53879) = -0.444312
[11] = f(0.269256) = 1.22147
[12] = f(-0.022888) = 1.01508
[13] = f(1.29065) = 1.37372
[14] = f(0.955597) = 0.0591375
[15] = f(1.46631) = 2.27802
[16] = f(0.884491) = 1.4141
[17] = f(1.46631) = 2.27802
[18] = f(1.24966) = 2.24959
[19] = f(1.29065) = 1.37372

GENERATION N: 6
Structures:
?2-[0]-{-0.827027, -0.401642, 1.46631, -0.348816, 0.160522}
?1-[1]-{0.697174, -0.611939, 1.98966, 1.29065, 1.97852}
?2-[2]-{-0.246643, -0.75528, 0.707672, -0.247955, 1.97852}
?1-[3]-{1.66873, 0.506866, 0.041412, 1.60007, -0.091796}
?3-[4]-{-0.388489, -0.554718, 1.94092, 1.97852, -0.502502}
?2-[5]-{-0.13208, -0.401642, -0.950196, 0.635193, 1.60007}
?4-[6]-{1.88129, 1.72037, 1.46631, 0.635193, 1.67365}
?3-[7]-{0.158081, 1.43738, 1.46631, -0.348816, 1.76785}
?3-[8]-{-0.958619, -0.341308, 1.5303, 0.38797, -0.022888}
?2-[9]-{-0.827027, -0.401642, 1.46631, -0.348816, 0.160522}
?3-[10]-{1.29065, 1.3107, -0.145813, -0.022888, -0.731751}
?2-[11]-{1.3107, 0.748627, 1.46631, 0.635193, 0.96988}
?4-[12]-{-0.850556, 0.506866, 1.9292, 0.431702, 0.748627}
?2-[13]-{1.49783, -0.401642, 1.46631, -0.348816, 0.763977}
?1-[14]-{0.697174, 1.05301, 0.763977, 1.29065, -0.091796}
?2-[15]-{-0.699127, -0.853363, 1.46631, 1.74533, 0.160522}
?0-[16]-{1.67365, 0.431702, 0.121093, -0.13208, 1.94092}
?2-[17]-{1.68832, 0.742951, 1.5303, 0.635193, 0.748627}

```
?3-[18]-{-0.958619, 0.316925, -0.82428, -0.416962, -0.022888}
?3-[19]-{-0.958619, 0.316925, 1.66873, -0.416962, 1.47864}

Outputs:
[ 0] = f(1.46631) = 2.27802
[ 1] = f(-0.611939) = 1.22418
[ 2] = f(0.707672) = 0.831081
[ 3] = f(0.506866) = 0.891514
[ 4] = f(1.97852) = -0.236276
[ 5] = f(-0.950196) = 0.049822
[ 6] = f(1.67365) = 2.23274
[ 7] = f(-0.348816) = 0.651425
[ 8] = f(0.38797) = 0.856839
[ 9] = f(1.46631) = 2.27802
[10] = f(-0.022888) = 1.01508
[11] = f(1.46631) = 2.27802
[12] = f(0.748627) = 0.252069
[13] = f(1.46631) = 2.27802
[14] = f(1.05301) = 2.04831
[15] = f(1.46631) = 2.27802
[16] = f(1.67365) = 2.23274
[17] = f(1.5303) = -0.246576
[18] = f(-0.416962) = 1.21182
[19] = f(-0.416962) = 1.21182
```

It is worth noticing that, except for the clones created by elitism, the arrays of random numerical constants of all these best-of-generation individuals are all different as a very high mutation rate (55%) was used to create new numerical constants. Note, however, that, in all cases, the gene sequence was kept unchanged and they all are expressing constant $c_2 = 1.46631$.

In the next generation a new individual was created (chromosome 2) that is better than all its ancestors (this individual is shown in bold):

```
GENERATION N: 7
Structures:
?2-[ 0]-{-0.699127, -0.853363, 1.46631, 1.74533, 0.160522}
?0-[ 1]-{1.67365, 0.431702, 0.121093, -0.13208, 1.94092}
?2-[ 2]-{0.697174, -0.611939, 1.44742, 1.29065, 0.037323}
?2-[ 3]-{1.49399, -0.401642, -0.434051, 0.041168, 0.160522}
?3-[ 4]-{1.55374, 0.249176, -0.582947, -0.411377, 0.160522}
?2-[ 5]-{1.49783, -0.401642, 1.46631, -0.348816, 0.763977}
?1-[ 6]-{0.818024, -0.75528, 0.568451, -0.247955, 1.97852}
?1-[ 7]-{0.19223, 0.748627, 0.05075, -0.582947, 0.059295}
?2-[ 8]-{-0.827027, 0.249176, -0.686157, -0.348816, 0.160522}
?4-[ 9]-{1.88129, 1.72037, 1.46631, 0.041168, 1.67365}
?3-[10]-{-0.958619, 0.316925, 0.033294, 0.626953, 1.47864}
```

```
?0-[11]-{1.49783,  -0.401642,  1.46631,  -0.348816,  0.041168}
?3-[12]-{0.19223, 1.3107, -0.145813, -0.022888, 0.255859}
?2-[13]-{-0.958619, -0.341308, 1.5303, 0.38797, -0.022888}
?2-[14]-{-0.839569, 1.44742, 1.98966, 0.346924, 0.033294}
?2-[15]-{0.697174, 0.556641, 0.626953, 1.29065, -0.686157}
?2-[16]-{1.3107, 0.19223, 0.568451, 0.635193, 0.96988}
?3-[17]-{1.29065, 1.3107, -0.145813, -0.022888, 0.027526}
?2-[18]-{-0.589783, -0.853363, 0.037323, 0.818024, 1.93112}
?0-[19]-{1.52893, -0.401642, 0.05075, 1.51398, 0.160522}


Outputs:
[ 0] = f(1.46631)   = 2.27802
[ 1] = f(1.67365)   = 2.23274
[ 2] = f(1.44742)   = 2.44266
[ 3] = f(-0.434051) = 1.3807
[ 4] = f(-0.411377) = 1.14392
[ 5] = f(1.46631)   = 2.27802
[ 6] = f(-0.75528)  = 0.255087
[ 7] = f(0.748627)  = 0.252069
[ 8] = f(-0.686157) = 1.28909
[ 9] = f(1.67365)   = 2.23274
[10] = f(0.626953)  = 1.46967
[11] = f(1.49783)   = 1.10189
[12] = f(-0.022888) = 1.01508
[13] = f(1.5303)    = -0.246576
[14] = f(1.98966)   = 0.36469
[15] = f(0.626953)  = 1.46967
[16] = f(0.568451)  = 0.524405
[17] = f(-0.022888) = 1.01508
[18] = f(0.037323)  = 1.0344
[19] = f(1.52893)   = -0.206058
```

The parameter value x_0 = 1.44742 encoded by the best individual of this generation (chromosome 2) results in a considerably higher output for function (8.4), giving $f(x_0)$ = 2.44266.

And finally, by generation 8, a perfect solution encoding a parameter value that maps into the global maximum was found:

```
GENERATION N: 8
Structures:
?2-[ 0]-{0.697174, -0.611939, 1.44742, 1.29065, 0.037323}
?0-[ 1]-{1.79529, 0.431702, 0.121093, 1.2038, 1.94092}
?2-[ 2]-{1.07755, -0.611939, 1.44742, -0.301575, 0.037323}
?0-[ 3]-{1.67365, 0.431702, 0.90503, 1.8645, -0.579163}
?2-[ 4]-{0.697174, -0.611939, 1.44742, 1.90302, -0.654175}
?2-[ 5]-{-0.253937, -0.401642, 1.46631, -0.348816, 0.763977}
```

```
?2-[ 6]-{0.494751, 1.90302, 1.46631, 1.78217, 0.763977}
?2-[ 7]-{1.49399, 1.53351, 1.73672, 0.041168, 0.160522}
?2-[ 8]-{1.49399, -0.401642, -0.434051, -0.301575, 1.142}
?0-[ 9]-{-0.662018, 1.27142, 1.35562, 1.90302, 0.430389}
?2-[10]-{0.406036, 0.150848, 1.98966, 0.346924, -0.538483}
?3-[11]-{-0.958619, 1.142, -0.654175, 0.626953, -0.654175}
?3-[12]-{-0.978241, -0.853363, 0.037323, 0.818024, 1.92236}
?1-[13]-{-0.958619, 0.316925, 1.05463, 0.19992, 1.47864}
?3-[14]-{0.19223, 1.3107, -0.662018, 1.85046, 1.69431}
?2-[15]-{-0.321991, 1.72037, 1.46631, 0.041168, -0.42337}
?3-[16]-{0.19223, -0.42337, -0.145813, -0.022888, 0.990174}
?2-[17]-{-0.253937, 0.249176, -0.582947, -0.932801, 1.53351}
?2-[18]-{0.430389, 1.3107, -0.387329, -0.022888, 1.05463}
?2-[19]-{0.697174, -0.611939, 1.44742, 1.29065, 0.037323}

Outputs:
[ 0] = f(1.44742)  = 2.44266
[ 1] = f(1.79529)  = 0.73521
[ 2] = f(1.44742)  = 2.44266
[ 3] = f(1.67365)  = 2.23274
[ 4] = f(1.44742)  = 2.44266
[ 5] = f(1.46631)  = 2.27802
[ 6] = f(1.46631)  = 2.27802
[ 7] = f(1.73672)  = -0.587869
[ 8] = f(-0.434051) = 1.3807
[ 9] = f(-0.662018) = 1.61539
[10] = f(1.98966)  = 0.36469
[11] = f(0.626953) = 1.46967
[12] = f(0.818024) = 1.43884
[13] = f(0.316925) = 0.839315
[14] = f(1.85046)  = 2.85027
[15] = f(1.46631)  = 2.27802
[16] = f(-0.022888) = 1.01508
[17] = f(-0.582947) = 0.702421
[18] = f(-0.387329) = 0.849855
[19] = f(1.44742)  = 2.44266
```

As you can see, the parameter value $x_0 = 1.85046$ encoded by the best of this generation corresponds to the output $f(x_0) = 2.85027$ and, therefore, this individual is a perfect solution to the problem at hand.

It is also interesting to see how close the HZero algorithm can get to the global maximum by letting it run for a few thousands of generations until no further improvement is observed. Consider, for instance, the evolutionary history presented below (the generation by which these solutions were discovered is shown in square brackets):

```
[0]   =  f(1.648345947)   =  2.64612100283072
[3]   =  f(1.843414307)  =  2.80410042875014
[13]  =  f(1.852813721)  =  2.84557968213877
[119] =  f(1.848297119)  =  2.84565284514179
[152] =  f(1.850646973)  =  2.85026471989876
[1400] =  f(1.850463868)  =  2.85026738184542
[2692] =  f(1.850585938)  =  2.85027241426006
[6416] =  f(1.85055542)   =  2.85027370872288
```

It was generated using the same settings of Table 8.1, with the difference that the number of generations was increased to 50,000. And as you can see, although a pretty good solution with an output higher that 2.85 was discovered early on in generation 152, the algorithm continued its search for the global optimum and three better approximations were discovered. And the best of all, discovered in generation 6416, is indeed a very good approximation to the global maximum of function (8.4).

8.2 The GEP-PO Algorithm

The GEP-PO algorithm is considerably more complex than either the HZero algorithm or the GA as it explores the GEP-RNC algorithm in all its complexity, making good use of its complex genes for fine-tuning the multiple parameters that optimize a function. Indeed, thanks to these sharp tools, the GEP-PO algorithm is rarely stuck in local optima for long stretches of time and is, therefore, always on the move to find the elusive global optimum.

8.2.1 The Architecture

Similarly to the HZero algorithm, the GEP-PO algorithm also uses N different genes to encode the values of the N parameters that optimize a function. But here the genes are much more complex than the simple structures of the HZero algorithm and the parameters are designed rather than just found.

Consider, for instance, the chromosome below composed of two genes:

```
Gene 1: +/+??+????????4374796
     C₁: {-0.698, 0.781, -0.059, -0.316, -0.912,
          0.398, 0.157, 0.473, 0.103, -0.756}

Gene 2: +////+????????4562174
     C₂: {0.104, 0.722, -0.547, -0.052, -0.876,
          -0.248, -0.889, 0.404, 0.981, -0.149}
```
(8.5a)

As you can see in Figure 8.2, its expression results in the following set of parameter values:

$$f(2.92008, 0.170599) \tag{8.5b}$$

which are then used to evaluate the output of the function at hand. For instance, for the function below:

$$f(p_1, p_2) = \cos\sqrt{p_1^2 + p_2^2} + 1.07 \cdot \sin(2p_1) \tag{8.6}$$

they give:

$$f(2.92008, 0.170599) = -1.4353299735$$

Note that, although the random numerical constants ranged, in this particular case, over the interval [-1, 1], a complete new range of constants can be created by performing a wide range of mathematical operations with the original random constants. Note also that this also means that sometimes the

a.   0123456789012345678901234567890123456789
      +/+??+??????**4374796**+////+???????**4562174**

C_1 = {-0.698, 0.781, -0.059, -0.316, -0.912, 0.398, 0.157, 0.473, 0.103, -0.756}
C_2 = {0.104, 0.722, -0.547, -0.052, -0.876, -0.248, -0.889, 0.404, 0.981, -0.149}

b.

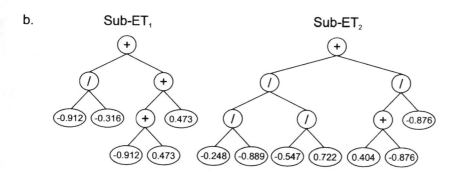

c. $$f(p_1, p_2) = (2.92008, 0.170599)$$

Figure 8.2. Expression of chromosomes encoding parameter values in a function optimization task. **a)** A two-genic chromosome with its arrays of random numerical constants. **b)** The sub-ETs codified by each gene. **c)** The values of the different parameters expressed in each sub-ET. The fitness of this program is the function value at point (p_1, p_2).

range of a parameter will fall below or above the imposed constraints; when this happens, the individual is made unviable and, therefore, will not be chosen to reproduce with modification, putting pressure on the selection of individuals that satisfy the chosen constraints and contributing to their dissemination in the population.

Let's now see how the GEP-PO algorithm fine-tunes its solutions by solving a simple optimization problem.

8.2.2 Optimization of a Simple Function

The function we are going to optimize with the GEP-PO algorithm is the same simple function of section 8.1.2, subjected to the same constraints. For the sake of simplicity, we are also going to use a small set of five random constants per gene and, again, represent them by the numerals 0-4, thus giving $R = \{0, 1, 2, 3, 4\}$. The ephemeral random constants "?" will be drawn from the rational interval [-2, 2], as, in this case, the constraints are imposed internally after the expression of each parameter value. The complete list of the parameters used per run is shown in Table 8.2.

The evolutionary dynamics of the successful run we are going to analyze is shown in Figure 8.3. And as you can see, in this run, a perfect solution

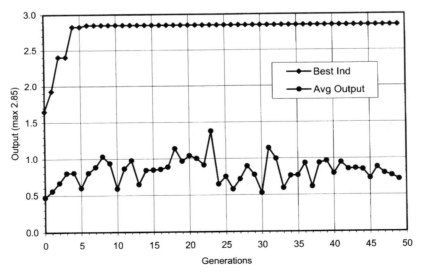

Figure 8.3. Progression of average output of the population and the output of the best individual for a successful run of the experiment summarized in Table 8.2.

Table 8.2
Settings for the simple optimization problem using the GEP-PO algorithm.

Number of generations	50
Population size	20
Function set	+ - * /
Terminal set	?
Random constants array length	5
Random constants type	Rational
Random constants range	[-2, 2]
Head length	5
Gene length	17
Number of genes	1
Chromosome length	17
Mutation rate	0.044
Inversion rate	0.1
IS transposition rate	0.1
RIS transposition rate	0.1
One-point recombination rate	0.3
Two-point recombination rate	0.3
Random constants mutation rate	0.55
Dc-specific mutation rate	0.044
Dc-specific inversion rate	0.1
Dc-specific transposition rate	0.1

encoding a parameter value that corresponds to the global maximum of function (8.4) was created in generation 6.

The initial population of this run, including both the structures and outputs of all its individuals, is shown below (a dash is used to flag the outputs that fall outside the chosen domain [-1, 2]; these individuals are all unviable and therefore have zero fitness):

```
GENERATION N: 0
Structures:
+-?+???????000141-[  0]-{-0.633606, 1.44528, -1.92072, 0.232788, -0.41391}
**//+??????211123-[  1]-{1.20468, -0.64856, -1.20969, -0.706726, -1.00031}
/?+//??????243140-[  2]-{0.801392, 0.965424, -0.887329, -1.9809, -1.3096}
**/-*??????140233-[  3]-{0.344513, -1.6423, 1.91223, -0.810486, -0.187408}
-+?*/??????124212-[  4]-{-1.44888, 0.859101, -0.359344, 0.763397, -1.00958}
*?**???????210103-[  5]-{-1.08029, 0.391937, -0.400848, -1.36026, -1.73004}
-*+/+??????420324-[  6]-{-0.908265, 1.15857, -0.871125, -0.459991, -0.345245}
/----??????022212-[  7]-{1.88175, -0.858857, -0.51532, 0.108032, 0.486572}
```

```
-/?/+??????331040-[  8]-{-1.02887, -0.418549, 0.227905, -0.816986, -0.466431}
-+/+-??????240300-[  9]-{1.83539, 1.11624, -0.917511, 0.395264, -1.71542}
+*/**??????021420-[10]-{-0.88263, -0.065887, -0.422607, -1.30252, -1.18634}
++**+??????032403-[11]-{1.40912, 0.489258, 1.1077, -1.08975, -0.546967}
*+/*+??????132410-[12]-{-0.243469, -0.438812, -0.882447, -0.906128, -1.05746}
-/-*+??????000234-[13]-{-1.58066, -0.344665, -0.249664, -0.405212, 0.219116}
/*-?*??????124240-[14]-{-0.419189, -1.16443, 1.82031, 0.203216, 1.75998}
***//??????010200-[15]-{-1.87112, -0.15274, -0.238952, -1.12656, -0.200226}
+*+?/??????112430-[16]-{-1.34305, -0.608338, 0.092956, -1.07773, -1.43314}
+-/-*??????340131-[17]-{1.5076, -0.039703, 1.36438, 1.03183, -1.09851}
/+/+*??????323432-[18]-{1.89792, -0.39209, 1.95444, -0.865998, -1.61389}
/*+*/??????021334-[19]-{-0.041564, 0.254913, 1.37637, -1.80347, -1.7045}
```

```
Outputs:
[ 0] = f(0.811676) = 1.2911
[ 1] = f(-3.57447) = --
[ 2] = f(11.664) = --
[ 3] = f(-9.02453) = --
[ 4] = f(-0.914593) = 0.59524
[ 5] = f(0.0665201) = 1.05776
[ 6] = f(-1.18539) = --
[ 7] = f(0.143316) = 0.859833
[ 8] = f(-0.488403) = 1.17403
[ 9] = f(1.69579) = 1.22345
[10] = f(2.11769) = --
[11] = f(-1.8221) = --
[12] = f(0.121492) = 0.924063
[13] = f(-2.12059) = --
[14] = f(-61.8317) = --
[15] = f(2.23794) = --
[16] = f(-1.32433) = --
[17] = f(0.64897) = 1.64863
[18] = f(9.4166) = --
[19] = f(-0.364412) = 0.672306
```

Note that, in this initial population, 11 out of 20 individuals have zero fitness but, as the system starts learning, this rate will start to decrease, being kept to a minimum in later generations.

The expression of the best individual of this generation (chromosome 17) is shown in Figure 8.4. As you can see, it corresponds to the parameter value $x_0 = 0.64897$, which gives $f(x_0) = 1.64863$. Note that some numerical constants are used more than once (c_1 and c_3 are both used twice) to create new ones. Note also that, for this small set of just five RNCs per gene, the algorithm is using most of the random constants at its disposal. In this particular case, only constant c_2 has no expression whatsoever in the final solution.

a.   01234567890123456
     +-/-*??????340131

     C = {1.5076, -0.039703, 1.36438, 1.03183, -1.09851}

b.   ET

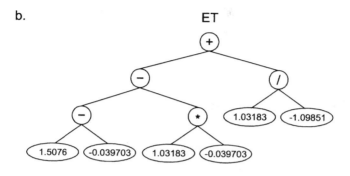

c. $f(0.64897) = (1.64863)$

Figure 8.4. Best individual of generation 0 (chromosome 17) created to find the global maximum of function (8.4). **a)** The chromosome of the individual with its random numerical constants. **b)** The fully expressed individual. **c)** The parameter value encoded in the chromosome and corresponding function value.

In the next generation a new individual was created, chromosome 9, considerably better than the best individual of the previous generation:

```
GENERATION N: 1
Structures:
+-/-*??????340131-[  0]-{1.5076, -0.039703, 1.36438, 1.03183, -1.09851}
/----??????022210-[  1]-{1.73029, -0.858857, -0.51532, -0.156433, 0.486572}
-/?/+??????331040-[  2]-{0.050384, 0.723206, 0.227905, -0.816986, -0.466431}
-+/**??????214212-[  3]-{-1.44888, -0.513245, -0.359344, 0.763397, -1.00958}
+-?+??????000141-[  4]-{-0.633606, -1.64124, -1.92072, -0.767426, -0.41391}
/?*++??????021334-[  5]-{-0.041564, 1.73029, -0.513245, -1.80347, -1.7045}
+-/-*??????340131-[  6]-{1.80936, 1.80936, -0.892914, 1.03183, 0.296997}
/?+?+??????210102-[  7]-{1.83539, 1.11624, 1.1489, -1.64124, -0.513245}
+-+?*??????120300-[  8]-{-1.44888, 0.859101, -0.359344, 0.763397, -1.00958}
-/?/+??????210143-[  9]-{-0.892914, 0.250061, 0.227905, -0.816986, -0.466431}
/*/+-??????021334-[10]-{-0.041564, 0.254913, 1.37637, -1.80347, 0.863404}
-?**??????244213-[11]-{0.723206, 0.391937, 0.623047, 1.68677, -1.73004}
*?*??????331040-[12]-{-1.08029, 0.604004, -0.400848, -1.36026, -1.73004}
-++?/??????240300-[13]-{0.723206, 1.11624, -0.917511, -0.17807, -1.32593}
-/?/+??????331040-[14]-{-1.02887, 0.808167, 0.494354, 1.60129, -0.466431}
/+?*/??????134212-[15]-{-1.44888, -0.502533, -0.156433, -0.502533, -1.00958}
```

```
+?/-???????340131-[16]-{1.5076, -0.039703, 1.36438, 1.80936, 1.33295}
-+++-??????302410-[17]-{1.83539, -0.208893, -0.917511, -0.502533, -1.71542}
+-/-*??????340131-[18]-{1.5076, -0.039703, 1.36438, 1.03183, -1.09851}
+-/-*??????340131-[19]-{-0.747101, -0.039703, -0.747101, 1.03183, -1.09851}
```

```
Outputs:
[ 0]  =  f(0.64897)    =  1.64863
[ 1]  =  f(1.15298)    =  -0.147927
[ 2]  =  f(3.53224)    =  --
[ 3]  =  f(-0.152923)  =  0.847722
[ 4]  =  f(-2.27484)   =  --
[ 5]  =  f(0.00946834) =  1.00278
[ 6]  =  f(1.60726)    =  1.3634
[ 7]  =  f(0.282432)   =  1.14809
[ 8]  =  f(0.156947)   =  0.846776
[ 9]  =  f(1.06641)    =  1.92784
[10]  =  f(-136.756)   =  --
[11]  =  f(-1.24176)   =  --
[12]  =  f(1.11759)    =  0.413278
[13]  =  f(-0.561012)  =  0.472227
[14]  =  f(-2.92636)   =  --
[15]  =  f(-1.62902)   =  --
[16]  =  f(2.97017)    =  --
[17]  =  f(-6.01007)   =  --
[18]  =  f(0.64897)    =  1.64863
[19]  =  f(-1.60573)   =  --
```

The expression of the best individual of this generation (chromosome 9) is shown in Figure 8.5. It encodes the parameter value $x_0 = 1.06641$, which corresponds to the output $f(x_0) = 1.92784$, thus a higher point in the fitness landscape than the best of the previous generation. Note that this individual is not related to the best of the previous generation, as they use a completely different set of random constants and show little homology in their sequences. Note again that almost all the constants are used to design this new parameter value and that constant c_1 is used twice.

In the next generation a new individual was created, chromosome 16, considerably better than the best individual of the previous generation:

```
GENERATION N: 2
Structures:
-/?/+??????210143-[  0]-{-0.892914, 0.250061, 0.227905, -0.816986, -0.466431}
+-/**??????340131-[  1]-{1.5076, -1.20621, 1.36438, 1.03183, -1.09851}
/?*++??????021334-[  2]-{-0.041564, 1.73029, -0.513245, 1.42899, -1.7045}
//+*+??????313334-[  3]-{1.42899, 1.11624, -0.489014, -1.64124, -0.513245}
*?*?????????103140-[  4]-{-1.08029, 0.604004, -0.400848, -1.36026, -0.963868}
```

a.   0123456789 0123456
     -/?/+??????210143

C = {-0.892914, 0.250061, 0.227905, -0.816986, -0.466431}

b.   ET

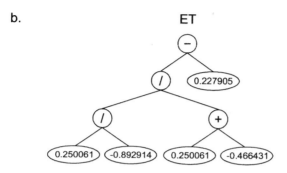

c. $f(1.06641) = (1.92784)$

Figure 8.5. Best individual of generation 1 (chromosome 9) created to find the global maximum of function (8.4). **a)** The chromosome of the individual with its random numerical constants. **b)** The fully expressed individual. **c)** The parameter value encoded in the chromosome and corresponding function value.

```
/-*+-??????340041-[  5]-{0.082397, -0.039703, 1.36438, -1.34146, -1.09851}
+*+?*??????123300-[  6]-{-1.44888, 0.859101, 0.1427, 1.66345, 1.79471}
+-/-*??????340431-[  7]-{1.5076, -0.039703, -0.761444, -1.20621, -1.09851}
/?*?+??????210102-[  8]-{1.76926, 1.73029, -0.513245, -0.674439, -1.7045}
?/*++??????021334-[  9]-{-1.09741, 1.73029, -0.513245, 1.79117, -1.34912}
-+/?*??????340431-[10]-{-0.892914, 1.79471, 0.227905, 0.082397, -1.65738}
+--?*??????120300-[11]-{-1.44888, 0.859101, 1.12442, 0.1427, -1.00958}
+/?*+??????021334-[12]-{-1.95117, 1.73029, -1.65738, -1.80347, -1.7045}
-/?/+??????210143-[13]-{-0.892914, 0.250061, 1.46106, -0.203644, -0.466431}
+??-*??????340132-[14]-{1.5076, -0.039703, 1.36438, 1.03183, -1.09851}
+/--/??????210143-[15]-{1.79117, 1.60388, -0.963868, 0.1427, -1.09851}
*-?????????331130-[16]-{1.79117, -0.287872, -0.400848, -1.36026, 0.1427}
-++?/??????240300-[17]-{0.723206, 1.11624, -0.917511, 1.25125, -1.95117}
-+/**??????214212-[18]-{-1.44888, -0.513245, -0.359344, 0.763397, -1.00958}
/-/+-??????210102-[19]-{-1.65738, 1.11624, 1.1489, -0.948121, -0.513245}

Outputs:
[ 0] = f(1.06641) = 1.92784
[ 1] = f(-1.51318) = --
[ 2] = f(-0.0119496) = 1.00438
[ 3] = f(2.38147) = --
```

```
[  4]  =  f(0.88757)   =  1.33786
[  5]  =  f(0.830343)  =  1.67699
[  6]  =  f(-0.264408) =  1.23778
[  7]  =  f(3.65626)   =  --
[  8]  =  f(-0.0847609)=  1.03905
[  9]  =  f(-1.09741)  =  --
[10]   =  f(-1.91031)  =  --
[11]   =  f(3.63916)   =  --
[12]   =  f(-1.15611)  =  --
[13]   =  f(-0.166747) =  0.855804
[14]   =  f(-0.066681) =  1.05773
[15]   =  f(-2.59208)  =  --
[16]   =  f(1.45873)   =  2.40425
[17]   =  f(2.0406)    =  --
[18]   =  f(-0.152923) =  0.847722
[19]   =  f(2.20076)   =  --
```

The expression of the best individual of this generation (chromosome 16) is shown in Figure 8.6. It encodes the parameter value $x_0 = 1.45873$, which corresponds to the output $f(x_0) = 2.40425$, thus considerably closer to the global maximum than the best of the previous generation. Note again that this individual is not related to the best of the previous generations, as neither the

a.   01234567890123456
      *-??????????331130

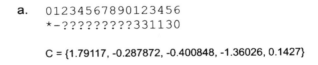

b.   **ET**

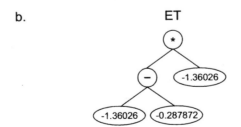

c. $f(1.45873) = (2.40425)$

Figure 8.6. Best individual of generation 2 (chromosome 16) created to find the global maximum of function (8.4). **a)** The chromosome of the individual with its random numerical constants. **b)** The fully expressed individual. **c)** The parameter value encoded in the chromosome and corresponding function value.

random constants nor the gene sequences show homology. Note also that this individual makes use of just two different constants (c_1 and c_3) to design this new parameter value.

In the next generation there was no improvement in best output and the best individual of this generation is the clone of the best of the previous generation and was, as usual, created by elitism:

```
GENERATION N: 3
Structures:
*-?????????331130-[  0]-{1.79117, -0.287872, -0.400848, -1.36026, 0.1427}
*-?????????340041-[  1]-{1.79117, -0.287872, -0.400848, 1.42456, 0.1427}
/?*?/??????210102-[  2]-{1.76926, -0.552368, 1.18143, -0.674439, -1.7045}
/?*?-??????340041-[  3]-{1.76926, 1.73029, 0.805542, -0.674439, 1.90161}
-/?/-??????211143-[  4]-{-1.55948, 0.308075, 0.227905, -0.816986, 1.90161}
/--+*??????210102-[  5]-{-1.80786, -0.039703, 1.36438, -1.34146, -1.09851}
/-*+/??????340041-[  6]-{-1.58328, 1.44235, 1.36438, -0.221069, -1.09851}
*?*?????????140212-[  7]-{-1.08029, 1.42456, 1.18143, -1.60004, 0.103332}
/?*?*??????210102-[  8]-{-1.85358, 1.76041, -1.58328, 1.76041, -1.7045}
**+/*??????331130-[  9]-{-1.44888, -0.513245, -0.359344, -0.496002, -1.00958}
/*??+??????213132-[10]-{1.76926, -1.44598, 0.465607, 1.44235, -1.7045}
*-?/??????214310-[11]-{1.18143, -0.287872, -0.400848, -1.36026, -0.132354}
*?-?????????331130-[12]-{1.79117, -0.287872, 1.07327, 1.18143, -1.75626}
+*+?*??????123300-[13]-{-1.44888, 0.859101, 0.1427, 1.66345, 1.79471}
-/?+*??????133143-[14]-{1.42456, 0.805542, 0.227905, -0.816986, 0.162994}
+?/+/??????210412-[15]-{1.5076, -0.039703, 0.426208, 1.18143, -0.132354}
*?*?????????103140-[16]-{-1.80786, 0.604004, -0.400848, 0.393646, -0.963868}
*-?????????331130-[17]-{1.79117, -0.287872, -0.400848, -1.86426, 0.1427}
/?*?+??????031130-[18]-{0.082397, -0.039703, 0.365814, -1.34146, -1.09851}
/-*+-??????210102-[19]-{0.162994, -0.259796, -0.152893, -0.674439, -1.05676}

Outputs:
[  0] = f(1.45873)  = 2.40425
[  1] = f(-2.34834) = --
[  2] = f(0.667753) = 1.56656
[  3] = f(2.38147)  = --
[  4] = f(-0.85544) = 1.84298
[  5] = f(0.44089)  = 1.42295
[  6] = f(-9.90319) = --
[  7] = f(-0.159022) = 0.847323
[  8] = f(0.275627) = 1.19102
[  9] = f(-0.712903) = 0.718877
[10] = f(0.0112795) = 1.00391
[11] = f(-0.154396) = 0.847074
[12] = f(1.73587)  = -0.567619
[13] = f(-0.264408) = 1.23778
```

```
[14] = f(-13.2503) = --
[15] = f(0.866541) = 1.75215
[16] = f(-0.429844) = 1.34651
[17] = f(2.93879) = --
[18] = f(0.773535) = 0.428444
[19] = f(-10.3897) = --
```

Note that the percentage of unviable individuals in this generation dropped down to 30%, considerably lower than the 55% of the initial population, showing that learning is under way.

In the next generation a new individual was created, chromosome 15, considerably better than the best individual of the previous generation:

```
GENERATION N: 4
Structures:
*-??????????331130-[  0]-{1.79117, -0.287872, -0.400848, -1.36026, 0.1427}
*?*???????210112-[  1]-{-1.54187, 0.743225, -1.20395, -0.35907, -0.963868}
/--+*??????102102-[  2]-{-1.80786, 0.699249, 1.36438, -1.34146, -1.09851}
-?*+/??????103140-[  3]-{-1.84558, 0.633179, -1.67188, 1.18143, -0.132354}
-/?/??????211143-[  4]-{-1.55948, -1.26907, -1.74191, -0.147216, -1.26907}
/--+*??????210102-[  5]-{-1.80786, -0.039703, 1.36438, -1.34146, -1.09851}
+?+/??????210412-[  6]-{-1.20395, -0.554138, -0.928986, 1.18143, -0.132354}
-/?/-??????211143-[  7]-{-1.55948, 0.308075, 0.227905, -0.816986, 1.34183}
/+-+*??????210402-[  8]-{-1.80786, -0.039703, 1.36438, -1.34146, -1.09851}
*?*???????101132-[  9]-{1.29898, 0.604004, -0.400848, 0.393646, -0.963868}
*?*???????140214-[ 10]-{-1.08029, 1.42456, 1.18143, 0.505127, 0.103332}
/?*+??????031130-[ 11]-{0.082397, -0.039703, 0.365814, 0.793213, -1.09851}
+?**+??????142230-[ 12]-{-1.44888, 1.41498, 0.793213, -0.496002, -0.055114}
+*+?*??????143300-[ 13]-{-1.44888, -0.35907, -0.737458, 1.66345, 1.79471}
/???+??????103130-[ 14]-{-1.52936, -1.84558, -0.994172, -0.147216, 0.103332}
+?/+/??????210412-[ 15]-{1.5076, -0.039703, 1.41498, 0.660401, -0.132354}
??/+*??????210412-[ 16]-{0.81247, -0.593659, 0.426208, 0.660401, -1.67188}
*??*??????140212-[ 17]-{-1.08029, -0.593659, 0.869568, -1.60004, 0.103332}
/*??+??????213142-[ 18]-{1.76926, -0.32016, 0.465607, 1.44235, -1.7045}
/--+/??????210102-[ 19]-{-0.41037, -0.039703, -0.985657, -1.6427, -1.71796}
```

```
Outputs:
[  0] = f(1.45873) = 2.40425
[  1] = f(1.37967) = 0.177624
[  2] = f(1.80696) = 1.39175
[  3] = f(-2.54412) = --
[  4] = f(0.953936) = 0.0533483
[  5] = f(0.44089) = 1.42295
[  6] = f(7.61331) = --
[  7] = f(-1.19526) = --
[  8] = f(-3.82668) = --
```

```
[ 9] = f(0.473895) = 1.34652
[10] = f(-0.159022) = 0.847323
[11] = f(-1.30818) = --
[12] = f(1.40198) = 1.08736
[13] = f(4.32357) = --
[14] = f(1.20677) = 1.25467
[15] = f(1.85531) = 2.82954
[16] = f(0.426208) = 1.31259
[17] = f(-0.061344) = 1.05749
[18] = f(-0.77164) = 0.399917
[19] = f(0.915916) = 0.560869
```

The expression of the best individual of this generation (chromosome 15) is shown in Figure 8.7. It encodes the parameter value $x_0 = 1.85531$, which corresponds to the output $f(x_0) = 2.82954$, reaching a new height in the solution space. Note again that this individual is not related to the best of the previous generations; it is rather a direct descendant of chromosome 15 of generation 3, the third best of its generation. Indeed, in the GEP-PO system, all newly created individuals have a past history that can be easily traced back to its ancestors in the initial population.

a. 01234567890123456
    +?/+/??????210412

C = {1.5076, -0.039703, 1.41498, 0.660401, -0.132354}

b.   ET

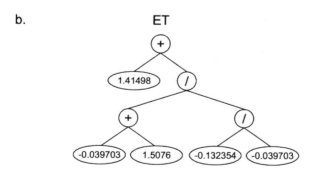

c. $f(1.85531) = (2.82954)$

Figure 8.7. Best individual of generation 4 (chromosome 15) created to find the global maximum of function (8.4). **a)** The chromosome of the individual with its random numerical constants. **b)** The fully expressed individual. **c)** The parameter value encoded in the chromosome and corresponding function value.

In the next generation there was no improvement in best output and, as usual, the best of this generation is the clone created by elitism:

```
GENERATION N: 5
Structures:
+?/+/??????210412-[  0]-{1.5076, -0.039703, 1.41498, 0.660401, -0.132354}
??/+*??????140412-[  1]-{0.81247, 1.5148, -0.777252, 0.660401, -1.96195}
--?/??????142230-[  2]-{-0.41037, -0.039703, -1.6095, 0.58136, -1.71796}
**?*??????140212-[  3]-{-1.08029, 0.609223, 0.869568, -1.60004, 0.103332}
/--/*??????210102-[  4]-{1.01096, -1.09009, 1.36438, -0.018035, 1.56934}
+?/?+??????100412-[  5]-{1.5076, -0.77179, -1.20801, 0.660401, -0.132354}
/--+*??????102100-[  6]-{-1.80786, 1.15878, 0.609223, -1.34146, -1.09851}
/---*??????210102-[  7]-{-1.80786, 0.699249, 1.36438, -1.34146, -1.09851}
*-????????331130-[  8]-{-0.018035, 0.318085, -1.65381, -1.36026, 0.1427}
/--+*??????210102-[  9]-{-1.80786, -0.039703, -0.112304, -1.34146, 0.818177}
-+*/-??????210302-[10]-{-1.80786, 1.22555, 0.852845, 0.445862, -1.09851}
/-+/*??????331130-[11]-{1.79117, 1.1149, 1.1149, -1.36026, 0.1427}
/?*?+??????103132-[12]-{-1.52936, -1.84558, -0.531647, -0.147216, -0.649262}
??/+*??????110402-[13]-{0.81247, -1.56635, -1.96341, 0.660401, 1.13275}
??/+*??????104120-[14]-{0.81247, -0.018035, 0.426208, 0.660401, -1.67188}
+?*+-??????210101-[15]-{-1.44888, -1.65381, 1.15878, -0.496002, -0.018035}
*?*????????140214-[16]-{-1.61643, 1.42456, 1.5148, 0.505127, -1.27096}
/??+/??????213130-[17]-{1.13275, -1.84558, -0.994172, -0.147216, 0.103332}
+?//+??????210412-[18]-{0.486664, -0.039703, 1.41498, -0.077697, -0.132354}
*?*????????101131-[19]-{1.29898, -1.96341, -0.875, -0.649262, -0.963868}

Outputs:
[ 0] = f(1.85531) = 2.82954
[ 1] = f(1.5148) = 0.320746
[ 2] = f(2.75766) = --
[ 3] = f(-0.0591365) = 1.05672
[ 4] = f(-0.939811) = 0.10793
[ 5] = f(0.32445) = 0.774575
[ 6] = f(-0.505743) = 0.909252
[ 7] = f(-0.0608906) = 1.05736
[ 8] = f(2.28299) = --
[ 9] = f(28.2447) = --
[10] = f(-7.76067) = --
[11] = f(-1.26316) = --
[12] = f(-0.605565) = 1.10534
[13] = f(-1.56635) = --
[14] = f(-0.018035) = 1.00968
[15] = f(1.79461) = 0.697303
[16] = f(2.92664) = --
[17] = f(0.538677) = 0.495047
[18] = f(1.88913) = 1.6324
[19] = f(5.00754) = --
```

And finally, in the next generation a parameter value that maps very closely to the global maximum of function (8.4) was discovered:

```
GENERATION N: 6
Structures:
+?/+/??????210412-[  0]-{1.5076, -0.039703, 1.41498, 0.660401, -0.132354}
+?/*+??????210012-[  1]-{-1.08066, -0.849427, -0.907471, -0.077697, -0.132354}
+?/?+??????100412-[  2]-{1.5076, -0.77179, -1.20801, -0.849427, 0.522186}
+?/?+??????100412-[  3]-{0.919007, -0.77179, -0.427673, 0.660401, -0.132354}
?/?*???????140212-[  4]-{-0.907471, 0.609223, 0.869568, 0.132232, 0.522186}
**?*???????140212-[  5]-{1.96539, -0.849427, 0.869568, 0.132232, 0.103332}
/--+*??????213130-[  6]-{-1.80786, 1.15878, 0.609223, -0.292877, 0.919007}
/?/+*??????104120-[  7]-{0.402527, 1.26566, 1.56201, -0.915558, 1.02359}
+--++??????100412-[  8]-{1.5076, 0.754364, 1.41498, 0.660401, -0.132354}
/?/+/??????202300-[  9]-{0.489838, 1.15878, 0.609223, 1.61295, -1.60925}
+?+-/??????210412-[10]-{1.5076, -1.32907, 0.782532, 0.660401, -0.132354}
+?/+/??????200412-[11]-{1.5076, -0.039703, 1.02359, 0.660401, -0.132354}
+?/+/??????210412-[12]-{1.49054, -0.039703, 1.41498, 0.660401, -0.132354}
+?//+??????210412-[13]-{0.486664, -0.039703, 1.02359, -0.077697, 1.49054}
+?/-/??????210412-[14]-{1.5076, -1.95691, 1.41498, 0.660401, -0.132354}
/---*??????102300-[15]-{1.61295, 1.15878, 0.609223, -1.34146, -1.09851}
/+*???????104120-[16]-{0.81247, 1.96539, 0.426208, 0.660401, -1.20557}
+?/+/??????102110-[17]-{0.402527, -0.039703, 1.41498, 0.522186, -0.132354}
+/??+??????041200-[18]-{1.13275, -1.84558, -0.994172, 0.132232, 0.103332}
+?*+-??????210101-[19]-{-1.44888, -0.476166, 1.15878, -0.496002, -0.018035}

Outputs:
[  0] = f(1.85531)   = 2.82954
[  1] = f(-1.38307)  = --
[  2] = f(-0.0290517) = 1.02298
[  3] = f(0.39646)   = 0.955994
[  4] = f(0.609223)  = 1.17406
[  5] = f(-0.150008) = 0.849992
[  6] = f(-0.612171) = 1.22841
[  7] = f(1.75453)   = -0.736818
[  8] = f(-1.54733)  = --
[  9] = f(1.82525)   = 2.30066
[10] = f(-1.95455)  = --
[11] = f(1.92808)   = -0.488558
[12] = f(1.85019)   = 2.85016
[13] = f(0.967359)  = 0.172961
[14] = f(-49.8093)  = --
[15] = f(1.43321)   = 2.2384
[16] = f(-1.17239)  = --
[17] = f(1.7778)    = -0.141712
[18] = f(1.09636)   = 1.12499
[19] = f(-0.713745) = 0.701288
```

The expression of this perfect solution is shown in Figure 8.8. It encodes the parameter value $x_0 = 1.85019$, giving $f(x_0) = 2.85016$, thus very close to the global maximum of function (8.4). Note that this individual is a direct descendant of the best individual of the previous generation. As you can see by comparing both solutions, this new individual was created thanks to a mutation in the array of random constants (the constant 1.5076 at position 0 was replaced by 1.49054). It is worth pointing out that this kind of learning is

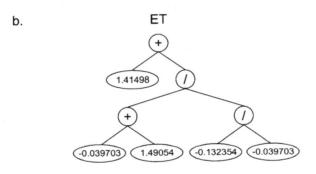

a.   01234567890123456
       +?/+/??????210412

       C = {1.49054, -0.039703, 1.41498, 0.660401, -0.132354}

b.

c. $f(1.85019) = (2.85016)$

Figure 8.8. Perfect solution designed in generation 6 (chromosome 12) encoding the parameter value for the global maximum of function (8.4). **a)** The chromosome of the individual with its random numerical constants. **b)** The expression tree encoded in the chromosome. **c)** The parameter value represented by the expression tree and corresponding function value.

not possible with the HZero algorithm as each parameter is expressed using just one node.

As we did before for the HZero algorithm, let's now see how close the GEP-PO algorithm can get to the global maximum of function (8.4) by letting it run for a few thousands of generations until no further improvement is observed. Consider, for instance, the evolutionary history presented below (the number in square brackets indicates the generation by which these solutions were discovered):

```
[0] = f(0.8662827398) = 1.75539246128292
[1] = f(1.243607008) = 2.2186092614651
[2] = f(1.442756154) = 2.405557613831
[3] = f(1.643576457) = 2.61022348179365
[7] = f(1.84369) = 2.80758281206205
[35] = f(1.851014) = 2.85007488485697
[164] = f(1.850769) = 2.85022892534645
[226] = f(1.850495) = 2.85027125170459
[526] = f(1.850586) = 2.85027240989802
[1894] = f(1.850525) = 2.85027330541754
[14604] = f(1.850565877) = 2.85027345685206
[14616] = f(1.850542097) = 2.85027374018891
[14636] = f(1.850545429) = 2.85027376273599
[14651] = f(1.850548264) = 2.85027376594376
[14784] = f(1.850547275) = 2.85027376649316
[14952] = f(1.850547484) = 2.85027376652619
[28863] = f(1.850547459) = 2.85027376652646
[41270] = f(1.85054747) = 2.85027376652649
[49986] = f(1.850547466) = 2.8502737665265
```

It was generated using the same settings of Table 8.2, with the difference that the number of generations was increased a thousand times. And as you can see, although a very good solution was discovered early on in generation 35, the algorithm continued its search for the global maximum and a total of 13 better approximations were made, considerably more than the additional three approximations achieved by the HZero algorithm (see page 312). This clearly shows the advantages of a hierarchical learning process where the parameter values are designed rather than just discovered; that is, in the GEP-PO system, from generation to generation very good solutions are selected to reproduce with modification and, by changing slightly their structures, it is possible to create even better solutions that will get closer and closer to the global maximum. As a comparison, this kind of learning is totally absent in the HZero system as it cannot learn and change at the same time with just one node to express the solution.

It is also worth noticing that the best approximation to the global maximum of function (8.4) designed by the GEP-PO algorithm in this experiment, $f(1.850547466) = 2.8502737665265$, is closer to the global maximum than the best solution discovered by the HZero algorithm in similar circumstances, $f(1.85055542) = 2.85027370872288$.

Let's now see how both these algorithms perform in two complex optimization problems of two and five parameters.

8.3 Maximum Seeking with GEP

In this section we are going to seek the maximum of two different functions. The first, is the following well-studied two-parameter function:

$$f(x, y) = -x\sin(4x) - 1.1y\sin(2y) \qquad (8.7a)$$

subjected to the constraints:

$$0 \le x \le 10 \qquad \text{and} \qquad 0 \le y \le 10 \qquad (8.7b)$$

For this function the maximum value is known, which, for simplicity, will be considered 18.5.

Although simple, this kind of functions with known global optima are very useful as they can be used to measure accurately the performance of different algorithms. For instance, the global minimum of the function (8.7) above could not be found by traditional minimum seeking algorithms such as the Nelder-Mead or the Broyden-Fletcher-Goldfarb-Shanno algorithms (Haupt and Haupt 1998).

On the other hand, less conventional methods such as the GA or GEP have no problems at all in finding the global maximum (or minimum) of this function (Table 8.3). Indeed, considering 18.5 the maximum output for function (8.7), both the HZero and the GEP-PO algorithms could find the exact parameters for which the function evaluation returns output values equal to or greater than 18.5 in virtually all runs.

As shown in Table 8.3, the HZero algorithm considerably outperforms the GEP-PO system at this simple task (note that in the GEP-PO experiment populations evolved 10 times longer). And taking into account the greater complexity of the GEP-PO algorithm, it is obviously advisable to use the simpler HZero system to solve function optimization problems with just one or two dimensions. However, the GEP-PO algorithm can find much more precise parameter values as it designs its solutions with as great a precision as necessary and, therefore, for those cases where such a precision is needed, the GEP-PO algorithm should be used instead. Furthermore, the HZero algorithm might sometimes converge prematurely, becoming stuck in some local maximum. In those cases, the higher flexibility of the GEP-PO algorithm becomes indispensable for navigating such treacherous landscapes.

By analyzing the best-of-experiment solutions discovered with both algorithms, the fine-tuning capabilities of the GEP-PO algorithm become clear.

Table 8.3
Performance and settings for the two-parameter function optimization problem using the HZero algorithm and the GEP-PO algorithm.

	HZero	GEP-PO
Number of runs	100	100
Number of generations	100	1000
Population size	30	30
Function set	--	+ - * /
Terminal set	?	?
Random constants array length	10	10
Random constants type	Rational	Rational
Random constants range	[0, 10]	[-1, 1]
Head length	0	6
Gene length	2	20
Number of genes	2	2
Chromosome length	4	40
Mutation rate	--	0.025
Inversion rate	--	0.1
IS transposition rate	--	0.1
RIS transposition rate	--	0.1
One-point recombination rate	0.3	0.3
Two-point recombination rate	0.3	0.3
Gene recombination rate	0.3	0.3
Gene transposition rate	0.1	0.1
Random constants mutation rate	0.35	0.01
Dc-specific mutation rate	0.01	0.025
Dc-specific inversion rate	--	0.1
Dc-specific transposition rate	--	0.1
Average best-of-run output	18.5148	18.5387
Success rate	80%	93%

For instance, the following set of parameter values was discovered by the HZero algorithm in generation 73 of run 77:

$$f(9.038421, 8.668366) = 18.5546969067$$

And the set below was designed by the GEP-PO algorithm in generation 155 of run 20:

$$f(9.03893887031, 8.66878467558) = 18.5547140746$$

330 GENE EXPRESSION PROGRAMMING

As you can see, the GEP-PO algorithm can approximate the parameter values with much more precision and, therefore, can find much better solutions than either the HZero algorithm or the GA.

It is also interesting to find out how close can these systems get to the global maximum by allowing evolution to take its course for several thousands of generations. For instance, running the HZero algorithm for 100,000 generations (three orders of magnitude longer than in the experiment summarized in the first column of Table 8.3), the following solutions were discovered (the number in square brackets indicates the generation by which they were created):

```
[0]   = f(9.126373291, 9.266235351) = 11.6629777173885
[3]   = f(8.857421875, 5.656158447) = 12.694624653527
[4]   = f(5.900817871, 8.469024658) = 14.6762045169257
[5]   = f(9.068511962, 5.491333008) = 15.0123472145664
[7]   = f(9.068511962, 8.772583008) = 18.2818098478718
[14]  = f(9.068511962, 8.751739502) = 18.3572449335579
[21]  = f(9.050445556, 8.751739502) = 18.4108930891067
[31]  = f(9.050445556, 8.708007813) = 18.5147538223188
[75]  = f(9.050445556, 8.693634033) = 18.5327871603454
[92]  = f(9.03729248, 8.693634033) = 18.5420822284372
[100] = f(9.03729248, 8.655212402) = 18.5512882545018
[256] = f(9.03729248, 8.66192627) = 18.5537607892668
[914] = f(9.037872314, 8.66192627) = 18.5538790786118
[946] = f(9.037872314, 8.663665772) = 18.5542384233189
[1148] = f(9.037872314, 8.66381836) = 18.5542644183508
[1724] = f(9.037872314, 8.664001465) = 18.5542944354892
[1833] = f(9.037872314, 8.668609619) = 18.554626996419
[2247] = f(9.038848876, 8.668609619) = 18.5547162111929
[5005] = f(9.038848876, 8.667785645) = 18.554716485223
[18408] = f(9.038848876, 8.668457031) = 18.5547182253551
[21950] = f(9.038848876, 8.668334961) = 18.554719194082
[41957] = f(9.038848876, 8.66809082) = 18.5547194180098
[63082] = f(9.038848876, 8.668273926) = 18.5547194642522
[63489] = f(9.039001464, 8.668273926) = 18.5547209319945
[72426] = f(9.039001464, 8.668243408) = 18.5547210135332
```

As you can see, the HZero algorithm is able to find the global maximum of function (8.7) with great precision.

And running the GEP-PO algorithm for 1,000,000 generations (also three orders of magnitude longer than in the experiment summarized in the second column of Table 8.3), the following solutions were designed (the generation by which they were created is in square brackets):

```
[0]    = f(0.5373683084, 8.549400615) = 8.80259358641701
[1]    = f(1.169346791, 8.654789962) = 10.6843782636572
[2]    = f(2.56474401, 8.549400615) = 11.1523261017334
[3]    = f(2.798309803, 8.654789962) = 12.2595682909331
[5]    = f(5.673876702, 8.654789962) = 13.1898377299992
[15]   = f(9.105922305, 8.654789962) = 18.2273236993991
[40]   = f(9.073935287, 8.654789962) = 18.4628024958807
[43]   = f(9.070846391, 8.654789962) = 18.4777504127046
[94]   = f(9.070846391, 8.66669448) = 18.4811446480484
[247]  = f(9.058678611, 8.651249962) = 18.5211448849144
[406]  = f(9.058678611, 8.664493) = 18.5263745426905
[1182] = f(9.058678611, 8.666354) = 18.5265717435938
[1520] = f(9.023396997, 8.666354) = 18.5370766419725
[1750] = f(9.048491662, 8.666354) = 18.5481190571652
[1905] = f(9.043386688, 8.666354) = 18.5532577056288
[2147] = f(9.042537043, 8.666354) = 18.5537463206678
[2576] = f(9.042537043, 8.668468023) = 18.5538093494319
[6747] = f(9.035891, 8.668468023) = 18.5540237663948
[6788] = f(9.036623, 8.668468023) = 18.5543135025517
[7108] = f(9.036837229, 8.668468023) = 18.5543836313701
[7865] = f(9.037875229, 8.668468023) = 18.5546293670704
[42316]  = f(9.039186229, 8.668468023) = 18.5547168426169
[43243]  = f(9.039186229, 8.668372756) = 18.5547176877183
[110851] = f(9.039186229, 8.66822349) = 18.5547183122798
[135051] = f(9.038857816, 8.66822349) = 18.5547197587122
[148895] = f(9.038901166, 8.66822349) = 18.5547204624021
[176618] = f(9.038940108, 8.66822349) = 18.5547208625485
[198737] = f(9.038978087, 8.66822349) = 18.554721041306
[465844] = f(9.039004378, 8.66822349) = 18.5547210427225
[569578] = f(9.039004378, 8.668181223) = 18.5547210644227
[819512] = f(9.039004378, 8.668181699) = 18.5547210645594
```

As you can see, the GEP-PO algorithm can approximate the global maximum of function (8.7) even better than the HZero algorithm. Indeed, the four best solutions designed with the GEP-PO algorithm are better than the best discovered with the HZero algorithm, $f(9.039001464, 8.668243408) = 18.5547210135332$, emphasizing again the fine-tuning capabilities of the GEP-PO algorithm.

Let's now see how both algorithms perform at a much more difficult task, the optimization of the following five-parameter function:

$$f(p_1, p_2, p_3, p_4, p_5) = 0.5 + \frac{\sin\sqrt{p_1^2 + p_2^2} - 1.25 \cdot \cos(p_3 p_4 p_5)}{\sqrt{1 + 0.001 \cdot (p_3^2 + p_4^2)}} \tag{8.8a}$$

subjected to the constraints:

$$-10 \leq p_1,..., p_5 \leq 10 \qquad (8.8\text{b})$$

The global maximum of this function is not known and, consequently, we won't be able to compare the performance of the algorithms in terms of success rate; consequently, we will use the average best-of-run output and the best-of-experiment output for that purpose.

As shown in Table 8.4, at this task, the GEP-PO algorithm performs slightly better than the simpler HZero system, with average best-of-run output of

Table 8.4
Performance and settings for the five-parameter function optimization problem using the HZero algorithm and the GEP-PO algorithm.

	HZero	GEP-PO
Number of runs	100	100
Number of generations	20,000	20,000
Population size	30	30
Function set	--	+ - * /
Terminal set	?	?
Random constants array length	10	10
Random constants type	Rational	Rational
Random constants range	[-10, 10]	[-2, 2]
Head length	0	6
Gene length	2	20
Number of genes	5	5
Chromosome length	10	100
Mutation rate	--	0.025
Inversion rate	--	0.1
IS transposition rate	--	0.1
RIS transposition rate	--	0.1
One-point recombination rate	0.3	0.3
Two-point recombination rate	0.3	0.3
Gene recombination rate	0.3	0.3
Gene transposition rate	0.1	0.1
Random constants mutation rate	0.35	0.01
Dc-specific mutation rate	0.01	0.025
Dc-specific inversion rate	--	0.1
Dc-specific transposition rate	--	0.1
Average best-of-run output	2.748712187	2.748801579
Best-of-experiment output	2.749120718	2.749128143

2.748801579 against 2.748712187. And the reason for this superiority resides obviously in the fine-tuning capabilities of the GEP-PO algorithm.

Consider, for instance, the best-of-experiment solution created with the HZero algorithm in generation 10399 of run 97:

$$f(p_1, p_2, p_3, p_4, p_5) = (1.21496582, 7.75982666, 0.558441162, 0.572570801, 9.826141358)$$

for which the function value at this point is:

$$f(p_1, p_2, p_3, p_4, p_5) = 2.74912071753426$$

Consider now the parameter values for the best-of-experiment solution designed by the GEP-PO algorithm in generation 14834 of run 10:

$$f(p_1, p_2, p_3, p_4, p_5) = (0, 1.571259, -0.548713482, -0.5760533165, -9.9444807)$$

And the function value at this point is:

$$f(p_1, p_2, p_3, p_4, p_5) = 2.7491281426491$$

which is slightly higher than the best solution discovered with the HZero algorithm.

As we did for the two-parameter function (8.7) above, let's see what kind of heights can be reached by letting these systems evolve for a considerable amount of time. For instance, running the HZero algorithm for 2,000,000 generations (two orders of magnitude longer than in the experiment summarized in the first column of Table 8.4), the following solutions were discovered (the number in square brackets indicates the generation in which they were discovered):

[0] = f(-1.037597656, -1.38204956, -2.631988526, -9.757141114, -7.959350586) = 2.58443976787767
[12] = f(7.128662109, -3.721313477, -7.474395752, -3.179992675, 8.333435058) = 2.63238917644785
[17] = f(1.474212646, 7.459075928, -2.71975708, -0.615753174, 5.488342285) = 2.67447123876539
[24] = f(1.474212646, 7.459075928, -2.71975708, -0.615753174, 9.441009521) = 2.70166476215306
[29] = f(7.459075928, 2.636352539, -2.71975708, -0.615753174, 9.441009521) = 2.73115091447637
[51] = f(-6.700866699, 4.013244628, -2.71975708, -0.615753174, 9.441009521) = 2.73185409502974
[111] = f(4.85131836, -6.269958496, 1.468261719, -2.069793701, -3.105651855) = 2.73837389558143
[247] = f(-6.269958496, 4.85131836, 1.248565674, -2.069793701, 3.654327393) = 2.73907137151942
[398] = f(1.624511719, -7.610443115, 1.248565674, -2.069793701, 3.654327393) = 2.73918633590436
[431] = f(1.902191163, -7.610443115, 1.248565674, -2.069793701, 3.654327393) = 2.74173170877181
[1004] = f(1.902191163, -7.610443115, 1.248565674, 0.693542481, 3.654327393) = 2.74683021639669
[1966] = f(1.902191163, -7.610443115, 0.593597412, -0.666992188, 8.001739501) = 2.74842158975813
[2200] = f(-7.610443115, 1.902191163, 0.593597412, -0.666992188, -7.920471192) = 2.74883995593118
[2811] = f(0.782867432, -7.819946289, 0.593597412, -0.666992188, -7.920471192) = 2.74887152265203
[2846] = f(1.358184814, 0.782867432, 0.593597412, -0.666992188, -7.920471192) = 2.74887936230107

[4571] = f(0.782867432, 1.358184814, 0.593597412, -0.666992188, 7.934143067) = 2.74889948309538
[6383] = f(0.782867432, 7.8152771, 0.593597412, -0.666992188, 7.934143067) = 2.74890432665425
[16617] = f(7.8152771, -0.778625489, 0.593597412, -0.666992188, 7.934143067) = 2.74890440978767
[23059] = f(-0.618377686, 7.8152771, 0.593597412, 0.54586792, 9.699584961) = 2.74900323784641
[23366] = f(7.8152771, -0.625030518, 0.593597412, 0.54586792, 9.699584961) = 2.74901062881376
[23583] = f(-0.868469239, 7.8152771, 0.593597412, 0.54586792, 9.699584961) = 2.74906095924613
[23729] = f(0.868041993, 7.8152771, 0.593597412, 0.54586792, 9.699584961) = 2.74906140150651
[24211] = f(0.74005127, 7.8152771, 0.593597412, 0.54586792, 9.699584961) = 2.7490981328368
[25711] = f(-0.795654297, 7.8152771, 0.593597412, 0.54586792, 9.699584961) = 2.7491037066131
[26735] = f(-0.795654297, 7.8152771, 0.593597412, 0.54586792, -9.696472168) = 2.74910473843031
[26806] = f(0.766784668, 7.8152771, 0.593597412, 0.54586792, -9.696472168) = 2.74910547649924
[30478] = f(7.8152771, -0.770874024, 0.593597412, 0.54586792, -9.696472168) = 2.7491058681066
[32709] = f(-0.77331543, 7.8152771, 0.593597412, 0.54586792, -9.696472168) = 2.74910602605621
[33647] = f(7.8152771, -0.779815674, 0.593597412, 0.54586792, -9.696472168) = 2.74910616541251
[57829] = f(-0.779815674, 7.8152771, -0.593475342, 0.54586792, -9.696472168) = 2.74910635743232
[58711] = f(-3.471252441, 7.045257568, -0.593475342, 0.54586792, -9.696472168) = 2.74910636279002
[103093] = f(-3.471252441, 7.045257568, -0.593475342, 0.54586792, 9.69732666) = 2.74910642967471
[165968] = f(-5.648040772, -5.457550049, -0.593475342, 0.54586792, 9.69732666) = 2.74910642970012
[238801] = f(-5.457550049, -5.648040772, 0.579986572, -0.558349609, 9.69732666) = 2.74910825622295
[240343] = f(-5.648040772, -5.457550049, 0.579986572, -0.558349609, 9.702819824) = 2.74910907148842
[306636] = f(-5.457550049, -5.648040772, 0.579986572, -0.558349609, 9.700164795) = 2.74910917117391
[321486] = f(-5.457550049, -5.648040772, 0.579986572, -0.558349609, 9.701019287) = 2.74910923989985
[485692] = f(-5.648040772, -5.457550049, 0.579986572, -0.558349609, 9.701293946) = 2.74910924166994
[725906] = f(-5.457550049, -5.648040772, 0.579986572, -0.558349609, -9.701263428) = 2.74910924196145
[1076601] = f(-5.648040772, -5.457550049, 0.579986572, 0.549194336, -9.868469239) = 2.74912122495307
[1079682] = f(-5.457550049, -5.648040772, 0.579986572, 0.549194336, 9.864898682) = 2.74912292499527
[1089992] = f(-5.648040772, -5.457550049, 0.579986572, 0.549194336, 9.863464356) = 2.74912315285548
[1189610] = f(-5.457550049, -5.648040772, 0.579986572, 0.549194336, 9.862487793) = 2.74912315874698
[1272149] = f(-5.648040772, -5.457550049, 0.579986572, 0.549194336, 9.863342286) = 2.74912316020413
[1364503] = f(-5.648040772, -5.457550049, 0.579986572, 0.549194336, 9.862701416) = 2.74912316778983
[1912653] = f(-5.457550049, -5.648040772, 0.579986572, 0.549194336, 9.863006592) = 2.74912317067176

As you can see, there are several peaks around 2.749 in the solution landscape and the HZero algorithm is able to jump from one to another with ease.

And running the GEP-PO algorithm for 2,000,000 generations (also two orders of magnitude longer than in the experiment summarized in the second column of Table 8.4), the following solutions were designed:

[0] = f(-2.437001231, 6.804853195, -3.920903837, -2.550900767, -1.5506605) = 2.50858529291059
[5] = f(0.759144582, -1.101565785, 1.910883268, 1.549652, -5.254417068) = 2.70105120412015
[7] = f(-0.4012764856, 1.377644253, 0.4893855291, 1.693482, -3.720840739) = 2.73444077987404
[13] = f(0.5232246315, -1.344425788, -2.110365863, -0.702179, -2.107452) = 2.73482739664025
[36] = f(1.528014249, -0.06285768887, -2.110365863, -0.702179, -2.107452) = 2.74214799527475
[38] = f(1.528014249, -0.07836708239, -2.110365863, -0.702179, -2.107452) = 2.74217736249844
[62] = f(1.528014249, -0.4777453751, -2.110365863, -0.702179, -2.107452) = 2.7425527268006

[65] = f(1.528014249, -0.2177268505, -2.110365863, -0.702179, -2.107452) = 2.74263343681217
[67] = f(1.528014249, -0.4481920585, -2.110365863, -0.702179, -2.107452) = 2.74277390586193
[78] = f(1.528014249, 0.3573702597, -2.114799175, -0.702179, -2.107452) = 2.74310545889365
[112] = f(-0.7304964568, -1.38086, 0.8695696511, -1.086248706, 3.362999512) = 2.74654017260428
[793] = f(0.5705395687, -1.479919, 1.158447, 0.5210973367, -5.207993049) = 2.74766398148367
[1029] = f(-0.4861430794, -1.479919, 1.158447, 0.5210973367, -5.207993049) = 2.74769539568901
[1112] = f(-0.4878039849, -1.479919, 1.158447, 0.5210973367, -5.207993049) = 2.74770204326874
[1146] = f(0.5309064376, -1.479919, 1.158447, 0.5210973367, -5.207993049) = 2.74777972524168
[1275] = f(0.52652, -1.479919, 1.158447, 0.5210973367, -5.207993049) = 2.74778080503404
[1888] = f(-1.479919, 0.52652, 0.643433, 0.9353127612, -5.207993049) = 2.74819571230492
[2491] = f(-1.479919, 0.52652, 0.99588, -0.3981750678, -7.900037418) = 2.74836958656445
[5138] = f(0.52652, -1.479919, 0.99588, 0.3997547218, -7.900037418) = 2.74841040697496
[8084] = f(-1.479919, 0.52652, 0.99588, 0.3997547218, -7.900037418) = 2.74841040697496
[10971] = f(-1.479919, 0.52652, -0.502015, 0.735291, -8.53958656) = 2.74884052359586
[11474] = f(0.52652, -1.479919, -0.502015, 0.735291, 8.52387084) = 2.74889633436337
[11849] = f(0.52652, -1.479919, 0.5013076707, 0.735291, 8.52387084) = 2.74891161461594
[12766] = f(0.52652, -1.479919, 0.5013076707, 0.7352766953, 8.52387084) = 2.74891166912289
[13077] = f(0.52652, -1.479919, 0.5013076707, -0.735138, 8.52387084) = 2.7489119555087
[25270] = f(-1.479919, 0.52652, -0.4549386616, -0.735138, 9.370843443) = 2.74893698445293
[25856] = f(0.52652, -1.479919, -0.4549386616, -0.735138, 9.376242465) = 2.74895205522925
[26522] = f(-1.479919, 0.52652, -0.441136548, -0.735138, 9.682325574) = 2.7489881887117
[27651] = f(0.52652, -1.479919, -0.441136548, -0.735138, 9.690788151) = 2.74898914439402
[40606] = f(-1.479919, 0.52652, 0.441136548, -0.735138, 9.68629455) = 2.74898980879368
[44810] = f(-1.479919, 0.52652, 0.441136548, -0.735138, 9.686785193) = 2.74898986529771
[49251] = f(-1.479919, 0.52652, 0.5472683993, -0.594269, 9.686785193) = 2.7490547749973
[49853] = f(0.52652, -1.479919, 0.5472683993, -0.594269, -9.662630922) = 2.74910248985673
[56095] = f(-1.479919, 0.52652, 0.5472306256, -0.594269, -9.662630922) = 2.74910277014878
[65836] = f(0.52652, -1.479919, 0.5472306256, 0.5940842608, -9.662630922) = 2.74910334969014
[761999] = f(0.52652, -1.479919, 0.5472306256, -0.5922689043, 9.686785193) = 2.74910377289617
[769174] = f(0.52652, -1.479919, 0.5472306256, 0.5922689043, 9.691515193) = 2.74910619550949
[822447] = f(-1.479919, 0.52652, 0.5472306256, 0.5922689043, 9.693133193) = 2.7491063501043
[983558] = f(-1.479919, -0.5265262738, 0.5472306256, 0.5922689043, 9.693133193) = 2.74910635011383
[1188173] = f(-0.5265262738, -1.479919, -0.5321242826, 0.5922689043, 9.954059193) = 2.74911629381814
[1195047] = f(-1.479919, -0.5265262738, -0.5321242826, 0.5922689043, 9.971637193) = 2.74912802623392
[1207956] = f(-0.5265262738, -1.479919, -0.5317327538, 0.5922689043, 9.971637193) = 2.74912836437082
[1254128] = f(-1.479919, -0.5265262738, -0.5317327538, 0.5922689043, 9.971881193) = 2.7491284794774
[1264991] = f(-1.479919, -0.5265262738, -0.5320337812, 0.5922689043, 9.971881193) = 2.74912864250163
[1271869] = f(0.5265282799, -1.479919, -0.5320337812, 0.5922689043, 9.971881193) = 2.74912864250375
[1290439] = f(-1.479919, 0.5265282799, -0.5320337812, 0.5922689043, 9.968860193) = 2.74912881093002
[1293212] = f(-1.479919, 0.5265282799, -0.5320337812, 0.5922689043, 9.970721193) = 2.7491288411013
[1305443] = f(0.5265282799, -1.479919, -0.5319474273, 0.5922689043, 9.970721193) = 2.74912896544059
[1595922] = f(0.5265282799, -1.479919, -0.5319474273, 0.5922689043, 9.971545193) = 2.74912900700103
[1613969] = f(-1.479919, 0.5265282799, -0.5318966927, -0.5922689043, 9.971545193) = 2.74912902568351
[1730118] = f(0.5265282799, -1.479919, -0.5318966927, -0.5922689043, 9.972034193) = 2.74912906820965
[1897838] = f(0.5265282799, -1.479919, -0.5318966927, -0.5922379043, 9.972034193) = 2.74912907219359
[1912565] = f(0.5265389368, -1.479919, -0.5318966927, 0.5922379043, 9.972034193) = 2.74912907219724

As you can see, the GEP-PO algorithm can also jump from peak to peak, finding along the way higher and higher positions in the solution landscape. For instance, the thirteen best solutions designed with the GEP-PO algorithm are better than the best solution discovered with the HZero algorithm, f(-5.457550049, -5.648040772, 0.579986572, 0.549194336, 9.863006592) = 2.74912317067176, which again emphasizes that whenever great precision is needed the more computationally demanding GEP-PO algorithm should be used.

In the next chapter we will study yet another interesting application of gene expression programming – decision tree induction – where random numerical constants also play an important role.

9   Decision Tree Induction

Decision tree induction is extremely popular in data mining, with most currently available techniques being refinements of Quinlan's original work (Quinlan 1986). His divide-and-conquer approach to decision tree induction involves selecting an attribute to place at the root node and then make the same decision about every other node in the tree.

Gene expression programming can also be used to design decision trees, with the advantage that all the decisions concerning the growth of the tree are made by the algorithm itself without any kind of human input, that is, the growth of the tree is totally determined and refined by evolution.

There are basically two different types of decision trees. The first one is the simplest and is used to induce decision trees with nominal attributes. But inducing decision trees both with nominal and numeric attributes (mixed attributes) is considerably more complicated and more sophisticated methods are required to grow the trees. This aspect of decision tree induction carries also to gene expression programming, and I developed two different algorithms to deal with both types of decision trees. The first one – evolvable decision trees or EDT for short – induces decision trees with nominal attributes; and the second one – evolvable decision trees with random numerical constants or EDT-RNC for short – was developed for handling numeric attributes but, in fact, can handle all kinds of attributes: from decision trees with just numeric attributes and decision trees with just nominal attributes to decision trees with both nominal and numeric attributes.

How both these algorithms are implemented and how they work will be explained in this chapter. We will also analyze their performance by solving four real-world classification problems: the already familiar breast cancer and iris problems, both of them with numeric attributes, and two new challenging problems: the lymphography problem with mixed attributes and the postoperative patient problem with nominal attributes.

Cândida Ferreira: *Gene Expression Programming*, Studies in Computational Intelligence (SCI) **21**, 337–380 (2006)
www.springerlink.com

9.1 Decision Trees with Nominal Attributes

Describing data using a decision tree is both easy on the eye and an excellent way of understanding our data. Consider, for instance, the play tennis data presented in Table 9.1, a famous toy dataset in decision tree induction (Quinlan 1986). This dataset concerns the weather conditions that are suitable for playing tennis. As you can see, there are four different attributes – OUTLOOK, TEMPERATURE, HUMIDITY, and WINDY – and the decision is whether to play or not depending on the values of the different attributes. In this particular case, all four attributes have nominal values. For instance, OUT-LOOK can be "sunny", "overcast", or "rainy"; TEMPERATURE can be "hot", "mild", or "cool"; HUMIDITY can be "high" or "normal"; and WINDY can be "true" or "false".

Table 9.1
A small training set with nominal attributes.

OUTLOOK	TEMPERATURE	HUMIDITY	WINDY	Play
sunny	hot	high	false	No
sunny	hot	high	true	No
overcast	hot	high	false	Yes
rainy	mild	high	false	Yes
rainy	cool	normal	false	Yes
rainy	cool	normal	true	No
overcast	cool	normal	true	Yes
sunny	mild	high	false	No
sunny	cool	normal	false	Yes
rainy	mild	normal	false	Yes
sunny	mild	normal	true	Yes
overcast	mild	high	true	Yes
overcast	hot	normal	false	Yes
rainy	mild	high	true	No

A decision tree learned from this dataset is presented in Figure 9.1. And the classification rules from such a tree are inferred from running all the paths from the top node to all the leaf nodes. For instance, for the decision tree of Figure 9.1, there are a total of five different paths or classification rules. And starting at the most leftward path and continuing towards the right, they are as follow:

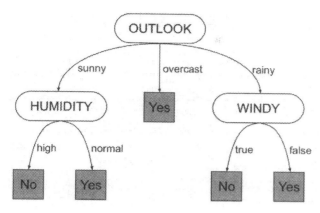

Figure 9.1. Decision tree for deciding whether to play tennis or not.

1. **IF** OUTLOOK = sunny **AND** HUMIDITY = high **THEN** PLAY = No;
2. **IF** OUTLOOK = sunny **AND** HUMIDITY = normal **THEN** PLAY = Yes;
3. **IF** OUTLOOK = overcast **THEN** PLAY = Yes;
4. **IF** OUTLOOK = rainy **AND** WINDY = true **THEN** PLAY = No;
5. **IF** OUTLOOK = rainy **AND** WINDY = false **THEN** PLAY = Yes.

As you can easily check, these rules classify correctly all the 14 instances of Table 9.1 and, therefore, are a perfect solution to the problem at hand.

Let's now see how such decision trees with nominal attributes can be induced with gene expression programming.

9.1.1 The Architecture

Gene expression programming can be used to induce decision trees by dealing with the attributes as if they were functions and the leaf nodes as if they were terminals. Thus, for the play tennis data of Table 9.1, the attribute set **A** will consist of OUTLOOK, TEMPERATURE, HUMIDITY, and WINDY, which will be respectively represented by "O", "T", "H", and "W", thus giving A = {O, T, H, W}. Furthermore, all these attribute nodes have associated with them a specific arity or number of branches n that will determine their growth and, ultimately, the growth of the tree. For instance, OUTLOOK is split into three branches (sunny, overcast, and rainy); HUMIDITY into two branches (high and normal); TEMPERATURE into three (hot, mild, and cool); and WINDY into two (true and false). The terminal set **T** will consist in this case of "Yes" and "No" (the two different outcomes of the class attribute "play"), which will be represented respectively by "a" and "b", thus giving $T = \{a, b\}$.

And now the rules for encoding a decision tree in a linear genome are very similar to the rules used to encode mathematical expressions with functions and variables in the nodes of the trees (this is explained in chapter 2, The Entities of Gene Expression Programming). So, for decision tree induction, the genes will also have a head and a tail, with the head containing attributes and terminals and the tail containing only terminals. The reason for this is again to ensure that all the decision trees created by GEP are always valid programs. Furthermore, the size of the tail t is also dictated by the head size h and the number of branches of the attribute with more branches n_{max} and is evaluated by the equation:

$$t = h\,(n_{max} - 1) + 1 \tag{9.1}$$

So, for the play tennis data of Table 9.1, both OUTLOOK and TEMPERATURE branch into three, whereas HUMIDITY and WINDY branch into two, thus giving maximum arity $n_{max} = 3$. For instance, if a head size of five is chosen, then the tail size will be $t = 5 \cdot (3 - 1) + 1 = 11$ and the gene size g will be $g = h + t = 5 + 11 = 16$. One such gene is shown below:

```
0123456789012345
HOTbWaaaaaabbbbb
```
(9.2)

The expression of this kind of gene is done in exactly the same manner as in all the GEP systems, giving the following decision tree:

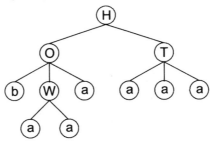

This simplified decision tree can also be represented as a conventional decision tree, that is, with all the branches clearly labeled (Figure 9.2). And as you can easily check, it solves correctly 12 of the 14 instances of Table 9.1.

The process of decision tree induction with GEP starts, as usual, with an initial population of randomly created chromosomes. Then these

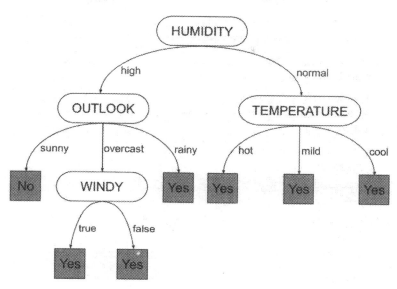

Figure 9.2. A more colorful rendition of the decision tree encoded in gene (9.2).

chromosomes are expressed and their fitnesses evaluated against a particular training set such as the data presented in Table 9.1. And according to fitness, they are selected to reproduce with modification. The modification mechanisms are exactly the same that apply in a conventional unigenic system, namely, mutation, inversion, IS and RIS transposition, and one- and two-point recombination (see section 3.3, Reproduction with Modification).

So, as you can see, nothing but the usual tricks of evolution will be used to design GEP decision trees. Let's now see how these decision trees learn by studying a design experiment in its entirety.

9.1.2 A Simple Problem with Nominal Attributes

For the simple problem of this section we are going to use the play tennis data presented in Table 9.1 and, therefore, the set of attributes will consist of A = {O, T, H, W} and the set of terminals will consist of T = {a, b}. The fitness function will consist of the number of hits, that is, the number of instances correctly classified. As usual for this kind of illustrative problem, we are going to use small populations of just 20 individuals so that we can analyze the complete evolutionary history of a successful run. The complete list of parameters used in this experiment is given in Table 9.2.

Table 9.2

Settings for the play tennis problem with nominal attributes.

Number of generations	50
Population size	20
Number of training instances	14
Number of attributes	4
Attribute set	O T H W
Terminal set / Classes	a b
Maximum arity	3
Head length	5
Gene length	16
Mutation rate	0.044
Inversion rate	0.1
IS transposition rate	0.1
RIS transposition rate	0.1
One-point recombination rate	0.3
Two-point recombination rate	0.3
Fitness function	Eq. (3.8)

The evolutionary dynamics of the successful run we are going to analyze is shown in Figure 9.3. And as you can see, in this run, a perfect solution was found in generation 5.

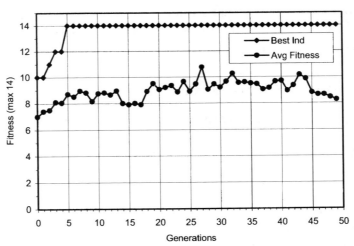

Figure 9.3. Progression of average fitness of the population and the fitness of the best individual for the illustrative run of the play tennis problem.

The initial population of this run and the fitness of each individual in the particular environment of Table 9.1 is shown below (the best of generation is shown in bold):

```
Generation N: 0
0123456789012345
WObbbbabbbabbabb-[  0] = 5
TaWbWaaaabaabbaa-[  1] = 7
ObOaabbbbbbabaab-[  2] = 6
TbHHWaababbbaaaa-[  3] = 7
HHbabbbbaaababbb-[  4] = 4
WHHWTbabbbbaaabb-[  5] = 9
HaaHabaaaaabaaba-[  6] = 9
WHaHWbbbbbaaaabb-[  7] = 9
OWWbabbabbbbbaba-[  8] = 7
WWbWabbbaabbbabb-[  9] = 5
TOWbaabbaabbbaaa-[10] = 6
OTWTOabaababbaab-[11] = 10
WTaaHbaabbabaaba-[12] = 9
HOWbWabbbabbabbaa-[13] = 6
WOaWHabababbaaaa-[14] = 8
Habaabbbbabaabbb-[15] = 4
OWTHHbbbaaababab-[16] = 8
HHHHTbbaaaabaaaa-[17] = 4
HbOHHaabaaabbabb-[18] = 8
HWaHbaaabbababaa-[19] = 9
```

The expression of the best individual of this generation (chromosome 11) is shown in Figure 9.4. As its fitness indicates, it explains 10 of the 14 instances of Table 9.1. As you can see in Figure 9.4, the decision tree it encodes is slightly redundant with a total of five neutral nodes. But we know already that these neutral blocks are essential for the design of good solutions as, without them, the system becomes less plastic and learns much more slowly (see a discussion of The Role of Neutrality in Evolution in chapter 12). Notwithstanding, we will see later in section 9.4, Pruning Trees with Parsimony Pressure, that there are ways of keeping these neutral blocks to a minimum by applying parsimony pressure.

In the next generation, despite no improvement in best fitness, there are now a total of three individuals with fitness 10:

```
Generation N: 1
0123456789012345
OTWTOabaababbaab-[  0] = 10
HWTWHbabbbbaaabb-[  1] = 7
```

a.   ```
0123456789012345
OTWTOabaababbaab
```

b.

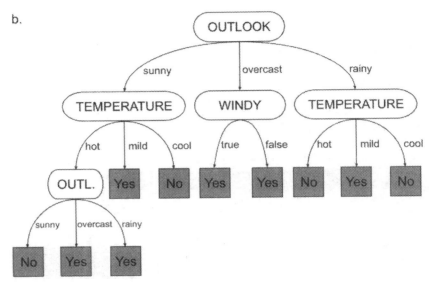

**Figure 9.4.** Best individual of the initial population (chromosome 11). It solves correctly 10 of the 14 instances presented in Table 9.1. **a)** The chromosome of the individual. **b)** The corresponding decision tree.

```
WWbWWbbbabbbbabb-[2] = 5
OTWTOabaabaabbaa-[3] = 10
TOabWababbababaa-[4] = 8
HHHHWbbbaaabaaaa-[5] = 5
OWWWabbabbbbbbab-[6] = 7
ObHHWbababbbaaaa-[7] = 6
WTaaHbaabbaaaaba-[8] = 9
TOWbaabbaabbbaba-[9] = 6
WHaHTbbabbaaaabb-[10] = 9
OaObObbbbbbabaab-[11] = 4
HWaHbaabaabaabba-[12] = 9
Wababbbbbbbbbabb-[13] = 5
HHTaaaaaaaabaaba-[14] = 9
OWWbbbbbababbbaba-[15] = 7
ObOTOabbababbaab-[16] = 10
ObOaabbbbbbababa-[17] = 6
TaWbWaaaababbbab-[18] = 7
HaaHabaaaaabaaba-[19] = 9
```
```

Note that two of the best individuals of this generation are the clone of the best individual of the initial population (chromosome 0) and a minor variant of this individual (chromosome 3) encoding a slightly different DT that solves a slightly different set of sample cases; the other one (chromosome 16) is totally unrelated to the best of the previous generation and encodes a completely different decision tree that also happens to have fitness 10.

In the next generation a new individual was created, chromosome 7, slightly better than the best individuals of the previous generations. Its expression is shown in Figure 9.5. The complete population is shown below (the best individual is shown in bold):

```
Generation N: 2
0123456789012345
ObOTOabbababbaab-[ 0] = 10
ObOTOabbababbaab-[ 1] = 10
HTWbaaababbbbabb-[ 2] = 6
ObHHWbbaabaabaab-[ 3] = 7
OHbWTabaababbaab-[ 4] = 8
OTWTOabaababaaaa-[ 5] = 8
OTOWbabbbbbbaabb-[ 6] = 5
HOWbaabababbbaba-[ 7] = 11
OTWTOabaabaabbaa-[ 8] = 10
OTWTOabbaabbbbba-[ 9] = 7
OTHTWabbbbabbaab-[10] = 5
Wabbabbaababababab-[11] = 5
WWbWWbbbabbbbabb-[12] = 5
HWaHbaabaabaabba-[13] = 9
WWbWWbbbabbbbabb-[14] = 5
WTaaHbabbbaaaaba-[15] = 8
OTWTOabaabaabbaa-[16] = 10
TOabHbaabbababaa-[17] = 5
TOObWababbbababaa-[18] = 6
OTWTOabaabaabbaa-[19] = 10
```

It is worth pointing out that the structure of the best DT of this generation is totally unrelated to the best-of-generation of the previous populations (compare Figures 9.4 and 9.5). As you can see, it encodes a very compact decision tree with just eight nodes that solves a total of 11 instances.

In the next generation a new individual was created, chromosome 11, that slightly outperforms the best decision tree of the previous generation. Its expression is shown in Figure 9.6. The complete population is shown below (the best of generation is shown in bold):

a. 0123456789012345
    HOWbaababababbbaba

b.

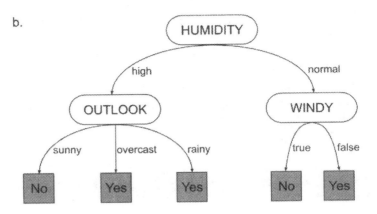

Figure 9.5. Best individual of generation 2 (chromosome 7). It solves correctly 11 of the 14 instances presented in Table 9.1. **a)** The chromosome of the individual. **b)** The corresponding decision tree.

```
Generation N: 3
0123456789012345
HOWbaababababbbaba-[ 0] = 11
OaaaTababbaaabaa-[ 1] = 9
OHbWTabaababbaab-[ 2] = 8
OTWTOabaababaaaa-[ 3] = 8
ObOTOabbababbaab-[ 4] = 10
OTWTOabaabaabbaa-[ 5] = 10
OOTObabbabaabaab-[ 6] = 6
HWWbaaaaabbbbbba-[ 7] = 10
ObOaOaaabbabbbab-[ 8] = 10
OTOWOaaaabaabbaa-[ 9] = 11
HTWaTabbbabbbaba-[10] = 4
HOWbaaaababbbaba-[11] = 12
OWTTOabbbabbbbba-[12] = 4
OTHTWabbbbabbaab-[13] = 5
TOaObabbababbaab-[14] = 10
OTWTOabbaabbbabb-[15] = 7
TOWbWababbababaa-[16] = 7
HOabWabbaaabbaab-[17] = 10
OTWTOabbbbbbaabb-[18] = 3
ObHHWbbaabaabaab-[19] = 7
```

a.   0123456789012345
      HOWbaaaababbbaba

b.

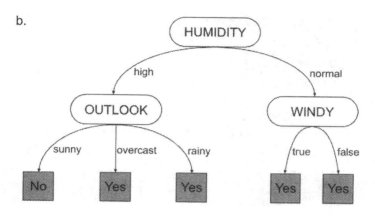

Figure 9.6. Best individual of generation 3 (chromosome 11). It solves correctly 12 of the 14 instances presented in Table 9.1. **a)** The chromosome of the individual. **b)** The corresponding decision tree.

Note that the best of this generation is a direct descendant of the best of the previous generation, as they differ only at position 6 (compare Figures 9.5 and 9.6). Note also that, thanks to this modification, a small neutral cluster was created in the tree.

In the next generation there was no improvement in best fitness, and the best individual of this generation is the clone of the previous generation created by elitism:

```
Generation N: 4
0123456789012345
HOWbaaaababbbaba-[ 0] = 12
OHHHWbbaababbaab-[ 1] = 6
OTWTObbaababbabaa-[ 2] = 8
OOWbaaaabbabbaba-[ 3] = 6
TOabWabbaaabbaab-[ 4] = 8
TOabbabbbbabbaaa-[ 5] = 9
TOaOTabaabaabbaa-[ 6] = 9
HWWbaaaababbbbba-[ 7] = 10
OTHTWabbbbabbaab-[ 8] = 5
HOaTbabbababbaab-[ 9] = 10
OaObTabbababbaab-[10] = 8
```

```
OHOWbabababbaaba-[11]  = 9
OOOTOabbabababbaab-[12]  = 8
OTHTWabbbabbbabb-[13]  = 4
OHbWTabaababbaab-[14]  = 8
TOWbWababbaabbaa-[15]  = 7
ObOTTabbababbaab-[16]  = 10
OTWOOabbaabbbabb-[17]  = 7
OTHbTababbaabaab-[18]  = 7
HOWbbaaaabbbbaba-[19]  = 10
```

Finally, in the next generation an individual with maximum fitness (chromosome 2), solving all the 14 instances of Table 9.1, was created. Its expression is shown in Figure 9.7. The complete population is shown below:

```
Generation N: 5
0123456789012345
HOWbaaaababbbaba-[ 0]  = 12
HOabaaaabbbababa-[ 1]  = 12
OHOWbababbabbaba-[ 2]  = 14
OaWOTababbaabaaa-[ 3]  = 8
HOWbbaaaaabbaaba-[ 4]  = 10
TOabbabbbbabbbaa-[ 5]  = 9
TOabbabbbbabbaaa-[ 6]  = 9
OOOTTabbababbaab-[ 7]  = 7
OOOTOabbababaaaa-[ 8]  = 8
HOaTbabbababbaab-[ 9]  = 10
OOWbaaaabaabbaba-[10]  = 6
OTOTOabbababbaab-[11]  = 7
TOabWabbaaabbaab-[12]  = 8
TbaaOabaabbbbabb-[13]  = 9
TabbOabbbaaabbaa-[14]  = 5
OaOOTabbababbaab-[15]  = 8
OOWbaaaabbabbaba-[16]  = 6
TOaOaabaabaabbaa-[17]  = 9
TaOOTabaabaabbaa-[18]  = 8
WbWaWaaabbabbaba-[19]  = 9
```

Note that the perfect solution created in this generation is not a direct descendant of the best of the previous generation; indeed, it was most probably created by an event of two-point recombination between chromosomes 3 and 11, both of which low profile individuals (chromosome 3 has fitness 6 and chromosome 11 is slightly better with fitness 9).

You probably noticed that all the chromosomes presented here encode decision trees with an attribute node in the root. Indeed, creating decision trees with a terminal at the root would be a total waste of resources and,

a.   0123456789012345
     OHOWbababbabbaba

b.

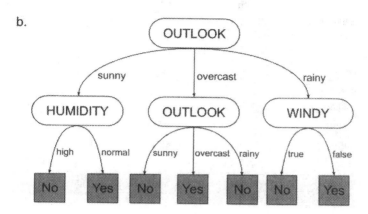

Figure 9.7. Perfect solution to the play tennis problem created in generation 5 (chromosome 2). It solves correctly all the 14 instances of Table 9.1. **a)** The chromosome of the individual. **b)** The corresponding decision tree.

therefore, both mutation and inversion (the only operators that could introduce a terminal at the root) were modified so that no terminals end up at the root. Thus, if the target point is the first position of a gene, the mutation operator can only replace an attribute by another (the mutation at the remaining positions in the gene obeys obviously the usual rules). As for the inversion operator, it was modified so as not to touch the first position of the head.

9.2 Decision Trees with Numeric/Mixed Attributes

Inducing conventional decision trees with numeric attributes is considerably more complex than inducing DTs with nominal attributes because there are many more ways of splitting the data and, no wonder, these trees can get messy very quickly. We will see that gene expression programming handles numeric attributes with aplomb and is not affected by the messiness that plagues the induction of conventional decision trees with numeric attributes.

Consider, for instance, a different version of the play tennis dataset, in which two of the attributes (TEMPERATURE and HUMIDITY) are numeric (Table 9.3). A decision tree that describes this dataset is presented in Figure 9.8. And the rules in this case are:

Table 9.3
A small training set with both nominal and numeric attributes.

OUTLOOK	TEMPERATURE	HUMIDITY	WINDY	**Play**
sunny	85	85	false	No
sunny	80	90	true	No
overcast	83	86	false	Yes
rainy	70	96	false	Yes
rainy	68	80	false	Yes
rainy	65	70	true	No
overcast	64	65	true	Yes
sunny	72	95	false	No
sunny	69	70	false	Yes
rainy	75	80	false	Yes
sunny	75	70	true	Yes
overcast	72	90	true	Yes
overcast	81	75	false	Yes
rainy	71	91	true	No

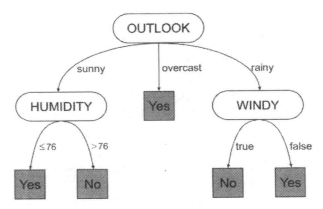

Figure 9.8. Decision tree with mixed attributes learned from the play tennis data.

1. **IF** OUTLOOK = sunny **AND** HUMIDITY ≤ 76 **THEN** PLAY = Yes;
2. **IF** OUTLOOK = sunny **AND** HUMIDITY > 76 **THEN** PLAY = No;
3. **IF** OUTLOOK = overcast **THEN** PLAY = Yes;
4. **IF** OUTLOOK = rainy **AND** WINDY = true **THEN** PLAY = No;
5. **IF** OUTLOOK = rainy **AND** WINDY = false **THEN** PLAY = Yes.

As you can easily check, these classification rules classify correctly all the 14 instances of Table 9.3 and, therefore, they are a perfect solution to the play tennis problem.

So, in order to implement such decision trees in gene expression programming one needs to find a way of encoding and fine-tuning the constants (such as the constant 76 in the decision tree presented in Figure 9.8) that are required by the numeric attributes to split the data. How this is accomplished is explained below.

9.2.1 The Architecture

The decision trees with numeric attributes of gene expression programming explore an architecture very similar to the one used for polynomial induction (see chapter 7 Polynomial Induction and Time Series Prediction). In this case, though, a Dc domain with the same size as the head is used to encode the constants used for splitting the numeric attributes.

Consider, for instance, the gene below with a head size of five (the Dc is shown in bold):

```
012345678901234567890
WOTHabababbbabba46336
```
(9.3)

Its expression results in the following decision tree:

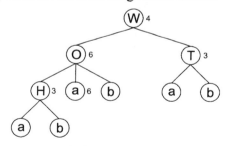

As you can see, every node in the head, irrespective of its type (numeric attribute, nominal attribute, or terminal), has associated with it a random numerical constant that, for simplicity, is represented by a numeral 0-9. These RNCs are encoded in the Dc domain and, as shown above, their expression follows a very simple path: from top to bottom and from left to right, the elements in Dc are assigned one-by-one to the elements in the tree. So, for the following array of random numerical constants:

$C = \{62, 51, 68, 83, 86, 41, 43, 44, 9, 67\}$

the chromosome (9.3) above results in the following decision tree:

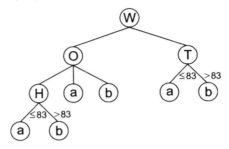

which can also be represented more colorfully as a conventional decision tree (Figure 9.9). As you can easily check, it classifies correctly 13 of the 14 instances of Table 9.3.

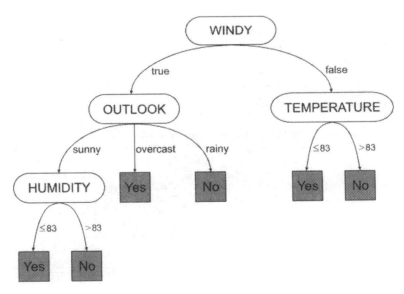

Figure 9.9. A more colorful rendition of the decision tree encoded in gene (9.3).

So, as you can see, every time a numeric attribute is expressed, it divides into two different branches and, for the scheme used above, just one numerical constant is needed to grow each branch. Let's now see how these complex structures grow and learn by solving the play tennis problem with both nominal and numeric attributes.

9.2.2 A Simple Problem with Mixed Attributes

The problem that we are going to solve here consists again in deciding whether or not to play tennis, with the difference that now we are going to use the dataset of Table 9.3 to train our decision trees. So, the attribute set will consist again of A = {O, T, H, W} and the set of terminals of T = {a, b}, with the difference that now the attributes TEMPERATURE and HUMIDITY will both branch off into two ("less than or equal to" and "greater than"). Furthermore, a set of 10 integer random constants ranging over the interval [60, 99] and represented by the numerals 0-9 will be used, giving R = {0, ..., 9}. As usual for this kind of illustrative problem, a small population size of 20 individuals evolving for just 50 generations will be used so that we can scrutinize a successful run in its entirety. The complete list of parameters used in this experiment is shown in Table 9.4.

The evolutionary dynamics of the successful run we are going to analyze here is shown in Figure 9.10. And as you can see, in this run, a perfect solution was found in generation 6.

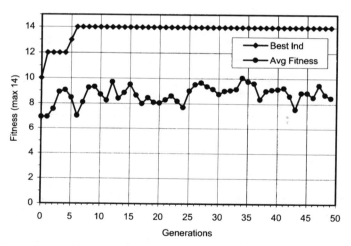

Figure 9.10. Progression of average and best fitnesses obtained in the illustrative run of the play tennis problem with mixed attributes.

The initial population of this run, together with the fitness of each individual evaluated against the training set of Table 9.3, is shown below. As you can see, we were lucky this time, and the best two individuals of this generation have both fitness 10 (these individuals are shown in bold):

Table 9.4
Settings for the play tennis problem with mixed attributes.

Number of generations	50
Population size	20
Number of training instances	14
Number of attributes	4
Attribute set	O T H W
Terminal set / Classes	a b
Random constants array length	10
Random constants type	Integer
Random constants range	[60, 99]
Maximum arity	3
Head length	5
Gene length	21
Mutation rate	0.044
Inversion rate	0.1
IS transposition rate	0.1
RIS transposition rate	0.1
One-point recombination rate	0.3
Two-point recombination rate	0.3
Dc-specific mutation rate	0.044
Dc-specific inversion rate	0.1
Dc-specific transposition rate	0.1
Random constants mutation rate	0.1
Fitness function	Eq. (3.8)

```
Generation N: 0
01234567890123456 7890
THWbaabaabbaabba28160-[  0]-{90, 82, 96, 78, 89, 82, 93, 61, 95, 88} = 6
WOWHHbbbaaabbbab15096-[  1]-{67, 82, 91, 72, 89, 62, 81, 79, 66, 68} = 6
OaHOHaabbabbbbbb15936-[  2]-{64, 99, 85, 85, 74, 97, 85, 81, 62, 92} = 6
WHHbObbababaaaaa14898-[  3]-{77, 96, 92, 93, 89, 93, 95, 63, 84, 85} = 3
TTWOWbababaabaaa06797-[  4]-{68, 87, 99, 63, 82, 73, 75, 79, 88, 71} = 9
TbTObaabbbbbaabb95446-[  5]-{91, 61, 72, 73, 61, 97, 90, 96, 86, 97} = 5
HaHWHababbbaaaba27509-[  6]-{82, 80, 80, 90, 77, 62, 60, 95, 91, 62} = 10
HOWWabbabbaabaaa07236-[  7]-{95, 71, 94, 62, 81, 90, 88, 90, 69, 62} = 10
WTaWabbabbaaaaab68604-[  8]-{60, 97, 72, 94, 76, 74, 77, 75, 75, 71} = 8
TWOWbbabbbaabbab19557-[  9]-{78, 64, 87, 80, 67, 69, 96, 78, 75, 99} = 8
TbaWbbbbbaaabbaba07452-[10]-{82, 77, 61, 61, 60, 81, 94, 92, 92, 67} = 5
WbabHaaaaaabbbab78976-[11]-{88, 92, 86, 98, 74, 99, 85, 82, 80, 69} = 9
OHbHbababaabbbba72885-[12]-{89, 86, 79, 90, 85, 80, 85, 61, 76, 94} = 4
WWTWbbbbbbbbaaaa30266-[13]-{79, 76, 82, 72, 99, 72, 84, 81, 98, 75} = 5
```

```
TaHHWbabaaaababa92576-[14]-{85, 76, 91, 65, 87, 82, 85, 88, 81, 85} = 9
WWWaabbbaaaaabab95018-[15]-{77, 63, 68, 70, 74, 78, 93, 73, 84, 71} = 5
WTWaObaabbbbbaab12420-[16]-{94, 91, 61, 76, 79, 98, 69, 98, 75, 65} = 9
TWWTHbbbbaababab68989-[17]-{96, 80, 73, 72, 95, 93, 95, 83, 93, 66} = 9
HOabbbabbabaabab06348-[18]-{95, 76, 98, 61, 95, 98, 85, 89, 62, 60} = 6
HTWHTbabaabbbbbb12933-[19]-{71, 78, 95, 78, 89, 60, 69, 91, 91, 69} = 7
```

Since it is the last of the two best individuals that is cloned unchanged into the next generation, we will focus our attention on the expression of chromosome 7 (Figure 9.11). As you can see, it classifies correctly 10 of the 14 samples presented in Table 9.3 and, therefore, has fitness 10.

In the next generation a new individual was created, chromosome 14, slightly better than the best individuals of the previous generation. Its expression is shown in Figure 9.12. The complete population is shown below (the best individual is shown in bold):

a.   012345678901234567890
     HOWWabbabbaabaaa07236

   C = {95, 71, 94, 62, 81, 90, 88, 90, 69, 62}

b.

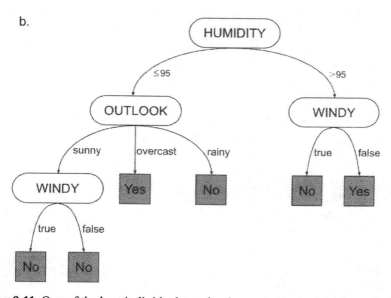

Figure 9.11. One of the best individuals randomly created in the initial population (chromosome 7). **a)** The chromosome of the individual with its random numerical constants. **b)** The corresponding decision tree. It classifies correctly 10 of the 14 sample cases presented in Table 9.3.

```
Generation N: 1
01234567890123456789O
HOWWabbabbaabaaa07236-[  0]-{95, 71, 94, 62, 81, 90, 88, 90, 69, 62} = 10
WWWTHbbbbaababab06348-[  1]-{67, 82, 91, 72, 89, 62, 81, 79, 67, 68} = 5
WTbaObaabbbbbaab12420-[  2]-{94, 91, 61, 76, 79, 98, 69, 98, 75, 65} = 5
TTWOWbababaabaaa06797-[  3]-{69, 87, 99, 63, 82, 66, 75, 79, 88, 71} = 8
HHbWHbbabbbaaaba27569-[  4]-{82, 80, 80, 90, 77, 62, 60, 95, 69, 62} = 5
HTWHTbabaabbabbb12933-[  5]-{71, 78, 95, 78, 89, 69, 69, 91, 86, 69} = 7
OWWOWabbaaabbbab15036-[  6]-{86, 82, 91, 72, 89, 62, 81, 79, 66, 68} = 5
TTWWObbaabbbaaa06797-[  7]-{68, 87, 99, 63, 82, 73, 75, 79, 88, 71} = 6
WTWaObabbbbbbaab82320-[  8]-{94, 91, 61, 76, 79, 98, 69, 98, 75, 65} = 9
TbaWbbbbaaabbaba07452-[  9]-{82, 77, 61, 61, 60, 81, 94, 92, 92, 67} = 5
WWTWbbbbbbbabbaa95957-[10]-{79, 76, 82, 72, 66, 72, 84, 82, 71, 75} = 5
HOWOWbbabbaabaaa07236-[11]-{95, 71, 94, 62, 81, 90, 88, 90, 69, 62} = 8
WTWaObaabbbbbaab12420-[12]-{94, 91, 61, 73, 79, 98, 69, 98, 75, 65} = 9
THHbObbbaaabbbab68989-[13]-{96, 80, 67, 72, 95, 93, 95, 82, 93, 66} = 6
OHOWWbbabbaabaaa07936-[14]-{95, 71, 94, 69, 81, 90, 88, 90, 69, 62} = 12
TWOWbbabbbabaaaa30266-[15]-{78, 64, 82, 80, 67, 69, 96, 78, 75, 99} = 7
WHTWbbbbbbbbaaaa26266-[16]-{79, 76, 82, 72, 99, 72, 84, 81, 98, 75} = 5
HOWWabbabbaabaaa07233-[17]-{67, 86, 94, 62, 81, 90, 88, 90, 69, 62} = 10
HOabWbabbabaabab15096-[18]-{95, 76, 98, 61, 95, 98, 73, 89, 62, 60} = 7
HTWHTbbbabaabbba12236-[19]-{71, 78, 95, 78, 89, 60, 69, 91, 91, 69} = 5
```

Note that despite having very different structures, the best of this generation (chromosome 14) is a direct descendant of the best of the previous generation, and was most probably created by a major event of root transposition (the one-element transposon "H" was inserted at the root) and some other minor modifications. Note also that the fact that they share practically the same set of random constants also supports this hypothesis.

For the next three generations there was no improvement in best fitness and all the best individuals of these generations are either the clone of the best or minor modifications of the best (all these individuals are highlighted):

```
Generation N: 2
01234567890123456789O
OHOWWbbabbaabaaa07936-[  0]-{95, 71, 94, 69, 81, 90, 88, 90, 69, 62} = 12
WObbWababbbabaaa07239-[  1]-{95, 71, 94, 62, 81, 90, 88, 86, 69, 62} = 3
WaHOWbbabbabbaaa07233-[  2]-{67, 86, 94, 72, 81, 90, 88, 90, 69, 62} = 7
WTbaObaabbbbbaab12426-[  3]-{94, 91, 61, 76, 79, 98, 69, 98, 75, 65} = 5
HTWWbbbabbaabaaa07233-[  4]-{71, 78, 95, 78, 89, 69, 69, 91, 86, 69} = 6
HTWHTbbbabbabbba12230-[  5]-{63, 78, 95, 78, 89, 60, 86, 91, 91, 69} = 5
HOWHTbabaabbabbb14933-[  6]-{67, 86, 94, 97, 81, 90, 88, 90, 72, 62} = 5
OHOOWbbabbaabaaa07936-[  7]-{95, 71, 94, 69, 81, 90, 88, 90, 69, 62} = 10
OHOWWbbabbaabaaa07936-[  8]-{95, 71, 94, 69, 81, 90, 75, 90, 69, 62} = 12
HOWWabbabbaabaaa07233-[  9]-{67, 86, 94, 97, 86, 90, 88, 90, 69, 63} = 10
```

a.
```
012345678901234567890
OHOWWbbabbaabaaa07936
```

C = {95, 71, 94, 69, 81, 90, 88, 90, 69, 62}

b.

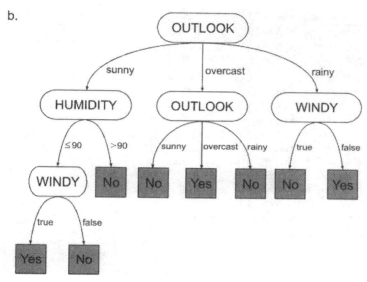

Figure 9.12. Best individual of generation 1 (chromosome 14). **a)** The chromosome of the individual with its random numerical constants. **b)** The corresponding decision tree. It classifies correctly 12 of the 14 sample cases presented in Table 9.3.

```
OOHOWbaabbaabaaa36793 - [10] - {95, 71, 94, 69, 81, 90, 88, 90, 69, 62} = 9
WbbWHbbabbbbaaba26262 - [11] - {79, 76, 82, 72, 99, 72, 84, 81, 98, 75} = 5
WOWWabbabbaabaaa07233 - [12] - {67, 86, 94, 62, 69, 90, 88, 90, 69, 62} = 11
OWOWaabbaaabbbab30363 - [13] - {86, 82, 91, 72, 89, 62, 81, 79, 66, 68} = 5
HHWTHbbbbbbaaaba27569 - [14] - {82, 80, 80, 90, 77, 62, 60, 95, 69, 86} = 5
WWWWHbbbbaaaabab06348 - [15] - {67, 82, 91, 72, 89, 62, 81, 79, 67, 68} = 5
HOWaWbbabbaabaaa07233 - [16] - {97, 86, 94, 62, 81, 90, 88, 90, 69, 62} = 4
HOWWabbabbaabaaa07236 - [17] - {95, 71, 94, 62, 66, 90, 88, 76, 69, 62} = 10
OHOWWbbabbaabaaa07936 - [18] - {95, 71, 94, 69, 69, 90, 88, 90, 69, 94} = 12
HOWWabbabbabbaaa07235 - [19] - {95, 71, 72, 62, 81, 90, 88, 90, 69, 62} = 10
```

```
Generation N: 3
012345678901234567890
OHOWWbbabbaabaaa07936 - [ 0] - {95, 71, 94, 69, 69, 90, 88, 90, 69, 94} = 12
OHOOWbbabbabbaaa07235 - [ 1] - {95, 71, 94, 69, 81, 90, 88, 90, 69, 62} = 9
OOHOWbaabbaabaaa36795 - [ 2] - {95, 71, 94, 69, 81, 90, 88, 94, 69, 62} = 9
WaHOWbbabbabbaaa07936 - [ 3] - {67, 86, 94, 81, 81, 90, 88, 90, 69, 62} = 5
```

```
OOWHTbabaabbabbb14933-[  4]-{67, 86, 94, 75, 81, 90, 88, 90, 72, 62} = 4
WWWWHbbbbaaaabab06348-[  5]-{67, 82, 91, 72, 89, 62, 81, 79, 67, 68} = 5
OHOWWbbabbaabaaa93796-[  6]-{95, 71, 94, 69, 69, 90, 88, 90, 69, 94} = 12
HOWWabbabbaabaaa07936-[  7]-{77, 61, 94, 62, 66, 90, 88, 85, 69, 62} = 9
HOWWabbaabaabaaa07936-[  8]-{95, 71, 72, 62, 81, 90, 88, 90, 69, 62} = 10
HOWWabbabbaabaaa07236-[  9]-{95, 71, 94, 62, 66, 90, 88, 76, 75, 62} = 10
WOWWObbabbaabbaa07233-[10]-{67, 86, 94, 65, 98, 90, 88, 90, 69, 62} = 11
OHOWWbbabbbabaaa07236-[11]-{81, 71, 94, 69, 81, 90, 75, 90, 69, 62} = 9
HTWWbbbabbbabaaa07273-[12]-{71, 78, 95, 78, 89, 69, 69, 91, 86, 69} = 6
HOWWabbabbaaaaaa07239-[13]-{67, 86, 94, 97, 86, 90, 88, 65, 98, 63} = 10
OHOWWbbabbaabaaa07936-[14]-{95, 91, 94, 69, 89, 90, 88, 90, 69, 94} = 12
HOWWabbabbaabaaa07233-[15]-{67, 86, 94, 97, 86, 90, 88, 90, 69, 89} = 10
OHOWWababbaabaaa95936-[16]-{95, 71, 94, 69, 81, 90, 88, 90, 69, 62} = 11
WHOWbbbabaaabaaa07233-[17]-{95, 76, 94, 69, 81, 90, 88, 90, 69, 62} = 8
HOHHHababbaaaaaa07936-[18]-{95, 71, 94, 69, 69, 90, 88, 60, 69, 94} = 10
OOHOWbaabbbbbaab12493-[19]-{95, 71, 94, 69, 81, 90, 89, 90, 69, 62} = 6

Generation N: 4
01234567890123456789 0
OHOWWbbabbaabaaa07936-[  0]-{95, 91, 94, 69, 89, 90, 88, 90, 69, 94} = 12
HOWWabbabbaabaaa07936-[  1]-{77, 61, 94, 62, 63, 90, 88, 85, 69, 62} = 9
WHOWWbbabbaabaaa07936-[  2]-{95, 91, 84, 69, 89, 90, 88, 90, 69, 84} = 7
HWWWabbaaaaabaaa23936-[  3]-{95, 71, 72, 62, 81, 90, 88, 90, 69, 62} = 7
OHOWWabbbbaabaaa95936-[  4]-{95, 71, 94, 69, 81, 90, 88, 90, 69, 62} = 7
OHOWWbbabbbabaaa36072-[  5]-{95, 71, 94, 69, 69, 90, 88, 74, 69, 97} = 10
WHOWbbbabbaaaaaa93233-[  6]-{95, 76, 94, 93, 81, 90, 88, 97, 69, 62} = 8
HOTWabbabbaabaaa07936-[  7]-{95, 71, 94, 62, 66, 90, 88, 76, 75, 62} = 10
HOWWabbabbaabaaa07286-[  8]-{93, 61, 94, 62, 66, 90, 88, 85, 69, 62} = 9
OHOWWbbabaaabaaa07796-[  9]-{95, 77, 94, 69, 69, 90, 88, 63, 69, 94} = 10
TOWWHbbaabaabaaa07936-[10]-{95, 71, 72, 62, 81, 90, 88, 90, 69, 62} = 9
HOHOWababbaabaaa04936-[11]-{95, 71, 94, 69, 81, 90, 88, 90, 69, 62} = 8
OHOWWbbabbbabaaa93796-[12]-{81, 71, 94, 69, 81, 90, 75, 90, 69, 62} = 9
HOHOTbbabbaabaaa93896-[13]-{95, 90, 94, 69, 69, 90, 88, 90, 69, 94} = 9
HOWWabbaabaabaaa07936-[14]-{95, 71, 72, 62, 81, 90, 88, 90, 69, 62} = 10
WOWWObbabbaabbab07233-[15]-{67, 86, 94, 65, 74, 90, 88, 93, 69, 62} = 11
OOHOWbaabbbbaaab12493-[16]-{95, 71, 94, 69, 81, 90, 89, 81, 69, 78} = 6
OHOWWbbabbaabaaa68796-[17]-{95, 71, 94, 69, 69, 90, 88, 90, 69, 94} = 12
OHOWWbbabbaabaaa07936-[18]-{95, 91, 94, 69, 89, 90, 88, 90, 69, 94} = 12
HWabObbabbaabaaa07936-[19]-{77, 61, 76, 62, 66, 90, 88, 85, 69, 62} = 6
```

It is worth noticing the appearance on generation 2 of another good solution with fitness 11 and that all these chromosomes with fitnesses higher than 10 are spreading rapidly into the populations.

In the next generation a new individual was created, chromosome 6, slightly better than the best individuals of the previous generations. Its expression is shown in Figure 9.13. The complete population is shown below:

```
Generation N: 5
01234567890123456789 0
```

Chromosome	C vector	Fitness
OHOWWbbabbaabaaa07936- [0]	{95, 91, 94, 69, 89, 90, 88, 90, 69, 94}	= 12
HOWWabbabbaabaaa07935- [1]	{77, 61, 94, 64, 63, 90, 88, 60, 69, 62}	= 9
HOHOWababbaabaaa04936- [2]	{95, 71, 94, 69, 81, 90, 88, 90, 69, 62}	= 8
TOWWHbbabbaabaaa99386- [3]	{95, 71, 72, 62, 81, 90, 88, 90, 69, 62}	= 9
OHWHOababbaabaaa68796- [4]	{95, 60, 94, 69, 69, 90, 88, 90, 69, 94}	= 6
OHOWWbbabbaabaaa68796- [5]	{95, 71, 86, 69, 70, 90, 88, 90, 69, 94}	= 12
OHOWWbbabbaabaaa07286- [6]	**{95, 71, 94, 69, 72, 90, 89, 81, 69, 78}**	**= 13**
OWHOWbbabbbabaaa97936- [7]	{95, 91, 94, 69, 89, 90, 88, 90, 69, 94}	= 5
THOWbbbabbaaaaaa93233- [8]	{95, 76, 94, 89, 81, 90, 88, 97, 69, 62}	= 6
OOHOWbaabbbbbaaa07936- [9]	{95, 91, 94, 69, 89, 94, 88, 90, 69, 94}	= 8
OOWWaabaabaaaaab12483- [10]	{93, 94, 94, 62, 66, 90, 88, 85, 69, 62}	= 11
OHOWWbbabaaabaaa07796- [11]	{95, 77, 94, 69, 69, 90, 88, 87, 69, 94}	= 11
OWHHWabbbbaabbaa04933- [12]	{60, 61, 94, 65, 70, 74, 88, 93, 69, 62}	= 4
HOWOWbaabbaabbab07236- [13]	{95, 71, 87, 69, 81, 90, 88, 90, 69, 62}	= 8
HOHOTbbabbabaaaa07936- [14]	{95, 90, 94, 69, 69, 90, 88, 90, 69, 60}	= 6

a. 01234567890123456789 0
OHOWWbbabbaabaaa07286

C = {95, 71, 94, 69, 72, 90, 89, 81, 69, 78}

b.

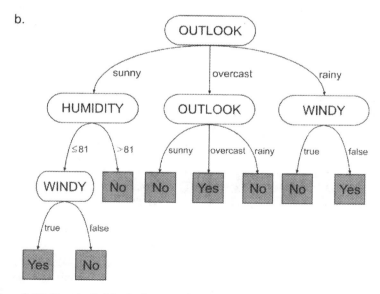

Figure 9.13. Best individual of generation 5 (chromosome 16). **a)** The chromosome of the individual with its random numerical constants. **b)** The corresponding decision tree. It classifies correctly 13 of the 14 sample cases presented in Table 9.3.

```
OHOWWbbabbabbaaa07933-[15]-{95, 91, 94, 69, 89, 90, 87, 90, 69, 94} = 12
WWWOObbabbabbbab07233-[16]-{67, 86, 94, 65, 74, 90, 88, 93, 69, 62} = 5
ObHOWbbabaaabaaa07796-[17]-{95, 77, 94, 69, 69, 90, 88, 63, 69, 94} = 5
OHOTWbbabbaabaaa07936-[18]-{95, 71, 94, 60, 66, 90, 88, 76, 75, 62} = 11
WHOWbbbaabaaaaaa93038-[19]-{95, 76, 94, 93, 81, 90, 88, 85, 69, 62} = 8
```

Note that the structure of the best of this generation (chromosome 6) matches exactly the structure of the best of the previous generations (compare Figures 9.12 and 9.13); they differ only on the numerical constant that is used to split the data under HUMIDITY (90 on the decision tree of Figure 9.12 and 81 on the one in Figure 9.13). Note also that, despite showing some variation on the neutral regions, these individuals managed to keep unchanged all the elements that really matter for their expression.

Finally, in the next generation an individual with maximum fitness (chromosome 9), classifying correctly all the 14 instances of Table 9.3, was created. Its expression is shown in Figure 9.14. The complete population is shown below (the perfect solution is shown in bold):

```
Generation N: 6
01234567890123456789O
OHOWWbbabbaabaaa07286-[ 0]-{95, 71, 94, 69, 72, 90, 89, 81, 69, 78} = 13
ObbaWbbababbbaaa07796-[ 1]-{95, 77, 94, 69, 68, 90, 67, 88, 69, 94} = 6
OWHHWabbbbaabbaa04933-[ 2]-{60, 61, 94, 65, 70, 74, 70, 93, 69, 62} = 4
OHOWWbbabbabbaaa07943-[ 3]-{95, 70, 94, 69, 89, 90, 87, 90, 69, 94} = 12
WWbOHbbabbaabaaa93936-[ 4]-{70, 91, 94, 69, 89, 90, 88, 68, 69, 94} = 3
HTHWTbbabbabaaaa07936-[ 5]-{95, 90, 94, 88, 69, 90, 88, 90, 69, 60} = 5
WHOWbbbaabbbaaaa93233-[ 6]-{95, 76, 94, 93, 81, 90, 88, 85, 69, 62} = 8
OOWaWabaabaaaaab12483-[ 7]-{93, 94, 94, 62, 66, 90, 88, 85, 69, 62} = 9
HOOWWbbabbaabaaa03936-[ 8]-{95, 71, 94, 69, 81, 90, 88, 90, 69, 62} = 6
```
OHOWHbbabbaabaaa07936-[9]-{95, 91, 94, 69, 89, 90, 78, 90, 69, 94} = 14
```
OHHOWbbabbbabaaa97936-[10]-{95, 91, 94, 71, 89, 68, 88, 90, 91, 94} = 5
OaOWWbbabbaabbaa93038-[11]-{95, 71, 94, 69, 72, 90, 89, 81, 69, 78} = 2
WHOWbbbaabaaaaaa07286-[12]-{95, 76, 94, 93, 81, 90, 88, 85, 85, 62} = 9
OHOTWbaabbaabaaa07286-[13]-{95, 71, 68, 69, 72, 90, 89, 81, 84, 78} = 10
OHbbOabbbbaabaaa07936-[14]-{95, 91, 94, 69, 89, 90, 88, 90, 69, 94} = 4
THOWbbbabbaaaaaa93038-[15]-{95, 76, 94, 89, 81, 90, 88, 97, 69, 62} = 6
OHHOWababbaabaaa70996-[16]-{95, 71, 86, 69, 70, 90, 88, 82, 69, 94} = 6
ObHOWbbabaaabaaa07796-[17]-{95, 67, 94, 69, 69, 90, 88, 63, 69, 94} = 5
TOWWWbbabbaabaaa99386-[18]-{95, 71, 72, 62, 81, 90, 88, 78, 69, 62} = 9
THOWbbbaabaaaaaa63032-[19]-{95, 76, 94, 93, 81, 90, 88, 85, 69, 62} = 5
```

Note that the perfect solution created in this generation is a direct descendant of the best of the previous generation. In this case, a point mutation at position 4, replacing WINDY by HUMIDITY, allowed a more precise

a.   012345678901234567890
      OHOWHbbabbaabaaa07936

C = {95, 91, 94, 69, 89, 90, 78, 90, 69, 94}

b.

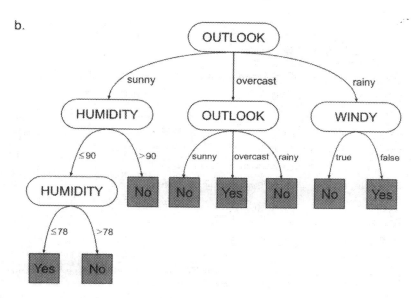

Figure 9.14. Perfect solution to the play tennis problem created in generation 6 (chromosome 9). **a)** The chromosome of the individual with its random numerical constants. **b)** The corresponding decision tree. It classifies correctly all the 14 sample cases presented in Table 9.3.

splitting of the data under HUMIDITY, resulting in a perfect solution to the problem at hand.

Let's now see how the decision trees of gene expression programming perform on more difficult real-world applications.

9.3 Solving Problems with GEP Decision Trees

In this section we are going to put the decision trees of gene expression programming to the test by solving four complex real-world problems. The first is the already familiar breast cancer problem with nine numeric attributes and two different outcomes: benign or malignant. The second is again the iris dataset, also with numeric attributes (four, as you may recall) and three different classes (*Iris setosa*, *Iris versicolor*, and *Iris virginica*). The third is the lymphography problem with 18 attributes (three numeric and 15 nominal) and four different classes (normal, metastases, malign lymph, and fibrosis). And the last one is the postoperative patient problem with eight nominal attributes and three different classes.

9.3.1 Diagnosis of Breast Cancer

The breast cancer problem is not a typical DT problem in the sense that all its attributes are numeric and has only two classes. But given the penchant for conventional decision trees with numeric attributes to overfit the data, it becomes mandatory that we test the performance of the EDT-RNC algorithm on a well-studied real-world problem with just numeric attributes in order to know not only how this algorithm compares with others but also to know how well these decision trees generalize.

As you would recall, we solved the breast cancer problem with five different algorithms (the basic GEA, the GEP-NC and the GEP-RNC algorithms, and the cellular systems with and without RNCs; see Tables 5.9 and 6.7) and concluded that all of them perform very efficiently, creating very good models that generalize extremely well. In this section, we will see that the EDT-RNC algorithm can also be used to create good DT models for the breast cancer problem and that these models also generalize extremely well.

So, preparing the data for decision tree induction means that the nine numeric attributes of the breast cancer problem are now the branching nodes of the decision trees. As usual, they are represented by capital letters, thus giving: CLUMP THICKNESS, UNIFORMITY OF CELL SIZE, UNIFORMITY OF CELL SHAPE, MARGINAL ADHESION, SINGLE EPITHELIAL CELL SIZE, BARE NUCLEI, BLAND CHROMATIN, NORMAL NUCLEOLI, and MITOSES, all branching off obviously into two. For simplicity and compactness, they will be respectively represented by A = {A, B, C, D, E, F, G, H, I}, and the two outcomes, benign or malignant, will be respectively represented by T = {a, b}. In this section, we will be using exactly the same 350 instances that were used for training and the 174 instances that were used for testing in the previous studies. And given that all the attributes are normalized between 0 and 1, the random numerical constants will be drawn from the real interval [0, 1]. The fitness function will consist of the number of hits and will be evaluated by equation (3.8), giving f_{max} = 350 on the training set and f_{max} = 174 on the testing set.

As you can see in Table 9.5, for this problem we are going to use small population sizes of just 30 individuals evolving for 1000 generations (exactly the same values used in the experiments summarized in Tables 5.9 and 6.7). However, as you can see by their performance, decision trees need more time than all the other GEP systems to design good solutions to this particular problem (average best-of-run number of hits of 337.40 versus 339.34 obtained with GEA-B; 339.05 with GEP-NC; 339.13 with GEP-RNC; 339.19 with MCS; and 339.06 with MCS-RNC), but they get there nonetheless. In fact, the best-of-experiment solution is an exceptional DT model with a fitness of 342 in the training set and 170 in the testing set, which corresponds to a training set accuracy of 97.714% and a testing set accuracy of 97.701% and, therefore, is as good a model as the best models evolved with the other

Table 9.5

Performance and settings used in the breast cancer problem.

Number of runs	100
Number of generations	1000
Population size	30
Number of fitness cases	350
Attribute set	A-I
Terminal set / Classes	a b
Random constants array length	10
Random constants type	Rational
Random constants range	[0, 1]
Head length	25
Gene length	76
Mutation rate	0.044
Inversion rate	0.1
IS transposition rate	0.1
RIS transposition rate	0.1
One-point recombination rate	0.3
Two-point recombination rate	0.3
Dc-specific mutation rate	0.044
Dc-specific inversion rate	0.1
Dc-specific transposition rate	0.1
Random constants mutation rate	0.01
Fitness function	Eq. (3.8)
Average best-of-run fitness	337.40

GEP systems. It was created in generation 837 of run 29 and looks like this:

```
CAAabCbBAFbDDHFGbHbDGGAI...
   ...aabaabaaabbabaaaaababbba...
   ...bbb245341719576825914725730 8
```

$$C = \{0.229217, 0.48703, 0.287506, 0.091735, 0.808899,$$
$$0.652405, 0.977082, 0.931214, 0.118988, 0.145019\} \tag{9.4}$$

The decision tree of this individual involves a total of 37 nodes and, given the bulkiness of the attribute nodes, we won't be able to show it here (you should be able to draw it easily on paper though). However, and because I believe it is important to see the structures of complex real-world models to make things a little bit more palpable, a much smaller model was created in another experiment using a head size of 10:

```
0123456789012345678901234567890
ABbBbaFDCIbababaaabbb3894830502
```

$$C = \{0.297699, 0.667999, 0.392853, 0.896088, 0.149322,$$
$$0.39212, 0.183807, 0.289062, 0.461792, 0.23587\} \tag{9.5}$$

This decision tree involves a total of just 15 nodes (see Figure 9.15) and has a fitness of 341 on the training set and a fitness of 171 on the testing set, again showing that GEP decision trees with numeric attributes do indeed generalize outstandingly well.

9.3.2 Classification of Irises

The iris dataset is also an interesting one to use with the EDT-RNC algorithm as it would allow us to compare the performance of this algorithm both with the cellular system with multiple outputs and the GEP-MO algorithm.

You know already that there are four numeric attributes in the iris dataset and three different classes of irises. The set of attributes will consist in this case of A = {S, T, P, Q}, corresponding, respectively, to SEPAL_LENGTH, SEPAL_WIDTH, PETAL_LENGTH, and PETAL_WIDTH, all of which will obviously divide into two different branches. The terminal set will consist of T = {a, b, c}, corresponding, respectively, to *Iris setosa*, *Iris versicolor*, and *Iris virginica*. Furthermore, a set of 10 random numerical constants drawn from the rational interval [0, 10] will be used and, as usual, they will be

a.
```
012345678901234567890123 4567890
ABbBbaFDCIbababaaabbb3894830502
```

C = {0.297699, 0.667999, 0.392853, 0.896088, 0.149322,
0.392120, 0.183807, 0.289062, 0.461792, 0.235870}

b.

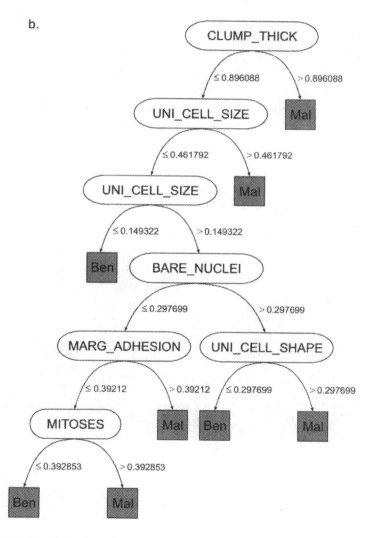

Figure 9.15. Model designed by the EDT-RNC algorithm to diagnose breast
cancer. **a)** The linear representation of the decision tree. **b)** The corresponding
decision tree. This model is extremely accurate and generalizes outstandingly well
(it has a training set accuracy of 97.43% and a testing set accuracy of 98.28%).

represented by the numerals 0-9. The fitness function will consist of the number of hits and, therefore, will be evaluated by equation (3.8). The population size and number of generations will be exactly the same used in the GEP-MO (see Table 4.7) and MCS-MO (see Table 6.8) experiments so that we can compare the performance of all these different algorithms. The complete list of parameters used in this experiment is shown in Table 9.6.

And as you can see by comparing Table 9.6 with Tables 4.7 and 6.8, the performance of the EDT-RNC algorithm is very similar to the performance obtained in the MCS-MO experiment and both of them are considerably better than the GEP-MO system (average best-of-run fitness of 147.52 for the decision trees, 147.20 for the multicellular system with multiple outputs, and 146.48 for the GEP-MO algorithm).

Table 9.6
Performance and settings used in the iris problem.

Number of runs	100
Number of generations	20,000
Population size	30
Number of fitness cases	150
Attribute set	S T P Q
Terminal set / Classes	a b c
Random constants array length	10
Random constants type	Rational
Random constants range	[0, 10]
Head length	10
Gene length	31
Mutation rate	0.044
Inversion rate	0.1
IS transposition rate	0.1
RIS transposition rate	0.1
One-point recombination rate	0.3
Two-point recombination rate	0.3
Dc-specific mutation rate	0.044
Dc-specific inversion rate	0.1
Dc-specific transposition rate	0.1
Random constants mutation rate	0.1
Fitness function	Eq. (3.8)
Average best-of-run fitness	147.52

Also important is the fact that the best-of-experiment solution created with decision trees classifies correctly 149 out of 150 instances, a feat only achieved by the three-step approach of the basic GEA (see Decomposing a Three-class Problem in section 4.2.3) and the multicellular system with multiple outputs (see section 6.6 Multiple Cells for Multiple Outputs: The Iris Problem). Its structure is shown below:

```
012345678901234567890123456789 0
PaQPPbTTQaccbbcaacaab5218839495
```

$$C = \{5.53, 1.69, 6.89, 9.67, 3.12, 2.53, 4.24, 3.96, 4.99, 1.79\} \qquad (9.6)$$

As you can see in Figure 9.16, it encodes a relatively compact decision tree with 15 nodes, with the PETAL_LENGTH ("P") at the root. It is also interesting to note that the attribute SEPAL_LENGTH ("S") is not used by this extremely accurate model, which was somewhat of a surprise since in all the previous best-of-experiment models analyzed thus far (see models (4.14) and (4.20), and the model (6.10) expressed in Figure 6.13) all the attributes were used in the trees. In other cases, though, it is the SEPAL_WIDTH ("T") that is not used, which most probably means that both these attributes are interchangeable.

It is also interesting to compare the structures of this extremely accurate decision tree with other DTs equally accurate in order to see, among other things, how these trees grow. For instance, the 30 decision trees presented below all have fitness equal to 149 and all of them are extremely parsimonious with just 13 nodes each:

```
QPSQccQabScbccaaaacbb4372201889- [1]
C = {9, 1.77, 0.68, 4.96, 1.59, 1.02, 6.99, 5.74, 1.59, 6.79}

SQQPTPcabcbbcccaccbbc5289076015- [2]
C = {3.12, 1.25, 1.62, 9.8, 1.25, 6.02, 7.88, 5.03, 1.77, 2.69}

PQQPTcQabcbbccaaaaacb6775813381- [3]
C = {3.64, 1.79, 3.64, 1.72, 0.56, 2.79, 4.9, 1.57, 3.19, 6.53}

PQPPTTcabcbcbabccbcba5740169053- [4]
C = {2.01, 3.1, 2.05, 8.27, 5, 4.97, 2.72, 1.64, 8.82, 4.89}

PQcaPTSQbcbbcbcccbcaa3520492834- [5]
C = {7.45, 4.58, 6.49, 5.04, 4.93, 0.64, 3.33, 4.44, 1.51, 3.09}

PQQQTcQabcbbcaacbbbac2008165834- [6]
C = {1.68, 3.11, 4.92, 0.67, 0.87, 1.76, 8.79, 6.8, 0.62, 8.19}
```

a. `01234567890123456789012345 67890`
`PaQPPbTTQaccbbcaacaab5218839495`

C = {5.53, 1.69, 6.89, 9.67, 3.12, 2.53, 4.24, 3.96, 4.99, 1.79}

b.

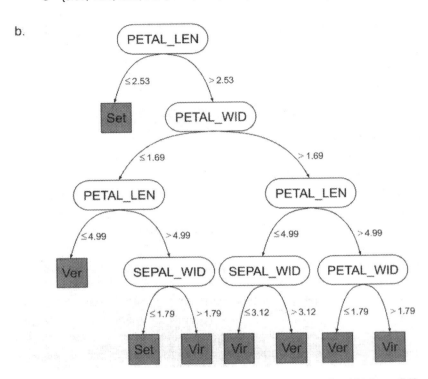

Figure 9.16. Model designed by the EDT-RNC algorithm to classify three different types of irises. **a)** The linear representation of the decision tree. **b)** The decision tree. This model classifies correctly 149 out of 150 sample cases.

```
PQcaQPSbcTbcbbaabcaab2541864899-[7]
C = {1.52, 2.46, 5.09, 5.91, 6.68, 0.98, 4.9, 4, 1.58, 3.1}

PQPPTTcabcbcbbcacaccc7819694588-[8]
C = {7.04, 5.06, 7.6, 4.06, 4.2, 5.25, 3.19, 4.92, 1.57, 2.65}

PQQQTQcabcbcbccaacbaa7134214051-[9]
C = {6.02, 1.67, 3.17, 1.79, 0.88, 0.61, 2.5, 4.96, 1.82, 8.03}

PPPaQScbTcbcbcbccaaba1605273480-[10]
C = {5.05, 4.92, 1.57, 1.47, 1.29, 7.18, 1.93, 6.37, 3.1, 8.98}

QPQQcScabSccbccabbaac3672713096-[11]
C = {2.55, 6.83, 0.88, 1.58, 4.06, 2.57, 4.93, 1.73, 0.62, 2.48}
```

PPcaPQSbTcbcbcbaaabaa1053859568-[12]
C = {2.58, 5.03, 8.72, 6.72, 7.75, 1.68, 3.16, 2.74, 4.94, 6.65}

PPQaQQcbTcbcbcccabccb8123556862-[13]
C = {5.91, 1.98, 1.73, 4.91, 8.08, 1.6, 3.14, 8.91, 4.96, 4.78}

PaPPcQTbTcbcbcbcaccbb0946017294-[14]
C = {2.95, 1.57, 0.8, 5.48, 5, 9.8, 4.95, 2.51, 8.39, 3.18}

PaPTPQbScbccbbcbabaac2170634423-[15]
C = {3.17, 6.37, 2.88, 1.63, 6.46, 3.53, 5.09, 4.91, 1.58, 3.38}

PPcaQPSbcTbcbcbbaccaa9562072924-[16]
C = {1.63, 5.09, 6.33, 6.98, 3.1, 2.63, 6.61, 4.91, 4.56, 5}

QPSaPQcbcSccbcbbcacab7901461971-[17]
C = {6.88, 5.59, 0.2, 1.35, 4.96, 8.26, 1.7, 1.56, 2.97, 1.99}

PQQaQcQbTbccbcbbabbcc5624231078-[18]
C = {6.27, 1.75, 1.6, 7.38, 2.28, 4.97, 0.63, 3.13, 0.03, 7.98}

PQTPTcPabcbbcbbcbcaca8159210916-[19]
C = {5.07, 1.67, 3.05, 5.02, 8.05, 2.79, 6.74, 9.18, 4.92, 2.14}

PQPaQScbTcbcbcbaccccb8325163196-[20]
C = {1.38, 1.55, 5, 0.64, 3.56, 8.49, 6.47, 1.98, 4.95, 3}

PQTQTcPabcbbccaaacbab1634962513-[21]
C = {5.87, 4.93, 5.02, 2.61, 0.67, 1.79, 1.55, 6.42, 5.08, 3.12}

PPPaQTcbTcbcbbcbaaaaa0834580398-[22]
C = {4.93, 4.59, 4.47, 5, 2.47, 1.62, 3.3, 3.43, 2.7, 3.1}

QPPPcTQabcbbcbbbbcccc6115824759-[23]
C = {7.84, 4.93, 3.05, 6.23, 1.73, 2.71, 1.64, 2.78, 3.96, 3.62}

PQQaQQcbTcbcbcccbbcba6987422730-[24]
C = {9.33, 2.07, 1.63, 3.18, 1.66, 1.71, 4.95, 7.98, 1.72, 0.9}

PQPPTTcabcbcbcabccbcc0134749756-[25]
C = {4.98, 1.66, 2.19, 5.06, 2.74, 2.99, 1.3, 3.13, 5.81, 3.93}

PaPTPQbScbccbbcbbbccb2501750368-[26]
C = {4.97, 3, 2.59, 6.58, 9.48, 1.69, 0.36, 5, 0.21, 6.21}

PQPPTScabcbcbacaabbbc0178359488-[27]
C = {4.98, 1.61, 3.75, 3, 6.36, 6.58, 6.56, 5, 2.46, 2.11}

PPQaQcQbTbccbcbbcbcba8064219247-[28]
C = {2.13, 6.37, 1.62, 6.11, 3.06, 7.81, 1.66, 1.82, 4.94, 1.71}

PQQPTcQabcbbcbccbcbbb4771335677-[29]
C = {8.79, 2.21, 7.27, 3.05, 4.97, 1.77, 1.64, 1.53, 1.18, 4.11}

```
PQQQTcQabcbbcabbbbbca2110399977-[30]
C = {0.73, 1.58, 4.93, 3.13, 7.95, 0.81, 5.06, 0.62, 5.84, 1.79}
```

They were obtained using exactly the same settings presented in Table 9.6, except that the number of generations was an order of magnitude higher and the trees were pruned by parsimony pressure (we'll talk more about this in section 9.4, Pruning Trees with Parsimony Pressure). And as you can see, 25 out of 30 decision trees have the attribute PETAL_LENGTH ("P") at the root; four of them have the PETAL_WIDTH ("Q"); and even the SEPAL_LENGTH ("S") can be found at the helm of one tree (DT 2). Note, however, that none of these accurate DTs starts with the SEPAL_WIDTH ("T") at the root. Thus, the decision trees of gene expression programming have no constraints whatsoever about the kind of attribute that is placed at the root (and of course at either of the other nodes): they explore all possibilities with equal freedom and, therefore, are able to search in all the crannies of the solution space, increasing the odds of finding the perfect solution.

Also interesting is that, in the same experiment, two decision trees were created that classify correctly all the 150 irises. Their structures are shown below:

```
PQQaQcQbTPccbbcaaabba5702028126-[1]
C = {1.5, 5.5, 3.03, 4.87, 4.1, 4.93, 5.63, 0.64, 1.74, 8.26}

PPQaQcQbTSccbbcabcabc8700094491-[2]
C = {1.53, 6.96, 8.5, 1.52, 1.72, 7.61, 0.03, 2.81, 4.96, 3.14}
```

As you can see, both these DTs involve a total of 15 nodes each and both of them start with the PETAL_LENGTH at the root. It is also worth knowing that the first of these perfect DTs is a descendant of DT 18 above and that the second is a descendant of DT 28. And by drawing these trees, it becomes clear that, in both cases, a perfect solution was created thanks to a more precise discrimination between *Iris versicolor* and *Iris virginica*. As an illustration, the DT 28 and its fitter descendant are shown in Figure 9.17.

Although we know already that, as a rule, GEP decision trees generalize extremely well, it is also interesting to see how GEP decision trees generalize with the iris dataset. For that purpose, the original iris dataset was randomized and then 100 instances were selected for training with the remaining 50 used for testing (these datasets are available for download at the gene expression programming website). And as expected, the decision trees trained on this dataset generalize outstandingly well. For instance, the decision tree below classifies correctly 99 instances on the training set and

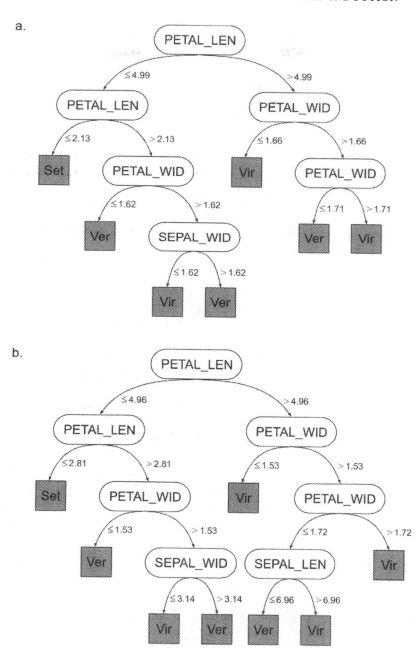

Figure 9.17. Fine-tuning solutions. The decision tree presented in **(a)** is a remote ancestor of the perfect solution presented in **(b)**. The first DT has an accuracy of 99.33% and the second one is obviously 100% accurate.

classifies correctly all the testing instances, which corresponds to a training accuracy of 99.0% and a testing set accuracy of 100%:

```
012345678901234567890123456 7890
PQQPTcQabcbbcccccbcbcc9336038121
```

$$C = \{3.16, 2.61, 1.76, 1.61, 3.11, 5.64, 2.25, 1.58, 1.74, 4.91\} \tag{9.7}$$

As you can see by its expression in Figure 9.18, it encodes a very compact decision tree with just 13 nodes with the PETAL_LENGTH ("P") at the root. Note again that this time it was the attribute SEPAL_LENGTH ("S") that was not used to distinguish between the three kinds of iris plants.

Let's now see how the EDT-RNC algorithm deals with complex problems with mixed attributes.

a.
```
012345678901234567890123456 7890
PQQPTcQabcbbcccccbcbcc9336038121
```

$$C = \{3.16, 2.61, 1.76, 1.61, 3.11, 5.64, 2.25, 1.58, 1.74, 4.91\}$$

b.

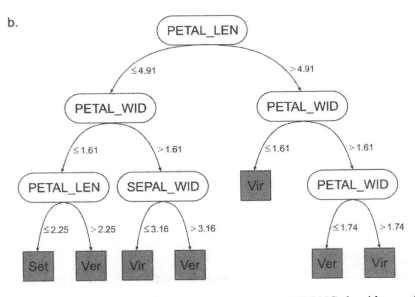

Figure 9.18. Testing the generalizing capabilities of the EDT-RNC algorithm on the iris data. **a)** The linear representation of the model. **b)** The decision tree model. This model has 99.0% accuracy on the training set and generalizes outstandingly well on the testing set, with 100% accuracy.

9.3.3 The Lymphography Problem

The lymphography problem is a complex real-world problem with four different classes (normal, metastases, malign lymph, and fibrosis) and 18 attributes, three of which are numeric and the remaining nominal (Table 9.7). The original dataset contains a total of 148 instances, 104 of which were selected at random for training with the remaining 44 used for testing. The class distribution in this dataset is pretty unbalanced with just two instances for class "normal", 81 for class "metastases", 61 for "malign lymph", and just four for "fibrosis".

As shown in Table 9.7, for this problem, the attribute set will be repre-. sented by A = {A, ..., R}. The terminal set will consist of T = {a, b, c, d}, representing, respectively, the four different classes: normal, metastases, malign lymph, and fibrosis. For the numeric attributes, a set of 10 random

Table 9.7
Organization of the lymphography dataset.

Attribute	Symbol	Branches	Arity
LYMPHATICS	A	normal, arched, deformed, displaced	4
BLOCK_OF_AFFERE	B	no, yes	2
BL_OF_LYMPH_C	C	no, yes	2
BL_OF_LYMPH_S	D	no, yes	2
BY_PASS	E	no, yes	2
EXTRAVASATES	F	no, yes	2
REGENERATION_OF	G	no, yes	2
EARLY_UPTAKE_IN	H	no, yes	2
LYM_NODES_DIMIN	I	numeric (integer)	2
LYM_NODES_ENLAR	J	numeric (integer)	2
CHANGES_IN_LYM	K	bean, oval, round	3
DEFECT_IN_NODE	L	no, lacunar, lac_margin, lac_central	4
CHANGES_IN_NODE	M	no, lacunar, lac_margin, lac_central	4
CHANGES_IN_STRU	N	no, grainy, drop_like, coarse, diluted, reticular, stripped, faint	8
SPECIAL_FORMS	O	no, chalices, vesicles	3
DISLOCATION_OF	P	no, yes	2
EXCLUSION_OF_NO	Q	no, yes	2
NO_OF_NODES_IN	R	numeric (integer)	2

constants, drawn from the integer interval [0, 10] will be used and, as usual, will be represented by the numerals 0-9, thus giving R = {0, ..., 9}. The number of hits will be again used to evaluate the fitness of the evolving decision trees. The complete list of parameters used per run in this experiment is given in Table 9.8.

And as you can see in Table 9.8, GEP decision trees are very good at learning from this dataset with an average best-of-run fitness of 90.73. Furthermore, the best-of-experiment solution is, as usual, considerably above average. It was created in generation 6861 of run 85 and encodes a decision tree with 28 nodes:

```
MBOFBPcGBIbBcREbbdcbcdcbbcadabdbcbd...
```

Table 9.8
Performance and settings used in the lymphography problem.

Number of runs	100
Number of generations	20,000
Population size	30
Number of training instances	104
Number of testing instances	44
Number of attributes	18
Attribute set	A-R
Terminal set / Classes	a b c d
Random constants array length	10
Random constants type	Integer
Random constants range	[0, 10]
Maximum arity	8
Head length	15
Gene length	136
Mutation rate	0.044
Inversion rate	0.1
IS transposition rate	0.1
RIS transposition rate	0.1
One-point recombination rate	0.3
Two-point recombination rate	0.3
Dc-specific mutation rate	0.044
Dc-specific inversion rate	0.1
Dc-specific transposition rate	0.1
Random constants mutation rate	0.01
Fitness function	Eq. (3.8)
Average best-of-run fitness	90.73

```
...acdcbddbbadbcbdaddcdcdbddcdbbaab...
...bdbacabccbaacabadbacadbaabdbcccb...
...dcbacbcdacdccbcabadabc158358772736301
```

$$C = \{1, 3, 0, 7, 2, 3, 8, 1, 2, 5\} \tag{9.8}$$

It has a fitness of 95 on the training set and 38 on the testing set, which corresponds to an accuracy of 91.35% on the training set and 86.36% on the testing set and, therefore, is better than the best results previously reported for this dataset (Cestnik et al. 1987, Clark and Niblett 1987, Michalski et al. 1986). For instance, a Bayesian classifier achieved 89% accuracy on the training set and 83% on the testing set, whereas the CN2 algorithm achieved 91% accuracy on training and 82% on testing (Clark and Niblett 1987).

But even better rules than the model (9.8) above can be learned from this dataset by letting the system evolve for a larger number of generations. For instance, the decision tree below, also with 28 nodes, was created after 146,833 generations and has a fitness of 95 on the training set (91.35% accuracy) and 39 on the testing set (88.64% accuracy):

```
MGOFRAdDBHbBBcabbcbdcbdccbcbbcadddb...
   ...dbcbddacabcbccbadcdcaccbabcbbcdd...
   ...dcbbaacbcdacadbbcbaaccbadbcdbaca...
   ...cadcdacaddbabcdcacdada458381830727089
```

$$C = \{3, 10, 6, 5, 2, 9, 8, 9, 1, 8\} \tag{9.9}$$

And obviously even better solutions can be created if (larger) populations are allowed to adapt for even longer periods of time. It all depends on how accurate one wishes the model to be and how much time one wants to spend on its design.

Let's now analyze other results obtained with a dataset containing only nominal attributes.

9.3.4 The Postoperative Patient Problem

The postoperative patient problem is a complex real-world problem with three different classes (admitted to general hospital floor "A", intensive care unit "I", and safe to go home "S") and eight nominal attributes (Table 9.9). The original dataset contains a total of 90 instances distributed very asymmetrically between the three classes (64 instances for class "A", two for class "I", and 24 for class "S"). Previous results on this dataset report very low accuracies of just 48% (Budihardjo et al. 1991).

Table 9.9
Organization of the postoperative patient dataset.

Attribute	Symbol	Branches	Arity
L-CORE	A	high, low, mid	3
L-SURF	B	high, low, mid	3
L-O2	C	excellent, good	2
L-BP	D	high, low, mid	3
SURF-STBL	E	stable, unstable	2
CORE-STBL	F	mod-stable, stable, unstable	3
BP-STBL	G	mod-stable, stable, unstable	3
COMFORT	H	05, 07, 10, 15, ?	5

For this experiment, a sub-set of 60 samples was randomly selected for training and the remaining 30 were used for testing (both these sets are available at the gene expression programming website). The fitness function was again based on the number of hits and was evaluated by equation (3.8). As shown in Table 9.9, the eight attributes were represented by A = {A, ..., H}, splitting respectively into 3, 3, 2, 3, 2, 3, 3, and 5 branches (note that "H" divides into five branches due to the presence of missing values, which, as you can see, are handled as just another branch). The terminal set consisted of T = {a, b, c}, representing respectively classes "A", "I", and "S". Both the performance and the parameters used per run are shown in Table 9.10.

And as you can see, the EDT algorithm performs quite well at this task with an average best-of-run fitness of 47.14. Indeed, several good solutions were designed in this experiment, and two of them are shown below:

```
GDAaHaBcBFcaaabaccaacaacab...
...acbcabbbcacaaabcaaaacccca
```
(9.10)

```
GaAEcBFBAaacaacacaacacbccc...
...bbbbcbbbcbbcacccbababcabc
```
(9.11)

As you can see by drawing the trees, the first one encodes a decision tree with a total of 24 nodes, whereas the second one encodes a DT with 21 nodes. These highly compact models are extremely accurate: the first one has a training fitness of 50 (83.33% accuracy) and a testing fitness of 23 (76.67% accuracy), whereas the second one has a training fitness of 49 (81.67% accuracy) and a testing fitness of 24 (80.00% accuracy) and are,

Table 9.10
Performance and settings used in the postoperative patient problem.

Number of runs	100
Number of generations	10,000
Population size	30
Number of training instances	60
Number of testing instances	30
Number of attributes	8
Attribute set	A-H
Terminal set / Classes	a b c
Maximum arity	5
Head length	10
Gene length	51
Mutation rate	0.044
Inversion rate	0.1
IS transposition rate	0.1
RIS transposition rate	0.1
One-point recombination rate	0.3
Two-point recombination rate	0.3
Fitness function	Eq. (3.8)
Average best-of-run fitness	47.14

therefore, very good solutions to the postoperative problem. Indeed, these results are considerably better than the 48% accuracy achieved by the rules induced with a learning system based on rough sets (Budihardjo et al. 1991).

Let's now see how the decision trees of gene expression programming can be pruned to create rules that are both accurate and compact.

9.4 Pruning Trees with Parsimony Pressure

We know already that all GEP systems learn better if slightly redundant configurations are used (see a discussion of The Role of Neutrality in Evolution in chapter 12). And the decision trees of gene expression programming are no exception: they also grow better when they are allowed to experiment with all kinds of configurations, which is only possible if neutral regions are around. So, it is a good strategy to maximize evolution by using slightly redundant organizations while simultaneously applying a little pressure on the size of the evolving solutions. As explained in section 4.3.1, Fitness Functions with Parsimony Pressure, this can be achieved by introducing a parsimony term on the fitness function.

Let's see, for instance, what happens when the play tennis problem with nominal attributes is solved with parsimony pressure. Thus, for the number of hits fitness function, the fitness will be evaluated by equation (4.21). And given that, for this problem, maximum raw fitness is equal to 14, f_{max} will be equal to 14.0028.

Let's now choose a fairly redundant configuration for the head size, for instance, $h = 10$, and observe how the parsimony pressure manifests itself on the size of the evolving decision trees. (Except for the head size and the number of generations, all the other settings used in this experiment remained exactly as shown in Table 9.2.)

Consider, for instance, the evolutionary history presented below:

```
TWTabaWHWHabababbaaaababaabbaaab-[0]   = 10.00093 (17)
OWaObaWHWHabababbaaaabbabaaabab-[7]    = 11.00103 (17)
OWaOWHbaWWababababaaaaaaabbaababb-[11] = 13.00121 (17)
OWaOWHbaWbbbabaababbbaabbabababab-[14] = 13.00139 (15)
OWaOHHbTWHabaaabbabaaabbbababaa-[39]   = 14.00103 (20)
OHaWWHbWTHbaaababbbaabbabbabbab-[78]   = 14.00112 (19)
OHaWWaWWHbbabababbbbabbbbaaaabab-[107] = 14.00140 (16)
OHaWbaHaHTbbaababbaababbbaabbbb-[108]  = 14.00149 (15)
OHaWbabaHaabaaaabbbaaaabaaaaabb-[126]  = 14.00215 (8)
```

As you can see, a perfect solution involving a total of 20 nodes was created early on in generation 39. Then, by generation 78, a slightly shorter solution with 19 nodes was created. Then an even more compact solution with just 16 nodes was created in generation 107. Then immediately in the next generation a slightly shorter solution was discovered using a total of 15 nodes. And finally, by generation 126 an extremely compact solution with just eight nodes was created. This is, as a matter of fact, the most compact configuration that can be achieved to describe the data presented in Table 9.1.

Decision trees with numeric/mixed attributes can also be efficiently pruned by applying parsimony pressure. Consider again the play tennis problem, but this time using the data presented in Table 9.3. Again, we will use the same basic settings presented in Table 9.4, with the difference that here we will allow for the creation of twice as big decision trees and increase the number of generations to 1000 to ensure that the most compact solution is indeed discovered. The fitness function will obviously also be changed in order to include parsimony pressure and will again be evaluated by equation (4.21).

Consider, for instance, the evolutionary history presented below:

```
WHHaHTOTHHbbaabaaabaaaaabaababa7904000659-[0]  = 11.00081 (20)
C = {72, 96, 60, 95, 61, 64, 94, 74, 67, 69}

OWTWbWOababababababaaaabbbaaabababa0047434150-[1]  = 11.00117 (15)
C = {79, 95, 96, 63, 82, 98, 62, 64, 98, 88}

ObaHbHOWaHababbaabbbbababbababa4730239023-[7]  = 12.00128 (15)
C = {71, 97, 76, 89, 85, 91, 79, 98, 65, 84}

ObaWOabObaaabbbaaaaababbaaababa8602318150-[22]  = 12.00152 (12)
C = {76, 76, 87, 63, 61, 86, 61, 73, 65, 62}

ObaWWabObaabbabaaaaababbaaababb8606328150-[36]  = 12.0016 (11)
C = {83, 76, 75, 69, 61, 86, 61, 86, 65, 62}

ObaWWababOabaaaaaaaabaaabbbaaaa0328132850-[48]  = 12.00184 (8)
C = {83, 76, 75, 69, 61, 86, 61, 86, 65, 62}

ObaWbaOabbabaaaaaaabbaabbbbaaab8338132620-[55]  = 12.002 (6)
C = {65, 73, 90, 87, 69, 64, 71, 78, 61, 64}

HaOOaWbaHbabbaabbbabbbbbbabaaaa0792852539-[63]  = 13.00156 (13)
C = {83, 76, 75, 95, 61, 71, 65, 62, 65, 77}

HaOOaWbabbaabbbbbaababababbbbaabb2283508339-[81]  = 13.00173 (11)
C = {83, 98, 70, 70, 70, 71, 65, 62, 65, 87}

HaObaWbabbaabababbabbababababaabab3543511337-[99]  = 13.00199 (8)
C = {88, 98, 75, 70, 70, 67, 65, 62, 77, 76}

OHWTOHaaWWabbbbbabbbbbbabaabaaa6503501843-[463]  = 14.00112 (19)
C = {80, 71, 86, 81, 97, 79, 83, 71, 85, 97}

OHWTOHaaWaaababbabbbbabababbaba3595962805-[473]  = 14.00131 (17)
C = {80, 71, 69, 97, 97, 79, 83, 71, 85, 97}

OOHWHbaaaWaabbbbaaababaaaabbbaa3690010008-[529]  = 14.00149 (15)
C = {76, 83, 64, 64, 75, 64, 61, 65, 87, 97}

OOHWHbbaabaabbbabbababbbaababba2522002090-[587]  = 14.00168 (13)
C = {78, 65, 95, 64, 75, 77, 80, 65, 80, 75}

OHaWHbbTabaabbbabbabababbababbbba5522002090-[607]  = 14.00177 (12)
C = {76, 70, 91, 67, 96, 75, 60, 69, 63, 75}

OHaWHbbaabaabaaaaaabaabbaabbbaa2006616086-[655]  = 14.00196 (10)
C = {78, 97, 79, 90, 90, 86, 70, 89, 96, 93}

OHaWabbaababbbaabbabbaaabbbbbba8603249940-[682]  = 14.00215 (8)
C = {84, 97, 72, 67, 82, 85, 70, 89, 96, 65}
```

It was chosen because it illustrates one of the dangers of applying parsimony pressure prematurely. As you can see, the system managed to create very compact decision trees that were far from perfect solutions (for instance, the decision trees discovered in generations 55 and 99 are both extremely compact but could only solve 12 and 13 instances respectively). And as you can see, it took a while for the system to find its way around those local optima. For instance, it took 364 more generations for the system to create a solution that surpassed the almost perfect and extremely compact solution discovered in generation 99. And not surprisingly, this first perfect solution is far from parsimonious and totally unrelated to the best of generation 99. As you can see, it went through six simplification events before ending up with just eight nodes, which, as we know, is the most parsimonious configuration that can be achieved to describe the data presented in Table 9.3.

So, no matter how redundant a configuration is chosen, for such simple problems, it is always possible to find the most parsimonious solutions by applying parsimony pressure. For more complex problems, this strategy might be useful to get rid of unnecessary neutral blocks that might be costly in terms of computational effort. However, as we learned here, it might be more productive to apply parsimony pressure just after the creation of the final model in order to avoid unnecessary local optima.

Let's now see in the next chapter yet another interesting application of gene expression programming – neural network induction – that relies even more heavily on the swift handling of large amounts of random numerical constants.

10   Design of Neural Networks

An artificial neural network (NN) is a computational device that consists of many simple connected units (neurons) that work in parallel. The connections between the units or nodes are usually weighted by real-valued weights. Weights are the primary means of learning in neural networks, and a learning algorithm is usually used to adjust the weights.

Structurally, a neural network has three different classes of units: input units, hidden units, and output units. An activation pattern is presented on its input units and spreads in a forward direction from the input units through one or more layers of hidden units to the output units. The activation coming into a unit from other units is multiplied by the weights on the links over which it spreads. All incoming activation is then added together and the unit becomes activated only if the incoming result is above the unit's threshold.

In summary, the basic elements of a neural network are the units, the connections between units, the weights, and the thresholds. And these are the elements one must encode in a linear chromosome in order to simulate fully a neural network so that it can adapt in a particular selection environment. For that I created a special chromosomal organization that explores some of the elements of the architecture used for polynomial induction (see chapter 7 Polynomial Induction and Time Series Prediction).

In this chapter we are going to learn how to modify the chromosomes of gene expression programming so that a complete neural network, including the architecture, the weights, and thresholds, can be totally encoded in a linear chromosome. Furthermore, we will see how this chromosomal organization allows also the training (evolution, in this case) of the neural network using the selection and modification mechanisms of gene expression programming, allowing the discovery of solutions in the form of neural networks. And as conventional neural networks, these evolvable neural networks can obviously be used to solve different kinds of problems: from symbolic

Cândida Ferreira: *Gene Expression Programming*, Studies in Computational Intelligence (SCI) **21**, 381–403 (2006)
www.springerlink.com

regression to classification problems to logic synthesis, for GEP-nets are as versatile as all the GEP systems we have studied so far. I will, however, restrict my presentation here to logic synthesis and the chapter closes with two illustrative Boolean problems: the simple exclusive-or function and the much more complex 6-multiplexer function.

10.1 Genes with Multiple Domains for NN Simulation

A neural network with all its elements is a rather complex structure, not easily constructed and/or trained to perform a certain task. Consequently, it is common to use sophisticated algorithms in which a genetic algorithm is used to evolve partial aspects of neural networks, such as the weights, the thresholds, and the neural network architecture (see Whitley and Schaffer 1992 for a collection of articles on neural networks and genetic algorithms).

Due to the simplicity and plasticity of gene expression programming, it is possible to fully encode complex neural networks of different sizes and shapes in linear chromosomes of fixed length. Indeed, by expressing them, these complex structures become fully functional and, therefore, they can grow and adapt in a particular training environment and then be selected according to fitness in that particular environment. And this means that populations of these entities can be used to explore a solution landscape and, therefore, evolve solutions virtually to all kinds of problems.

In GEP nets, the network architecture is encoded in the familiar structure of a head/tail domain. The head contains special functions (neurons) that activate the hidden and output units (in the GEP context, more appropriately called functional units) and terminals that represent the input units. The tail contains obviously only terminals. Besides the head and the tail, these genes (neural network genes or NN-genes) contain two additional domains, Dw and Dt, encoding, respectively, the weights and the thresholds of the neural network encoded in the head/tail domain. Structurally, the Dw comes after the tail and its length d_w depends on the head size h and maximum arity n_{max} and is evaluated by the expression $h \cdot n_{max}$. The Dt comes after Dw and has a length d_t equal to h. Both domains are composed of symbols representing the weights and thresholds of the neural network.

For each NN-gene, the weights and thresholds are created at the beginning of each run, but their circulation is guaranteed by the usual genetic operators of mutation, inversion, transposition, and recombination.

Furthermore, special mutation operators were created that allow the permanent introduction of variation in the set of weights and thresholds.

It is worth emphasizing that the basic genetic operators such as mutation, inversion, IS and RIS transposition, are not affected by Dw or Dt as long as the boundaries of each region are maintained and the alphabets of each domain are used correctly within the confines of the respective domains. Note also that this mixing of alphabets is not a problem for the recombinational operators and, consequently, their port to GEP-nets is straightforward. However, these operators pose other problems that will be addressed later in the next section.

Consider, for instance, the conventionally represented neural network with two input units (i_1 and i_2), two hidden units (h_1 and h_2), and one output unit (o_1) (for simplicity, the thresholds are all equal to one and are omitted):

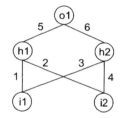

It can also be represented by a conventional tree:

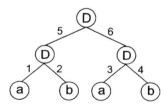

where a and b represent, respectively, the two inputs i_1 and i_2 and "D" represents a function with connectivity two. This function multiplies the values of the arguments by the respective weights and adds all the incoming activation in order to determine the forwarded output. This output (zero or one) depends on the threshold, that is, if the total incoming activation is equal to or greater than the threshold, then the output is one, zero otherwise.

We could linearize the above NN-tree as follows:

```
0123456789012
DDDabab123456
```

where the structure in bold (Dw) encodes the weights. The values of each weight are kept in an array and are retrieved as necessary. For simplicity, the number represented by the numeral in Dw indicates the order in the array.

Let's now analyze a simple neural network encoding a well-known Boolean function, the exclusive-or. Consider, for instance, the chromosome below with $h = 3$ and a domain encoding the weights with $d_w = 6$:

```
0123456789012
DDDabab752393
```

Its translation results in the following neural network:

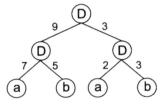

And for the set of weights:

$$W = \{-1.978, 0.514, -0.465, 1.22, -1.686, -1.797, 0.197, 1.606, 0, 1.753\}$$

the neural network above gives:

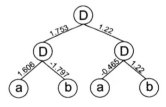

which is a perfect solution to the exclusive-or function.

10.2 Special Genetic Operators

The efficient evolution of such complex entities composed of different domains and different alphabets requires a special set of genetic operators. The operators of the basic gene expression algorithm are easily transposed to neural-net encoding chromosomes, and all of them can be used as long as the boundaries of each domain are maintained and the different alphabets are used within the confines of the corresponding domain.

The mutation operator was extended to all the domains and it continues to be the most important genetic operator, single-handedly allowing an efficient evolution. However, different mutation operators were implemented (head/tail mutation, Dw mutation, and Dt mutation) so that their roles could be understood in more detail but, for practical matters, one usually uses the same mutation rate for all of them, which is obviously equivalent to having just one operator controlling the mutation rate.

Inversion was also extended to GEP-nets, with the usual operator working in the heads of genes and two others working in Dw and Dt to further enhance the circulation of weights and thresholds in the genetic pool (see their description in section 10.2.1).

The IS and RIS transposition operators were also transposed to GEP-nets and their action is obviously restricted to the heads and tails of genes. However, special transposition operators were created that operate within Dw and Dt in order to help in the circulation of weights and thresholds in the population (see their description in section 10.2.2).

The extension of both recombination and gene transposition to GEP-nets is straightforward as their actions never result in mixed domains or alphabets. However, for them to work properly (i.e., to allow an efficient evolution), one must be careful in determining which weights and/or thresholds go to which region after the splitting of the chromosomes, otherwise the system is incapable of evolving efficiently. In the case of gene recombination and gene transposition, to keep track of a gene's weights and thresholds is not difficult and, in fact, these operators are easily implemented and work very well in GEP-nets. But in the case of one-point and two-point recombination, chromosomes are split anywhere and, therefore, it is impossible to keep track of the weights and thresholds that are attached to each neuron. In fact, if applied straightforwardly, these operators produce such evolutionary monsters that they are of little use in multigenic systems. Therefore, for multigenic systems, a special intragenic two-point recombination was created in order to restrict the exchange of sequences to a particular gene (see its description in section 10.2.3).

And finally, in addition to all these operators that work exclusively on the chromosome sequence, special mutation operators – direct mutation of weights and direct mutation of thresholds – were also created in order to introduce modification in the set of available weights and thresholds (see their description in section 10.2.4).

10.2.1 Domain-specific Inversion

Domain-specific inversion is similar to the inversion operator that works in the heads of genes, with the difference that it operates either within Dw (Dw-specific inversion) or Dt (Dt-specific inversion). Thus, these operators randomly choose the chromosome, the gene with its Dw/Dt to be modified, and the start and termination points of the sequence to be inverted.

Consider, for instance, the chromosome below with $h = 6$ (Dw and Dt are shown in different shades):

```
0123456789012345678901234567890123456789012
TDUDcTddccabdddadbd8069454390401674317 93 059
```

where "U" represents a function of one argument and "T" represents a function of three arguments. Suppose that the sequence "5439" in Dw and the sequence "93" in Dt were chosen to be inverted, giving:

```
0123456789012345678901234567890123456789012
TDUDcTddccabdddadbd8069493450401674317 39 059
```

Suppose now that the arrays below represent the weights and the thresholds of both chromosomes:

W = {0.701, 1.117, 0.148, -0.94, -0.044, 1.124, -1.575, 0.877, -1.22, 1.614}
T = {-1.756, -1.776, 0.825, 0.628, -0.263, 0.127, 0.965, -1.651, -0.894, -1.078}

As you can see by their expression in Figure 10.1, they encode very different neural networks because the weights and the thresholds are moved around by the inversion operator, creating new neurons with different strengths and responses.

10.2.2 Domain-specific Transposition

Domain-specific transposition is restricted to the NN-specific domains, that is, Dw and Dt. Its mechanism, however, is similar to IS transposition. This operator randomly chooses the chromosome, the gene with its respective Dw or Dt, the first and last positions of the transposon, and the target site (obviously also chosen within Dw or Dt). Then it moves the transposon from the place of origin to the target site.

Consider, for instance, the chromosome below with $h = 4$ (Dw and Dt are shown in different shades):

a. 01234567890123456789**9012345678901234567**89012
TDUDcTddccabdddadbd**8069454390401674317**93059

W = {0.701, 1.117, 0.148, -0.94, -0.044, 1.124, -1.575, 0.877, -1.22, 1.614}
T = {-1.756, -1.776, 0.825, 0.628, -0.263, 0.127, 0.965, -1.651, -0.894, -1.078}

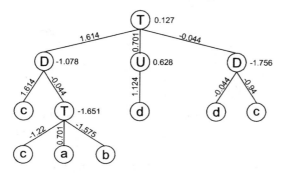

b. 01234567890123456789**9012345678901234567**89012
TDUDcTddccabdddadbd**8069493450401674317**39059

W = {0.701, 1.117, 0.148, -0.94, -0.044, 1.124, -1.575, 0.877, -1.22, 1.614}
T = {-1.756, -1.776, 0.825, 0.628, -0.263, 0.127, 0.965, -1.651, -0.894, -1.078}

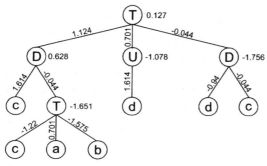

Figure 10.1. Illustration of Dw-specific and Dt-specific inversion. **a)** The neural network before inversion. **b)** The new neural network created by inversion. Note that the network architecture is the same before and after inversion and that the neural networks share the same arrays of weights/thresholds. However, the old and new neural networks are different because a different constellation of weights and thresholds is expressed in these individuals.

01234567890123456789**9012345678901234567**89012
DTQaababaabbaabba**8664826375471750**7682

where "Q" represents a function of four arguments. Suppose that the sequence "86648" was chosen as a transposon and that the insertion site was bond 7 in Dw (between positions 23 and 24). Then the following chromosome is obtained:

```
012345678901234567890123456789012345 6
DTQaababaabbaabba26375478664817507682
```

Note that the transposon is deleted at the place of origin, maintaining the domain length and gene integrity.

Suppose now that the arrays below represent the weights and the thresholds of both chromosomes:

W = {-1.64, -1.834, -0.295, 1.205, -0.807, 0.856, 1.702, -1.026, -0.417, -1.061}
T = {-1.14, 1.177, -1.179, -0.74, 0.393, 1.135, -0.625, 1.643, -0.029, -1.639}

As you can see by their expression in Figure 10.2, they encode very different solutions because the weights are moved around and new combinations of weights and thresholds are tested.

10.2.3 Intragenic Two-point Recombination

Intragenic two-point recombination was created in order to allow the modification of a particular gene without interfering with the other sub-NNs encoded in the other genes. The mechanism of this kind of recombination is exactly the same as in the already familiar two-point recombination of the basic gene expression algorithm, with the difference that now the recombination points are chosen within a particular gene.

Consider the following parent chromosomes composed of two genes, each with a Dw domain ($W_{i,j}$ represents the weights of gene j in chromosome i):

$W_{0,0}$ = {-0.78, -0.521, -1.224, 1.891, 0.554, 1.237, -0.444, 0.472, 1.012, 0.679}
$W_{0,1}$ = {-1.553, 1.425, -1.606, -0.487, 1.255, -0.253, -1.91, 1.427, -0.103, -1.625}

```
012345678901234560123456789012 3456
TTabahaab55239341QDbabbabb40396369- [0]
Qaabbbabb29127879QDbabbaaa36972318- [1]
```

$W_{1,0}$ = {-0.148, 1.83, -0.503, -1.786, 0.313, -0.302, 0.768, -0.947, 1.487, 0.075}
$W_{1,1}$ = {-0.256, -0.026, 1.874, 1.488, -0.8, -0.804, 0.039, -0.957, 0.462, 1.677}

Suppose that gene 0 was chosen to recombine and that point 1 (between positions 0 and 1) and point 12 (between positions 11 and 12) were chosen as recombination points. Then the following offspring is formed:

a. 012345678901234567**8901234567890123**456
DTQaababaabbaabba**8664826375471750**7682

W_m = {-1.64, -1.834, -0.295, 1.205, -0.807, 0.856, 1.702, -1.026, -0.417, -1.061}
T_m = {-1.14, 1.177, -1.179, -0.74, 0.393, 1.135, -0.625, 1.643, -0.029, -1.639}

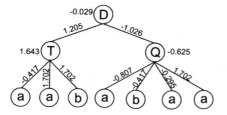

b. 012345678901234567**8901234567890123**456
DTQaababaabbaabba**2637547866481750**7682

W_d = {-1.64, -1.834, -0.295, 1.205, -0.807, 0.856, 1.702, -1.026, -0.417, -1.061}
T_d = {-1.14, 1.177, -1.179, -0.74, 0.393, 1.135, -0.625, 1.643, -0.029, -1.639}

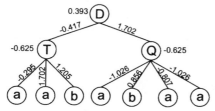

Figure 10.2. Illustration of Dw-specific transposition. **a)** The mother neural network. **b)** The daughter NN created by transposition. Note that the network architecture is the same for both mother and daughter and that $W_m = W_d$ and $T_m = T_d$. However, mother and daughter are different because different combinations of weights and thresholds are expressed in these individuals.

$W_{0,0}$ = {-0.78, -0.521, -1.224, 1.891, 0.554, 1.237, -0.444, 0.472, 1.012, 0.679}
$W_{0,1}$ = {-1.553, 1.425, -1.606, -0.487, 1.255, -0.253, -1.91, 1.427, -0.103, -1.625}

0123456789012345601234567890123456
Taabbbabb29139341QDbabbabb**40396369**- [0]
QTababaab55227879QDbabbaaa**36972318**- [1]

$W_{1,0}$ = {-0.148, 1.83, -0.503, -1.786, 0.313, -0.302, 0.768, -0.947, 1.487, 0.075}
$W_{1,1}$ = {-0.256, -0.026, 1.874, 1.488, -0.8, -0.804, 0.039, -0.957, 0.462, 1.677}

Note that the weights of the offspring are exactly the same as the weights of the parents. However, due to recombination, the weights expressed in the parents are different from those expressed in the offspring (compare their expressions in Figure 10.3).

It is worth emphasizing that this gene-restricted recombination allows a greater control of the modification mechanism and, consequently, permits a much finer tuning of all the neural network elements. If we were to use, in multigenic systems, two-point recombination as it is used in the basic GEA, that is, disrupting chromosomes anywhere, the fine adjustment of the weights and thresholds in multigenic systems would be an almost impossible task. The restriction of two-point recombination to only one gene, however, ensures that only this gene is modified and, consequently, that the weights and thresholds of the remaining genes are kept in place.

Remember, however, that intragenic two-point recombination is not the only source of recombination in multigenic neural networks: gene recombination is fully operational in those systems and it can be combined with gene transposition to propel evolution further through the creation of duplicated genes. And in unigenic systems, the old one-point and two-point recombination are also fully operational as no synchronization of weights/thresholds is necessary.

```
a. 012345678901234560123456789 0123456
   TTababaab55239341QDbabbabb40396369-[0]
   Qaabbbabb29127879QDbabbaaa36972318-[1]
                  ⇓
   012345678901234560123456789 0123456
   Taabbbabb29139341QDbabbabb40396369-[0]
   QTababaab55227879QDbabbaaa36972318-[1]
```

$W_{0,1}$ = {-0.78, -0.521, -1.224, 1.891, 0.554, 1.237, -0.444, 0.472, 1.012, 0.679}
$W_{0,2}$ = {-1.553, 1.425, -1.606, -0.487, 1.255, -0.253, -1.91, 1.427, -0.103, -1.625}

$W_{1,1}$ = {-0.148, 1.83, -0.503, -1.786, 0.313, -0.302, 0.768, -0.947, 1.487, 0.075}
$W_{1,2}$ = {-0.256, -0.026, 1.874, 1.488, -0.8, -0.804, 0.039, -0.957, 0.462, 1.677}

Figure 10.3. Intragenic two-point recombination in multigenic chromosomes encoding neural networks. **a)** An event of intragenic two-point recombination illustrated at the chromosome level. Note that the arrays of weights are not modified by recombination and that parents and offspring share the same weights. **b)** The sub-NNs codified by the parent chromosomes (facing page). **c)** The sub-NNs codified by the daughter chromosomes (facing page).

b.

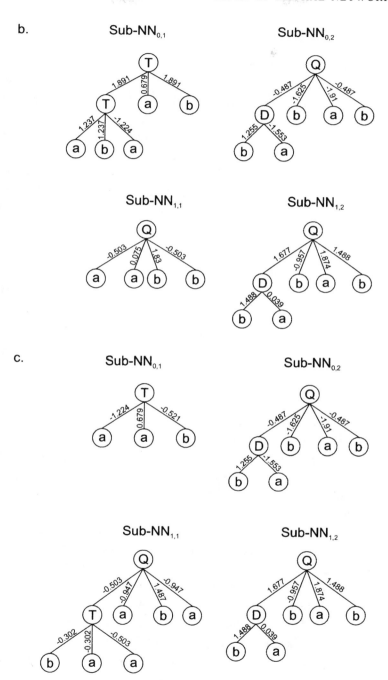

Figure 10.3. *Continued.*

10.2.4 Direct Mutation of Weights and Thresholds

We have already seen that all genetic operators contribute, directly or indirectly, to move the weights and thresholds around. And, in fact, this constant shuffling of weights and thresholds is more than sufficient to allow an efficient evolution of GEP-NNs as the appropriate number of random weights and thresholds can be easily created at the beginning of each run. However, we can also implement special mutation operators that replace the value of a particular weight or threshold by another in order to maintain a constant influx of novelty into the pool of weights and thresholds. These operators, called direct mutation of weights and direct mutation of thresholds, randomly select particular targets in the arrays where the weights/thresholds are kept, and randomly generate a new real-valued number.

Consider, for instance, the array:

$$W_{i,j} = \{-0.433, -1.823, 1.255, 0.028, -1.755, -0.036, -0.128, \mathbf{-1.163}, 1.806, 0.083\}$$

encoding the weights of gene j in chromosome i. Now suppose a mutation occurred at position 7, replacing the weight -1.163 occupying that position by -0.494, giving:

$$W_{i,j} = \{-0.433, -1.823, 1.255, 0.028, -1.755, -0.036, -0.128, \mathbf{-0.494}, 1.806, 0.083\}$$

The consequences of this kind of mutation are very diverse: they might be neutral in effect (for instance, when the gene itself is neutral or when the weight/threshold is not used in the sub-NN) or they might have manifold effects. The latter occurs whenever the weight/threshold modified happens to be used in more than one place in the sub-NN (Figure 10.4).

Interestingly, if all the other operators are being used, this kind of mutation seems to contribute very little to adaptation and, indeed, better results are obtained when this operator is switched off or used at very low rates. This suggests that, thanks to the constant restructuring of the neural network architecture achieved by the chromosomal operators (that is, those that operate directly on the genome sequence), a well dimensioned initial diversity of weights and thresholds is more than sufficient to allow their evolutionary tuning. Typically, I use a small mutation rate of 0.01 for the weights/thresholds and an array length of 10 weights/thresholds for domain lengths equal to or less than 20. For larger domains one could increase the number of elements but most of the times an array of length 10 works very well.

a.   0123456789012345678901234567890123456789012
     TDQDbaabababababaaaaa1523277252856897-[m]

   W_m = {**-0.202**, -1.934, **-0.17**, 0.013, 1.905, 1.167, 1.801, -1.719, 1.412, 0.434}

   W_d = {**1.49**, -1.934, **1.064**, 0.013, 1.905, 1.167, 1.801, -1.719, 1.412, 0.434}

   0123456789012345678901234567890123456789012
   TDQDbaabababababaaaaa1523277252856897-[d]

b.

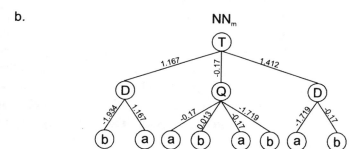

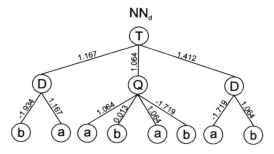

Figure 10.4. Illustration of direct mutation of weights. **a)** The mother and daughter chromosomes with their respective arrays of weights. In this case, weights at positions 0 and 2 were mutated. Note that the network architecture is the same for both mother and daughter. **b)** The mother and daughter neural networks encoded in the chromosomes. Note that the point mutation at position 2 (weight -0.17) has manifold effects as this weight appears four different times in the neural network. Note also that a mutation in the weights at positions 0, 4, 6, and 9 would have had a neutral effect as these weights have no expression on the neural network.

10.3 Solving Problems with GEP Neural Networks

The problems chosen to illustrate the evolution of linearly encoded neural networks are two well-known problems of logic synthesis. The first one, the exclusive-or problem, was chosen both for its historical importance in the neural network field and for its simplicity. The second one, the 6-bits multiplexer, is a rather complex problem and can be useful in evaluating the efficiency of the GEP-NN algorithm.

10.3.1 Neural Network for the Exclusive-or Problem

The XOR is a simple Boolean function of two activities and, therefore, can be easily solved using linearly encoded neural networks.

For this study, two different experiments were designed in order to illustrate the advantages of having the algorithm choosing and evolving the neural network architecture (Table 10.1). The first one uses a fairly redundant organization in order to give us an idea about the kinds of solutions that can be designed for this particular function; and the second uses a much more compact organization in order to force the system into designing only the most parsimonious solutions.

The functions used to solve this problem have connectivities 2, 3, and 4, and are represented, respectively, by "D", "T" and "Q". For the experiment summarized in the first column of Table 10.1, an $h = 4$ was chosen, which, together with the high connectivity level chosen for this problem, allows the discovery of hundreds of different perfect solutions to the XOR function. Most of them are more complicated than the conventional solution with seven nodes presented on page 384; others have the same degree of complexity evaluated in terms of total nodes; but others are, surprisingly, more parsimonious than the aforementioned conventional solution to the XOR function.

Consider, for instance, the first perfect solution found in the experiment summarized in the first column of Table 10.1:

```
01234567890123456789 0123456789012
TQaTaaababbbabaaa6305728327795806
```

$W = \{1.175, 0.315, -0.738, 1.694, -1.215, 1.956, -0.342, 1.088, -1.694, 1.288\}$

As you can see in Figure 10.5, it is a rather complicated solution to the XOR function, as it uses neurons with a relatively high number of connections. The reason for choosing this flamboyant architecture for this simple

Table 10.1
Parameters for the XOR problem using a redundant and a compact system.

	Redundant	Compact
Number of runs	100	100
Number of generations	50	50
Population size	30	30
Number of fitness cases	4	4
Function set	D T Q	D T Q
Terminal set	a b	a b
Weights array length	10	10
Weights range	[-2, 2]	[-2, 2]
Head length	4	2
Number of genes	1	1
Chromosome length	33	17
Mutation rate	0.061	0.118
Inversion rate	0.1	--
One-point recombination rate	0.3	0.3
Two-point recombination rate	0.3	0.3
IS transposition rate	0.1	--
RIS transposition rate	0.1	--
Dw mutation rate	0.061	0.118
Dw-specific inversion rate	0.1	0.1
Dw-specific transposition rate	0.1	0.1
Weights mutation rate	0.01	0.01
Fitness function	Eq. (3.8)	Eq. (3.8)
Success rate	77%	30%

problem, was to show the plasticity of GEP neural networks and that they also thrive in slightly redundant architectures (see, for instance, a discussion of The Role of Neutrality in Evolution in chapter 12). And as you can see in Table 10.1, the success rate for this problem using the redundant chromosomal organization is higher (77%) than the obtained with more compact organizations with $h = 2$ (30%). And it is thanks to this plasticity that the neural network architecture can evolve without any kind of human intervention, contrasting sharply with conventional neural networks where the structure is chosen a priori for each problem and maintained unchanged throughout the learning process.

On the other hand, as you already know, gene expression programming can be useful for searching parsimonious solutions, and very interesting

a. 012345678901234567890123456789012
    TQaTaaababbbabaaa6305728327795806

W = {1.175, 0.315, -0.738, 1.694, -1.215, 1.956, -0.342, 1.088, -1.694, 1.288}

b.

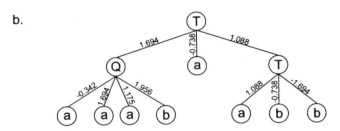

Figure 10.5. A perfect, slightly complicated solution to the exclusive-or problem designed with the GEP-NN algorithm. **a)** Its chromosome and respective weights. **b)** The fully expressed neural network encoded in the chromosome.

parsimonious solutions to the XOR function were found in our second experiment (they were also found among the perfect solutions of the first experiment, which was the reason why the second experiment was designed).

The parameters used per run in this experiment are summarized on the second column of Table 10.1 and, as you can see, despite keeping the functions "T" and "Q" with high connectivities, the smallest workable head size for this problem was chosen (that is, $h = 2$) in order to increase the odds of finding the most parsimonious solutions. One such solution is shown below:

    01234567890123456
    TDbabaabb73899388

W = {0.713, -0.774, -0.221, 0.773, -0.789, 1.792, -1.77, 0.443, -1.924, 1.161}

which, as you can see in Figure 10.6, is a perfect, extremely parsimonious solution to the XOR problem.

Curiously, in this experiment, several other solutions to the XOR function were found that use exactly the same kind of structure of this parsimonious solution. Indeed, the algorithm discovered not only one but several Boolean functions of three arguments and invented new, unexpected solutions to the XOR problem by using them as building blocks. This clearly shows that GEP is an astonishing invention machine, totally devoid of preconceptions.

a.   01234567890123456
     TDbabaabb73899388

W = {0.713, -0.774, -0.221, 0.773, -0.789, 1.792, -1.77, 0.443, -1.924, 1.161}

b.

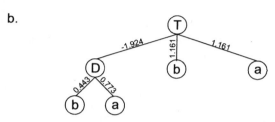

Figure 10.6. A perfect, extremely parsimonious solution to the exclusive-or problem designed with the GEP-NN algorithm. **a)** Its chromosome and respective array of weights. **b)** The fully expressed NN encoded in the chromosome.

10.3.2 Neural Network for the 6-Multiplexer

Multiplexers are logic circuits frequently used in communications and input/output operations for transmitting a number of separated signals simultaneously over a single channel. And in particular, the task of the 6-bit Boolean multiplexer is to decode a 2-bit binary address (00, 01, 10, 11) and return the value of the corresponding data register (d_0, d_1, d_2, d_3). Thus, the 6-multiplexer is a function of six activities: two, a_0 and a_1, determine the address, and four, d_0 through d_3, determine the answer.

There are therefore $2^6 = 64$ possible combinations for the six arguments of the 6-multiplexer function (Table 10.2) and, for this problem, the entire set of 64 combinations was used as the selection environment. The fitness was evaluated by equation (3.10), thus giving $f_{max} = 64$.

In order to simplify the analysis, I chose a rather compact (hence, less efficient) organization and used only neurons with connectivities one, two, and three, which were respectively represented by "U", "D", and "T", thus

Table 10.2
Lookup table for the 6-multiplexer. The output bits are given in lexicographic order starting with 000000 and finishing with 111111.

00000000 11111111 00001111 00001111 00110011 00110011 01010101 01010101

---

giving F = {U, D, T}. The terminal set consisted of T = {a, b, c, d, e, f}, representing, respectively, {a_0, a_1, d_0, d_1, d_2, d_3}. Furthermore, a set of 10 random weights, drawn from the interval [-2, 2] and represented as usual by the numerals 0-9 were used, giving W = {0, 1, 2, 3, 4, 5, 6, 7, 8, 9}.

For this study, we are going to use both a unigenic and a multigenic system in order to show the performance and evolvability of multigenic neural network systems on this difficult task. Both the performance and parameters used per run in both experiments are shown in Table 10.3.

Table 10.3
Parameters for the 6-multiplexer using a unigenic and a multigenic system.

	Unigenic	Multigenic
Number of runs	100	100
Number of generations	2000	2000
Population size	50	50
Number of fitness cases	64 (Table 10.2)	64 (Table 10.2)
Function set	3(U D T)	3(U D T)
Terminal set	a b c d e f	a b c d e f
Linking function	--	Or
Weights array length	10	10
Weights range	[-2, 2]	[-2, 2]
Head length	17	5
Number of genes	1	4
Chromosome length	103	124
Mutation rate	0.044	0.044
Inversion rate	0.1	0.1
One-point recombination rate	0.3	--
Intragenic two-point recombination rate	0.3	0.6
Gene recombination rate	--	0.1
Gene transposition rate	--	0.1
IS transposition rate	0.1	0.1
RIS transposition rate	0.1	0.1
Dw mutation rate	0.044	0.044
Dw-specific inversion rate	0.1	0.1
Dw-specific transposition rate	0.1	0.1
Weights mutation rate	0.002	0.002
Fitness function	Eq. (3.10)	Eq. (3.10)
Success rate	4%	6%

For the experiment summarized in the first column of Table 10.3, unigenic chromosomes were chosen in order to simulate more faithfully a neural network. One of the most parsimonious solutions found in this experiment has a total of 32 nodes and is shown in Figure 10.7.

Obviously, we could explore the multigenic nature of GEP systems and also evolve multigenic neural networks for the 6-multiplexer. The solutions found are, however, structurally more constrained as we have to choose some kind of linking function to link the sub-NNs encoded by each gene. In this case, the Boolean function OR was chosen to link the sub-NNs. (If the mixing of OR with "U", "D", and "T" functions is confusing, think of OR as a function with connectivity two and weights and thresholds all equal to one, and you have a neural network solution to the OR function.)

In the experiment summarized in the second column of Table 10.3, four genes posttranslationally linked by OR were used. The first solution found in this experiment is shown in Figure 10.8. Note that some of the weights in genes 1 and 2 have identical values, and that the same happens for genes 3 and 4. This most probably means that these genes share a common ancestor and were, therefore, created by an event of gene duplication, which, as you would recall, can only be achieved by the concerted action of gene transposition and gene recombination.

The fact that the problems chosen to illustrate the workings of GEP-nets are Boolean in nature and that the neurons we used belong to the simplest class of McCullouch-Pitts neurons, doesn't mean that only this kind of crisp Boolean problems with binary inputs and binary outputs can be solved by the GEP-NN algorithm. In fact, all kinds of neurons (linear neuron, tanh neuron, atan neuron, logistic neuron, limit neuron, radial basis and triangular basis neurons, all kinds of step neurons, and so on) can be implemented in GEP-nets as no restrictions whatsoever exist about the kind of functions the system can handle. And the exciting thing about this is that one can try different combinations of different neurons and let evolution work out which ones are the most appropriate to solve the problem at hand. So, GEP-nets can be used to solve not only Boolean problems but also classification and symbolic regression problems using not only unigenic and multigenic systems but also acellular and cellular systems. Furthermore, classification problems with multiple outputs can also be solved in one go by GEP-nets either by using a multigenic system or a multicellular system. All these different kinds of neural network systems work beautifully and can be used not only as an

a. `TbDTTTfTTaUDcUUTTafeefebabbdabffddfc...`
    `...feeeabcabfabdcfe9512593853547311...`
    `...15067183564799391795954974176167907`

W = {0.241, 1.432, 1.705, -1.95, 1.19, 1.344, 0.925, -0.163, -1.531, 1.423}

b.

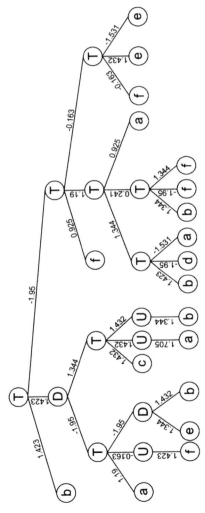

Figure 10.7. A perfect, extremely compact solution to the 6-multiplexer function designed with the GEP-NN algorithm. **a)** Its chromosome and corresponding array of weights. **b)** The fully expressed neural network encoded in the chromosome.

a.   012345678901234567890123 4567890
TecTDdfafabdddfa304107197476784
TDcbTbadddfceacc065920651207105
TfTTUbadbcdffdce813622123399395
TDTbaceaaeeacacd869940072636270

W₁ = {1.126, 0.042, 1.588, -0.03, -1.91, 1.83, -0.412, 0.607, -0.294, -0.659}
W₂ = {-1.961, 1.161, 1.588, -0.03, -1.91, 1.762, -0.412, -0.121, -0.294, -0.659}
W₃ = {1.558, -0.69, 0.921, 0.134, 0.468, -1.534, 0.966, 1.399, 0.023, 0.915}
W₄ = {1.558, 0.767, 0.076, 0.071, 0.468, -1.534, 1.387, -1.857, -1.88, 0.331}

b.

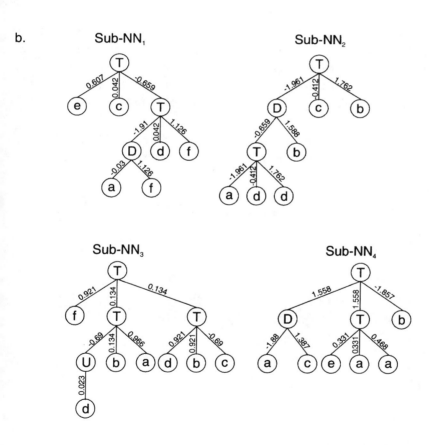

Figure 10.8. A perfect solution to the 6-multiplexer problem encoded in a four-genic chromosome. **a)** Its chromosome with each gene shown separately. W₁-W₄ are the arrays containing the weights of each gene. **b)** The sub-NNs codified by each gene. In this perfect solution the sub-NNs are linked by OR.

alternative modeling tool but also to shed light on the behavior of intricately connected networks of neurons.

10.4 Evolutionary Dynamics of GEP-nets

The neural networks described in this chapter are perhaps the most complex entities created with gene expression programming. Therefore, it would be interesting to see if the evolutionary dynamics of these systems exhibit the same kind of pattern observed in other less complex systems.

As shown in Figure 10.9, GEP-NN systems exhibit the same kind of dynamics found on other, less complex GEP systems. The particular dynamics shown in Figure 10.9 was obtained for a successful run of the experiment summarized in the second column of Table 10.3. Note the characteristic oscillatory pattern on average fitness and that the best fitness is considerably above average fitness.

The ubiquity of these dynamics suggests that, most probably, all healthy genotype/phenotype evolutionary systems are ruled by them.

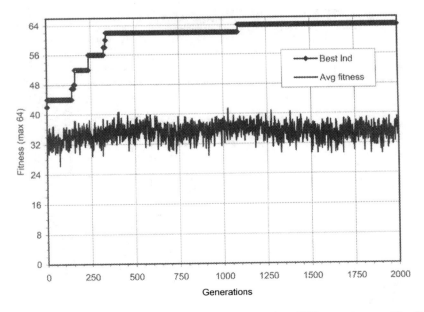

Figure 10.9. Evolutionary dynamics found in complex GEP systems, specifically, on run 4 of the experiment summarized in the second column of Table 10.3.

The dynamics of gene expression programming will be explored further in chapter 12 but, before that, let's analyze yet another modification to the basic gene expression algorithm in order to solve scheduling problems

11 Combinatorial Optimization

In gene expression programming the simplest chromosome will code for a single gene composed of only one terminal. This kind of gene is obtained when the head length h is zero. Then, applying equation (2.4) for determining tail length, we get a gene length $g = 1$ and genes exclusively composed of terminals. So, in its simplest representation, gene expression programming is equivalent to the canonical genetic algorithm in which each gene consists of just one terminal.

Although of little use in unigenic chromosomes, in multigenic chromosomes, one-element genes are extremely useful as they can be organized in multigene families (MGFs). These multigene families consist of clusters of related genes encoding, for instance, a particular class of terminals. Thus, in MGFs, each gene codes for a particular terminal or task. We will see in this chapter that such chromosomes composed of MGFs are very useful for evolving solutions to scheduling problems. We will see, however, that the evolution of such solutions requires special combinatorial-specific operators so that populations of candidate solutions can evolve efficiently. Indeed, in the GA community, several researchers created modifications to the basic genetic operators of mutation and recombination in order to create high performing combinatorial-specific operators (see, e.g., Larrañaga 1998 for a review). However, it is not known which operators perform better as no systematic comparisons have been done.

In this chapter, a new combinatorial optimization algorithm that explores a new chromosomal organization based on multigene families will be described. We will see that this new chromosomal organization together with several combinatorial-specific search operators, namely, inversion, gene deletion/insertion, sequence deletion/insertion, restricted permutation, and generalized permutation, allow the algorithm to perform with high efficiency, astoundingly outperforming the canonical genetic algorithm.

Cândida Ferreira: *Gene Expression Programming*, Studies in Computational Intelligence (SCI) **21**, 405–420 (2006)
www.springerlink.com

11.1 Multigene Families and Scheduling Problems

Multigene families are very useful for finding solutions to combinatorial problems as different classes of terminals/items can be organized into different multigene families. For instance, the different cities in the traveling salesperson problem can be encoded in a multigene family, where each gene codes for a city. Consider, for instance, the simple chromosome below, composed of one MGF with nine members:

$$\begin{array}{l} \texttt{012345678} \\ \texttt{GCDAHEIFB} \end{array} \qquad (11.1)$$

where each element represents one city. In this case, the expression consists of the spatial organization of all the elements, i.e., the orderly way in which they interact with one another (Figure 11.1). This kind of chromosomal structure is going to be used in section 11.3.1 to solve the traveling salesperson problem.

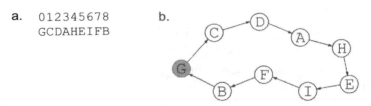

a.   012345678
      GCDAHEIFB   b.

Figure 11.1. Expression of one-element genes as a spatial organization of sub-ETs. If each symbol represented a city visited by a salesperson, the complete expression of the chromosome in (a) would correspond to the nine-city tour shown in (b). The starting and finishing city is shown in gray.

For combinatorial problems with N classes of terminals, multigenic chromosomes composed of N multigene families can be used. In the task assignment problem of section 11.3.2, a simple six-by-six optimization problem, two multigene families representing two different classes of elements will be used. For instance, the following chromosome:

$$\begin{array}{l} \texttt{012345012345} \\ \texttt{632451EDFCBA} \end{array} \qquad (11.2)$$

is composed of two multigene families. The first one is composed of the six elements of set **A** representing the assistants A = {1, 2, 3, 4, 5, 6} and the second is composed of the six elements of set **T** representing the tasks to be assigned T = {A, B, C, D, E, F}. Obviously, how the members of one MGF interact with the members of the other must be decided a priori and implicitly encoded in the chromosome. Figure 11.2 illustrates a very straightforward interaction between the members of two multigene families, in which the first member of MGF_1 interacts with the first member of MGF_2, the second member of MGF_1 with the second member of MGF_2, and so forth.

a. **012345**012345
   632451EDFCBA

b.
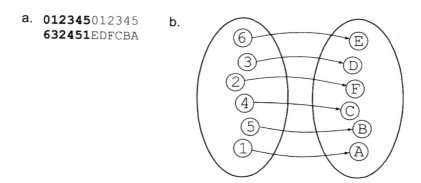

Figure 11.2. Expression of chromosomes composed of multigene families. **a)** The chromosome with two multigene families. **b)** A fully developed individual where the interactions between the sub-ETs are represented by the arrows.

Obviously, different combinatorial problems require different chromosomal organizations and the specification of the interactions between multigene families, but this kind of structure is the basis to encode virtually all kinds of scheduling problems.

11.2 Combinatorial-specific Operators: Performance and Mechanisms

In combinatorial optimization problems, the elements of a multigene family must all be present and cannot be represented more than once. Therefore, special modification mechanisms must be created in order to introduce genetic variation in the population. In this section, I will start by describing the three most efficient combinatorial-specific genetic operators: inversion, gene deletion/insertion, and restricted permutation, and finish by describing the less efficient ones: sequence deletion/insertion and generalized permutation. All these combinatorial-specific operators allow the introduction of genetic variation without disrupting the balance of multigene families and, therefore, always produce valid structures.

Before proceeding with the description of their mechanisms, it is useful to compare their performances (Figure 11.3). The problem chosen to make this analysis is the traveling salesperson problem (TSP) of section 11.3.1 with 19 cities using population sizes of 100 individuals and evolutionary times of 200 generations. As Figure 11.3 clearly demonstrates, the best operator is by far inversion, followed by gene deletion/insertion, whereas restricted permutation is the less efficient of all.

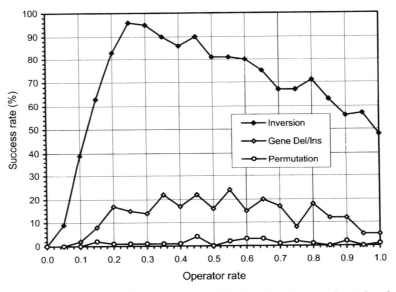

Figure 11.3. Comparison of inversion, gene deletion/insertion, and restricted permutation on the traveling salesperson problem with 19 cities. For this analysis, $P = 100$ and $G = 200$. The success rate was evaluated over 100 identical runs.

11.2.1 Inversion

The inversion operator randomly selects the chromosome, the multigene family to be modified, the inversion points in the MGF, and inverts the sequence between these points. Each chromosome can only be modified once by this operator.

Consider, for instance, the chromosome below composed of two multigene families:

```
012345678901234567801234567890123456 78
snpbqhfgkicmjlaedorPRMDCNLEBQFGKJOHIAS
```
(11.3)

Suppose, for instance, that genes 2 and 7 in MGF_2 were chosen as inversion points. Then the sequence between these points is inverted, giving:

```
012345678901234567801234567890123456 78
snpbqhfgkicmjlaedorPRELNCDMBQFGKJOHIAS
```
(11.4)

Note that with inversion the whole multigene family can be inverted. This happens whenever the first and last genes of a multigene family are chosen as inversion points. For instance, the inversion of MGF_2 in chromosome (11.3) above gives:

```
012345678901234567801234567890123456 78
snpbqhfgkicmjlaedorSAIHOJKGFKBELNCDMRP
```
(11.5)

Note also that this operator allows small adjustments like, for instance, the permutation of two adjacent genes. For instance, if genes 7 and 8 in MGF_1 of chromosome (11.5) were chosen as inversion points, these genes would be permuted, giving:

```
012345678901234567801234567890123456 78
snpbqhfkgicmjlaedorSAIHOJKGFKBELNCDMRP
```
(11.6)

As shown in Figure 11.3, inversion is the most powerful of the combinatorial-specific genetic operators, causing populations to evolve with great efficiency even if used as the only source of genetic modification. Indeed, this operator alone produces better results than when combined with gene deletion/insertion or permutation. Inversion rates between 20% and 60% produce good results for most problems.

11.2.2 Gene Deletion/Insertion

As you can see in Figure 11.3, gene deletion/insertion is the second in importance of the combinatorial-specific operators.

This operator randomly selects the chromosome, the multigene family to be modified, the gene to transpose, and the insertion site. Each chromosome can only be modified once by this operator.

Consider, for instance, the chromosome below composed of two multigene families:

```
0123456789012345678012345678901234567  8
rpifghasbdeocjknlqmQSKLHCIGDONPFEJMBRA
```
(11.7)

Suppose gene 5 in MGF_1 was selected to transpose to site 14 (between genes 13 and 14). Then gene 5 ("h") is deleted at the place of origin and inserted between genes "j" and "k", giving:

```
0123456789012345678012345678901234567  8
rpifgasbdeocjhknlqmQSKLHCIGDONPFEJMBRA
```
(11.8)

The deletion/insertion of genes when combined with the most powerful operator (inversion) might be useful for finer adjustments. However, for all the problems analyzed in this chapter, the performance was higher when inversion alone was used in the search.

11.2.3 Restricted Permutation

Restricted permutation allows two genes occupying any positions within a particular multigene family to exchange positions. This operator might also be useful for making finer adjustments when combined with inversion, but if used as the only source of genetic variation, its performance is very poor (see Figure 11.3 above).

The restricted permutation operator randomly chooses the chromosome, the multigene family to be modified and the genes to be exchanged. Furthermore, each chromosome is only modified once by this operator.

Consider, for instance, the following chromosome composed of two multigene families:

```
0123456789012345678012345678901234567  8
ikmosfghdeqprljncabLNJIHGCDPSRQOBKMFAE
```
(11.9)

Now suppose that genes 6 ("C") and 15 ("M") in MGF_2 were chosen to be exchanged. Then the following chromosome is formed:

```
0123456789012345678012345678901234567  8
ikmosfghdeqprljncabLNJIHGMDPSRQOBKCFAE
```
(11.10)

Restricted permutation, if used at small rates and in combination with inversion, might be useful for making finer adjustments. But, again, for all the problems analyzed in this chapter, when restricted permutation is used in conjunction with inversion the success rate decreases slightly.

11.2.4 Other Search Operators

In this section, we are going to analyze another set of combinatorial-specific operators: sequence deletion/insertion and generalized permutation. These operators are related, respectively, to gene deletion/insertion and restricted permutation. Compared to inversion, these operators are extremely ineffi-cient but, nevertheless, the analysis of their performance and mechanisms can give us some clues about the fundamental attributes that a good combi-natorial-specific operator must have.

Sequence Deletion/Insertion

We have seen that the gene deletion/insertion operator described in section 11.2.2 permits only the transposition of single genes, in other words, it al-lows the transposition of small sequences composed of only one element. A different operator can be easily implemented that deletes/inserts sequences of varied length (sequence deletion/insertion operator). This might appear more advantageous than the deletion/insertion of genes, but experience shows the opposite (see Figure 11.4 below). In fact, this operator produces results that are even worse than the restricted permutation operator on the traveling salesperson problem with 19 cities (compare with Figure 11.3). Indeed, an identical analysis done with this operator showed that sequence deletion/insertion is incapable of solving the 19 cities TSP using population sizes of 100 individuals for 200 generations. Thus, an easier version of the TSP with 13 cities was chosen in order to allow the comparison between gene dele-tion/insertion and sequence deletion/insertion (Figure 11.4). For this analy-sis, a population size of 100 individuals and an evolutionary time of 200 generations were used, that is, exactly the same values of P and G used in the much harder TSP with 19 cities.

Generalized Permutation

Generalized permutation is a variation of the restricted permutation operator described in section 11.2.3. Recall that during restricted permutation only a pair of genes are exchanged per chromosome, that is, the restricted permuta-tion rate p_{rp} is evaluated by $p_{rp} = N_C/P$, where N_C represents the number of

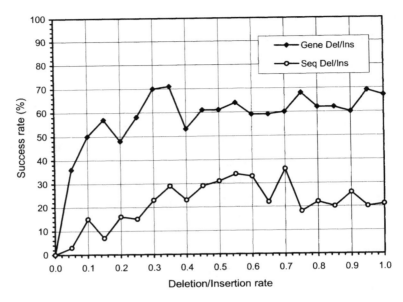

Figure 11.4. Comparison of gene deletion/insertion with sequence deletion/insertion on the traveling salesperson problem with 13 cities. For this analysis, $P = 100$ and $G = 200$. The success rate was evaluated over 100 identical runs.

chromosomes modified. A more generalized version of this operator can be easily implemented where a different number of genes in a chromosome can trade places with other genes according to a certain rate. More formally, the generalized permutation rate p_{gp} is evaluated by $p_{gp} = N_G/(C_L \cdot P)$, where N_G represents the number of genes modified and C_L the chromosome length. Again, this might appear more efficient than the restricted permutation described in section 11.2.3, but experience shows that restricted permutation is slightly better (see Figure 11.5 below). For instance, in the TSP with 19 cities (see Figure 11.3 above), generalized permutation performed worse than restricted permutation and, in fact, was incapable of finding a perfect solution to this problem.

The results obtained for a simpler version of the TSP with 13 cities are shown in Figure 11.5. In this analysis, both the restricted permutation and generalized permutation operators are compared using populations of 100 individuals evolving for 200 generations, that is, exactly the same values of P and G used to solve the much more complex TSP with 19 cities shown in Figure 11.3.

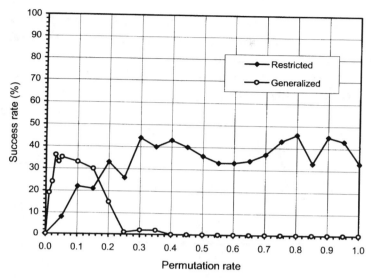

Figure 11.5. Comparison of restricted permutation with generalized permutation on the traveling salesperson problem with 13 cities. For this analysis, $P = 100$ and $G = 200$. The success rate was evaluated over 100 identical runs.

11.3 Two Scheduling Problems

The first problem of this section is the already explored traveling salesperson problem with 19 cities. And we have already seen that this problem requires only a multigene family consisting of the genes representing the 19 cities the salesperson should visit.

The second problem is a task assignment problem that requires two different multigene families: one containing the agents and the other the tasks assigned to the agents.

11.3.1 The Traveling Salesperson Problem

The TSP represents a classical optimization problem and good, traditional approximation algorithms have been developed to tackle it down (see, e.g., Papadimitriou and Steiglitz 1982 for a review). However, to evolutionary computists, the TSP serves as the simplest case of a variety of combinatorial problems which are of enormous relevance to industrial scheduling problems (Bonachea et al. 2000; Hsu and Hsu 2001; Johnson and McGeoch 1997; Katayama and Narihisa 1999; Merz and Freisleben 1997; Reinelt 1994).

Indeed, several evolution-inspired algorithms used the TSP as a battleground to develop combinatorial-specific search operators (see, e.g., Ferreira 2002b for a detailed list of the most common combinatorial-specific operators).

For the TSP with 19 cities there are $19! = 1.21645 \times 10^{17}$ combinations to search. If we fix the starting point (and consequently the ending point, for salespersons must return home), the number of possible combinations is halved to 6.0823×10^{16}. Furthermore, if we choose a configuration where all the cities lie in a rectangle as shown in Figure 11.6, we can rigorously evaluate the performance of the algorithm as we know beforehand the correct solution; for a 19 cities tour, the minimum distance will be obviously 20.

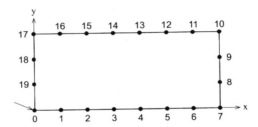

Figure 11.6. Configuration of 19 cities arranged in a rectangle. The arrow indicates the starting and finishing point. Obviously, the shortest tour is to trace the rectangle which has a distance of 20.

Obviously, we cannot use the tour length directly as a measure of fitness of the evolving solutions, as the shorter the tour the fitter the individual. Thus, for each generation, the fitness f_i of an individual program i in generation g is evaluated by the formula:

$$f_i = T_g - t_i + 1 \tag{11.11}$$

where t_i is the length of the tour encoded in i, and T_g is the length of the largest tour encoded in the chromosomes of the current population. This way, the fitness of the worst individual of the population is always equal to one. As usual, individuals are selected according to fitness by roulette-wheel selection and in each generation the best individual is cloned unchanged into the next generation. Both the performance and parameters used per run are summarized in Table 11.1.

Table 11.1
Parameters for the traveling salesperson problem with 19 cities.

Number of runs	100
Number of generations	200
Population size	100
Number of multigene families	1
Number of genes per multigene family	19
Chromosome length	19
Inversion rate	0.25
Success rate	96%

The results obtained by gene expression programming are astounding if we compare them with the performance the GA achieved on the 19-city TSP. For instance, Haupt and Haupt (1998) could not find the shortest route using population sizes of 800 for 200 generations. As shown in Figure 11.3 and Table 11.1, the algorithm described here not only is capable of finding the shortest route using populations of only 100 individuals and for the same 200 generations, but also is capable of finding the shortest route in practically all runs (in 96% of the runs, in fact). It is worth pointing out that, in this experiment, inversion is the only source of genetic variation. Indeed, the presence of other genetic operators, namely, gene deletion/insertion and restricted permutation, decreases slightly the success rate and, therefore, these operators were not used. Apparently, they are unnecessary for finer adjustments whenever inversion is doing the search.

As stated earlier in this chapter, the chromosomal architecture used to solve combinatorial problems corresponds exactly to the canonical genetic algorithm. So, why the stunning difference in performance between these two algorithms? Obviously, the answer lies in the set of genetic operators used by GEP and by the GA. As shown in Figures 11.3, 11.4, and 11.5 above, inversion performs significantly better than permutation (both restricted and generalized implementations). And different kinds of permutation and extremely complicated forms of crossover are exactly the kind of search operators favored by GA researchers to solve combinatorial problems (see Larrañaga 1998 for a review). Unfortunately, inversion was abandoned earlier in the history of genetic algorithms and maybe because of this it is seldom used today even in combinatorial optimization.

11.3.2 The Task Assignment Problem

The task assignment problem (TAP) of this section is the simple toy problem chosen by Tank and Hopfield (1987) in their *Scientific American* article to illustrate the workings of Hopfield networks on combinatorial cost-optimization problems.

In TAP there are *n* tasks that must be accomplished by using only *n* workers. Each worker performs better at some tasks and worse at others and obviously some workers are better than others at certain tasks. The goal is to minimize the total cost for accomplishing all tasks or, stated differently, to maximize the overall output of all the workers as a whole.

Suppose we had to shelve *n* book collections in a library using *n* shelving assistants. Each assistant is familiar with the subject areas to varying degrees and shelves the collections accordingly. The input data or fitness cases in this task assignment problem consist of the rates at which books are shelved per minute (Figure 11.7).

	A	B	C	D	E	F
1	10	6	1	5	3	7
2	5	4	8	3	2	6
3	4	9	3	7	5	4
4	6	7	6	2	6	1
5	5	3	4	1	8	3
6	1	2	6	4	7	2

Figure 11.7. The task assignment problem. Each assistant (1-6) should be assigned to one collection of books (A-F) based on the rates at which books are shelved per minute (fitness cases). Shaded squares show the best assignment with the largest sum of shelving rates, 44.

For this simple six-by-six problem there are already 6! = 720 possible assignments of assistants to book collections. The best solution has the highest sum of rates for the chosen assistants. For the particular set of fitness cases shown in Figure 11.7, f_{max} = 44.

This kind of toy problem is very useful for comparing the performance of different algorithms and different search operators and, here, the potentialities of inversion will be further tested in systems in which chromosomes

composed of more than one multigene family are used. Indeed, the task assignment problem is solved very efficiently using only inversion as the source
of genetic variation and two multigene families: one to encode the assistants
(represented by 1-6) and another to encode the book collections (represented
by A-F). Such a chromosome is shown below:

```
012345012345
652314DECFAB
```
(11.12)

It contains two different MGFs, the first encoding the assistants and the second the book collections. And its expression gives:

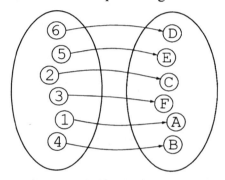

where the assignments are represented by the arrows. As you can easily
check in Figure 11.7, this individual has $f_i = 41$.

So, for the task assignment problem, we are going to use small populations
of 30 individuals and evolutionary times of 50 generations. The parameters
used per run and the performance of the algorithm expressed in terms of
success rate are shown in Table 11.2.

Table 11.2
Parameters for the task assignment problem.

Number of runs	100
Number of generations	50
Population size	30
Number of multigene families	2
Number of genes per multigene family	6
Chromosome length	12
Inversion rate	0.30
Success rate	69%

In the first successful run of this experiment, a solution with maximum fitness was discovered in generation 16:

```
012345012345
536241EDCFBA
```
(11.13)

which corresponds to the best assignment of 44 (see also Figure 11.7 above):

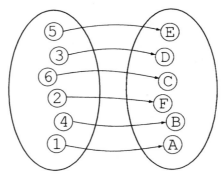

11.4 Evolutionary Dynamics of Simple GEP Systems

We concluded the last chapter by analyzing the evolutionary dynamics of one of the most complex GEP systems – multigenic neural networks – and observed that GEP-nets exhibit exactly the same kind of evolutionary dynamics found in less complex genotype/phenotype systems.

As stated earlier in this chapter, the simple chromosomal organization used for combinatorial optimization is very similar to the canonical GA. Thus, it would be interesting to see if the evolutionary dynamics of these simpler GEP systems are also of the kind observed in GA's populations.

Let's first analyze the simplest GEP system in which only one multigene family is used per chromosome. Its evolutionary dynamics is shown in Figure 11.8. And curiously enough, the evolutionary dynamics for this simple system is similar to the dynamics characteristic of GA's populations. As you can see, for this kind of dynamics the plot for average fitness closely accompanies the plot for best fitness and the oscillatory pattern on average fitness is less pronounced.

And finally, let's analyze a slightly more complex system where chromosomes composed of two multigene families are used. The evolutionary dynamics shown in Figure 11.9 was obtained for the first successful run of the experiment summarized in Table 11.2. It is worth noticing that this

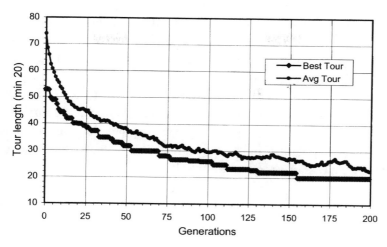

Figure 11.8. Evolutionary dynamics found in the simplest GEP systems. This dynamics was obtained for the first successful run of the experiment summarized in Table 11.1 (traveling salesperson problem with 19 cities).

evolutionary dynamics is no longer of the type expected for a GA. In fact, it has all the hallmarks of a GEP dynamics: the oscillatory pattern in average fitness and the considerable difference between the fitness of the best

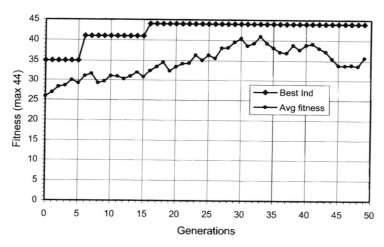

Figure 11.9. Evolutionary dynamics of simple GEP systems. This dynamics was obtained for a successful run of the experiment summarized in Table 11.2 (six-by-six task assignment problem). Note that, although simple, this system no longer exhibits a typical GA dynamics (compare with Figure 11.8).

individual and the fitness of the average individual. Indeed, one should expect this kind of dynamics because of the higher complexity required to express chromosomes composed of two multigene families encoding complex interactions between their members.

In the next chapter I will conduct a series of evolutionary studies, including the detailed analysis of the evolutionary dynamics typical of complex GEP systems. The aim of those studies is to discuss some controversial topics in evolutionary computation, most of them raised and intensified by the invention of gene expression programming itself.

12   Evolutionary Studies

The evolutionary studies of this chapter have both practical and theoretical interest. On the one hand, there are many controversial questions in evolution theory and the simple genotype/phenotype system of gene expression programming can give us some insights into natural evolutionary processes. On the other hand, evolutionary computation not only inherited all the controversial questions from evolution theory but also created some of its own, casting more shadow than light into evolution theory.

In this chapter we are going to discuss three fundamental questions of evolution theory: (1) genetic operators and their power; (2) the importance of the initial diversity; and (3), the importance of neutrality in evolution. Furthermore, related to these fundamental questions, are several other pertinent questions that have been around for a long time in evolutionary computation and are here clearly answered in an empirical way. Such questions include, among others, the role of sex, the role of building blocks, and the problem of premature convergence.

12.1 Genetic Operators and their Power

Everybody agrees that, by and large, evolution relies on genetic variation coupled with some kind of selection and, in fact, all evolutionary algorithms explore these fundamental processes. However, there is no agreement concerning the best way to create genetic variation, with researchers divided between mutation and recombination. This fact per se is extremely revealing, suggesting that existing artificial evolutionary systems are fundamentally different from one another. Indeed, artificial evolutionary systems are themselves still evolving and among them, camouflaged by different representations, can be found simple replicator systems, rudimentary genotype/

Cândida Ferreira: *Gene Expression Programming*, Studies in Computational Intelligence (SCI) **21**, 421–456 (2006)
www.springerlink.com

phenotype systems, and full-fledged genotype/phenotype systems. And the mechanisms of genetic modification of all these systems are intricately connected with their representation schemes.

We have already seen that gene expression programming uses not only mutation and recombination but also different kinds of transposition and, therefore, can be useful for conducting a rigorous analysis of the power of different search operators in order to gain some insights into their role in evolution. We will see that mutation is by far the single most important genetic operator, outperforming recombination considerably. In fact, all three kinds of genetic recombination analyzed here (one-point, two-point, and gene recombination) perform considerably worse than mutation and also considerably worse than simple intrachromosomal transposition mechanisms. In addition, we are also going to analyze with great detail the evolutionary dynamics produced by all these genetic operators in order to understand their importance in evolution.

12.1.1 Their Performances

For this analysis of the genetic operators and their power, the following relatively complex test sequence was chosen:

$$a_n = 5n^4 + 4n^3 + 3n^2 + 2n + 1 \tag{12.1}$$

where n consists of the nonnegative integers. This sequence was chosen for four reasons: (1) it can be exactly solved and therefore provide an accurate measure of performance in terms of success rate; (2) it requires relatively small populations and relatively short evolutionary times, making the task feasible; (3) it provides sufficient resolution to allow the comparison of dissimilarly performing operators such as mutation and gene recombination; and (4) it is appropriate to study all the genetic operators, including operators specific of multigenic systems like gene recombination.

In all the experiments of this section, the first 10 positive integers n and their corresponding term a_n were used as fitness cases (Table 12.1); the fitness function was evaluated by equation (3.3b) and a selection range of 20%

Table 12.1
Set of fitness cases for the sequence induction problem.

n	1	2	3	4	5	6	7	8	9	10
a_n	15	129	547	1593	3711	7465	13539	22737	35983	54321

and maximum precision (0% error) were chosen, giving $f_{max} = 200$; population sizes $P = 500$ and evolutionary times $G = 100$ generations were used and the success rate Ps of each experiment was evaluated over 100 independent runs; $F = \{+, -, *, /\}$ and the terminal set consisted of the independent variable; and finally, seven-genic chromosomes of length 91 (head length $h = 6$) linked by addition were used. In the experiments where transposition was switched on, three transposons with lengths 1, 2, and 3 were used.

In this section, we are going to study only six of the usual set of eight genetic operators of gene expression programming, leaving out the late-addition inversion and gene transposition as, alone, this operator contributes nothing in the particular conditions chosen for this study (that is, genes encoding sub-ETs linked by a commutative function).

The performance of all these genetic operators is shown in Figure 12.1 and, as you can see, mutation is the single most powerful operator, followed by RIS transposition and IS transposition, whereas recombination is the less powerful of all. Note also that the three recombinational operators show appreciable differences among themselves. As you can see, two-point recombination is the most efficient of the three whereas gene recombination is the most inefficient. It is worth pointing out that the finger-shaped plot

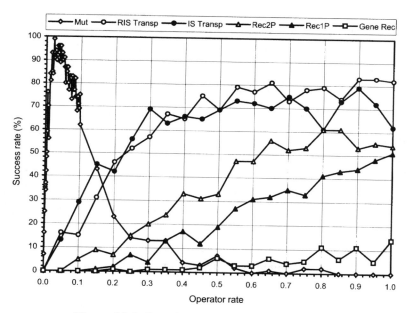

Figure 12.1. Genetic operators and their power.

observed for mutation, is very different from the plots obtained both for transposition and recombination. Note that only mutation can climb "Mount Everest" and, indeed, systems undergoing mutation can be easily tuned so that populations stay, albeit precariously, poised at the peak as small variations on mutation rate have dramatic effects within the finger region.

As Figure 12.1 emphasizes, mutation has a tremendous creative power and, indeed, this operator alone is more than sufficient to evolve solutions to virtually all problems. Nonetheless, other genetic operators can be and are regularly used in GEP both for practical and theoretical reasons. For instance, RIS transposition is interesting as it challenges our views on the importance of too drastic modifications in evolution. Moreover, it is known that mutation is not sufficiently innovative to account for all the wonders of nature. Consider, for instance, the creation of the eukaryotes by symbiosis (Margulis 1970) or the simpler phenomenon of gene duplication (Lynch and Conery 2000) also observed in gene expression programming.

Also interesting are the results obtained for IS and RIS transposition. Particularly interesting is the fact that these two kinds of simple intrachromosomal transposition far exceed all kinds of recombination in the first place and that RIS transposition is slightly more efficient than IS transposition. It is worth pointing out that with RIS transposition the first position of a gene is always the target. And this means that, at the phenotype level, the root of sub-ETs is modified. Indeed, this kind of modification is one of the most disruptive and is similar to a frameshift mutation occurring at the beginning of a protein gene. Note also that the transforming power of both kinds of transposition is slightly smaller than mutation (compare maximum performance obtained for each operator).

Also worth discussing are the results obtained for the three kinds of recombination. Recall that two-point recombination is the most disruptive of the recombinational operators and, as shown in Figure 12.1, it is also the most efficient kind of recombination. Not surprisingly, the most conservative of the recombinational operators – gene recombination – is also the less efficient. In addition, it is worth noticing that all the recombinational mechanisms analyzed here, even the most conservative, are more disruptive than the homologous recombination that occurs during sexual reproduction because, in GEP, the exchanged genes rarely are homologous.

One of the unsolved questions of biology is the role of sex in evolution (Margulis and Sagan 1986) and, most of the times, biological sex in its overwhelming diversity is confounded with the homologous recombination that occurs during sexual reproduction. Consequently, many erroneously assume that homologous recombination creates more diversity despite several facts against this hypothesis (Margulis and Sagan 1997). The comparison of GEP operators, especially recombination, suggests that a more conservative recombinational mechanism such as homologous recombination would only be useful for maintaining the status quo in periods of stasis. Interestingly, this hypothesis is further supported by the kind of evolutionary dynamics found in populations undergoing recombination alone (see the sudy on Recombination in the next section).

12.1.2 Evolutionary Dynamics of Different Types of Populations

The analysis of the evolutionary dynamics of some of the populations shown in Figure 12.1 can be helpful to gain some insights into the adaptive strength of different evolutionary systems. The most interesting are those with the highest performances and those are, obviously, the ones we are going to analyze first, comparing them with other, less efficient systems.

Mutation

Based on the analysis of Figure 12.1, the study of key populations undergoing mutation can be used as a reference against which other populations undergoing different genetic modifications could be compared. Populations on the ascending side of the mutation plot are healthily evolving under small mutation rates and, therefore, are generally called *healthy* populations. Populations on the descending side are evolving under excessive mutation and are generally called *unhealthy*. Obviously, populations at the peak have an ideal mutation rate. Figure 12.2 shows how populations behave in terms of average fitness and how the plot for average fitness relates to the plot for best fitness as populations move along the mutation curve of Figure 12.1.

In the first evolutionary dynamics, $p_m = 0.001$ and $Ps = 16\%$ (plot **a**), the plot for average fitness closely accompanies the plot for best fitness and only a small oscillation is observed in average fitness. These populations are called *moderately innovative* as they evolve very sluggishly and only a little bit of genetic diversity is introduced in the population.

In the second dynamics, $p_m = 0.0045$ and $Ps = 48\%$ (plot **b**), it can be observed that, although closely accompanying the plot for best fitness, the gap between both plots increases and the characteristic oscillatory pattern on average fitness is already evident even for such small variation rates. As

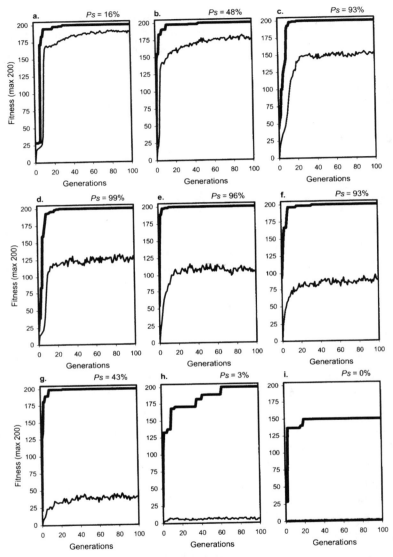

Figure 12.2. The evolutionary dynamics of mutation. The success rate above each plot was determined in the experiment shown in Figure 12.1. **a)** *Moderately Innovative* ($p_m = 0.001$). **b)** *Healthy But Weak* ($p_m = 0.0045$). **c)** *Healthy And Strong* ($p_m = 0.016$). **d)** *Healthy And Strong* ($p_m = 0.022$). **e)** *Healthy And Strong* ($p_m = 0.034$). **f)** *Unhealthy But Strong* ($p_m = 0.046$). **g)** *Unhealthy And Weak* ($p_m = 0.15$). **h)** *Almost Random* ($p_m = 0.45$). **i)** *Totally Random* ($p_m = 1.0$). Note that, except for the last dynamics, only successful runs were chosen so that the plot for best fitness reached its maximum value and the distance between average and best fitness could be better appreciated.

shown in Figure 12.1, this population is midway to the peak with a success rate of 48%. Such systems, although adaptively healthy, are not very efficient and are called *healthy but weak*.

As shown in Figure 12.1, the success rate increases abruptly with mutation rate until it reaches a plateau around $p_m = 0.022$. In terms of dynamics, this is reflected in a more pronounced oscillatory pattern in average fitness and an increase in the gap between average and best fitness. For obvious reasons, populations evolving with maximum performance are called *healthy and strong*. The next three dynamics (plots c, d, and e) are all drawn from the performance plateau or peak. Note, however, that from the peak onward, an increase in mutation rate results in a decrease in performance until populations are totally incapable of adaptation (see Figure 12.1). In these populations, the gap between average and best fitness continues to increase (see plots f through i) until the plot for average fitness reaches the bottom and populations become totally random and incapable of adaptation (see plot i). Note also that some populations, despite evolving under excessive mutation rates, are still very efficient. For instance, the evolutionary dynamics shown in plot f, $p_m = 0.046$ and $Ps = 93\%$, is extremely efficient. This kind of population is called *unhealthy but strong* (with "unhealthy" indicating the excessive mutation rate). The next population, $p_m = 0.15$ and $Ps = 43\%$ (plot g), is not as efficient as the previous one and therefore is called *unhealthy and weak*.

The last two dynamics were obtained for populations outside the finger region of the mutation plot. In the first, $p_m = 0.45$ and $Ps = 3\%$ (plot h), the plot for average fitness is not far from the minimum position obtained for totally random populations (plot i). Thus, populations of this kind are called, respectively, *almost random* and *totally random*. Note that oscillations on average fitness are less pronounced in these plots. It is worth noticing that totally random populations (the presence of elitism at $p_m = 1.0$ is, in this experiment with 500 individuals, insignificant in terms of average fitness) are unable to find a perfect solution to the problem at hand. And this tells us that every time a perfect solution was found, a powerful search mechanism was responsible for this and not a random search.

Transposition

The evolutionary dynamics of transposition are shown in Figure 12.3. As you can see, the dynamics obtained for both RIS and IS transposition are similar to the ones obtained for mutation. And this is worth pointing out, because not all operators display this kind of dynamics (see the dynamics of recombination below). Like with mutation, the gap between average and best fitness increases with transposition rate. Also worth pointing out is that populations are evolving more efficiently at $p_{ris} = 1.0$ ($Ps = 82\%$) than at $p_{is} = 1.0$ ($Ps = 62\%$). Consequently, the gap between average and best fitness is smaller in the last case (compare plots **c** and **f**). More generally, the plots for average fitness obtained for RIS transposition occupy lower positions than the corresponding IS dynamics. As previously shown, for healthy populations, this is an indicator of a more efficient evolution. Indeed, as shown in Figure 12.1, RIS transposition performs slightly better than IS transposition.

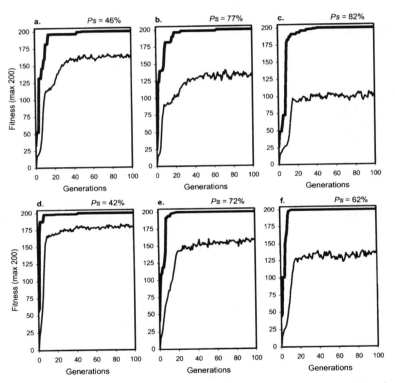

Figure 12.3. Evolutionary dynamics characteristic of populations evolving by RIS transposition (plots **a**, **b**, and **c**) and IS transposition (plots **d**, **e**, and **f**). The success rate above each plot was determined in the experiment of Figure 12.1. **a)** $p_{ris} = 0.2$. **b)** $p_{ris} = 0.6$. **c)** $p_{ris} = 1.0$. **d)** $p_{is} = 0.2$. **e)** $p_{is} = 0.6$. **f)** $p_{is} = 1.0$. Notice that the evolutionary behavior is similar to mutation (see Figure 12.2) and very different from recombination (see Figure 12.4 below). Note also that the plots for average fitness obtained for RIS transposition occupy lower positions than the corresponding IS dynamics, an indicator of a more tumultuous and, in this case, more efficient evolution.

The fact that RIS and IS transposition exhibit a behavior similar to muta-tion further reinforces the uniqueness of recombination. We will see in the next section that populations undergoing recombination alone behave very strangely: they evolve very smoothly without appreciable oscillations on average fitness and are constantly trying to close the gap between average and best fitness.

Recombination

The evolutionary dynamics of recombination show that all recombinational operators, from the most conservative (gene recombination) to the most dis-ruptive (two-point recombination), display a homogenizing effect (Figure 12.4). For obvious reasons, these recombination-specific dynamics are called *homogenizing* dynamics. Note that, in all cases, the plot for average fitness closely accompanies the plot for best fitness and, given enough time, the plots tend to overlap and populations lose all genetic diversity (see, for in-stance, plots **c**, **g**, and **h**). Note that this happens despite the recombination rate or the kind of recombinational operator involved and, in fact, by looking solely at these plots, not only is impossible to distinguish one kind of recom-bination from another but also more efficient systems from less efficient ones. Indeed, populations subjected to recombination alone become invari-ably less and less diverse with time, becoming totally incapable of adapta-tion. Obviously, if populations converge to this stage before finding a good solution, they become irrevocably stuck in that point if no other, non-ho-mogenizing operators are available. As the small success rates obtained for recombination emphasize (see Figure 12.1), when populations evolve exclu-sively by recombination, most of the times they converge before finding a good solution. This shows that recombination should never be used as the only source of genetic variation if an efficient adaptation is what is desired.

Note also that the oscillations on average fitness are less pronounced in homogenizing dynamics. Nevertheless, oscillations increase slightly with recombination rate showing that highly recombining populations are more resilient and will take more time before they become stagnant.

In summary, recombination is a homogenizing operator and, therefore, inappropriate to create genetic diversity; in the long run, populations exclu-sively subjected to recombination become stagnant. Consequently, recombi-nation is removed from the center of the evolutionary storm and new roles must be proposed for this operator. For one thing, it can play an important role in the duplication of genes if it is combined with other genetic operators

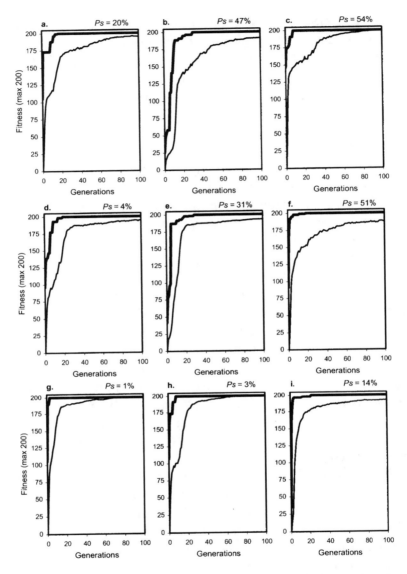

Figure 12.4. Evolutionary dynamics characteristic of populations undergoing two-point recombination (plots **a**, **b**, and **c**), one-point recombination (plots **d**, **e**, and **f**), and gene recombination (plots **g**, **h**, and **i**). The success rate above each plot was evaluated in the experiment shown in Figure 12.1. **a)** $p_{2r} = 0.3$. **b)** $p_{2r} = 0.6$. **c)** $p_{2r} = 1.0$. **d)** $p_{1r} = 0.3$. **e)** $p_{1r} = 0.6$. **f)** $p_{1r} = 1.0$. **g)** $p_{gr} = 0.3$. **h)** $p_{gr} = 0.6$. **i)** $p_{gr} = 1.0$. Note that all kinds of recombination generate the same type of homogenizing dynamics. Note also that these dynamics are very different from the dynamics of mutation and transposition (see Figures 12.2 and 12.3).

such as gene transposition. For another, it can play the role of a status quo agent, maintaining populations steady by permanently working at homogenizing their genetic makeup. Indeed, species are stable entities from their birth to their extinction (Eldredge and Gould 1972) and operators such as homologous recombination are fundamental to maintaining the status quo.

On the other hand, we also observed that the performance of a system not only changes dramatically with mutation rate but also that the performance peak is accessible to mutation alone. Therefore, mutation rates can be easily tuned so that systems could evolve with maximum efficiency. In fact, mutation rates are themselves tightly controlled and subjected to selection pressures in nature, another indication that mutation, and not recombination, is the center of the evolutionary storm.

And finally, we also observed that transposition operators display dynamics similar to those of mutation (i.e., non-homogenizing dynamics) and that populations undergoing transposition evolve significantly better than populations undergoing recombination alone, further emphasizing the unique, homogenizing effect of recombination.

12.2 The Founder Effect

We have seen in the previous section that the evolvability of a system will depend heavily on the kind of genetic operator used to create genetic modification. And the size and kind of initial populations is closely related to this question.

In all evolutionary algorithms, an evolutionary epoch or run starts with an initial population. Initial populations, though, are generated in many different ways, and the performance and the costs (in terms of CPU time) of different algorithms depend greatly on the characteristics of initial populations. The simplest and less time consuming initial population is the totally random initial population. However, few evolutionary algorithms are able to use this kind of initial population due not only to structural constraints but also to the kind of genetic operators available to create genetic modification. The initial populations of gene expression programming are totally random and consist of the linear genomes of the individuals of the population.

In artificial evolutionary systems, the question of the initial diversity is pertinent for two main reasons. First, for some complex problems, the random generation of viable individuals (i.e., individuals with positive fitness) can be a rare event and, in those cases, it would be advantageous if the evolutionary process could get started from one or a few founder individuals; whether this is possible or not, will depend on the modification mechanisms available to the system. And, second, because of this, the kind of mechanism used to create genetic modification becomes of paramount importance. If genetic modification is created by non-homogenizing operators such as point mutation, then populations will be able to adapt and evolve. However, if genetic variation is created by homogenizing operators such as recombination, then evolution is either altogether halted when only one founder individual is available or seriously compromised when the number of founder individuals is excessively small.

The importance of the initial diversity in evolution was stressed by Ernst Mayr in what he called founder effect speciation (Mayr 1954, 1963). This process may be thought of as the establishment of a new population due to a founder event initiated by genetic drift and followed by natural selection. An extreme case of a founder event is the colonization of a previously uninhabited area by a single pregnant female. In nature, besides recombination, other genetic operators are used to create modification and populations that pass through a bottleneck are capable of adaptation, sometimes even originating new species.

Similarly, in artificial evolutionary systems, the capability of founder populations to evolve depends greatly on the kind of mechanism used to create genetic modification. Indeed, if homogenizing operators are the only source of genetic modification, populations will not be able to evolve efficiently or not at all in the extreme case of only one founder individual.

In this section, we will analyze the importance of the initial diversity in evolution in two different systems. The first evolves under mutation and has a non-homogenizing dynamics characteristic of an efficient adaptation. The second evolves under recombination and has a homogenizing dynamics characteristic of poorly evolving systems.

12.2.1 Choosing the Population Types

In order to quantify accurately how different populations respond to the number of actual founders in initial populations, a simple, exactly solved problem must be chosen. This problem must allow the comparison of dissimilarly performing genetic operators, such as the high-performing point mutation and the less powerful recombination. In addition, the populations chosen to make the comparisons must follow different evolutionary dynamics so that the results discussed here could be useful not only theoretically but also for understanding the evolutionary strategies chosen by different artificial evolutionary systems.

For this analysis, we are going to use the already familiar test sequence (5.14) of section 5.6.1. In all the experiments, the first 10 positive integers n and their corresponding term were used as fitness cases (see Table 5.5); the fitness function was evaluated by equation (3.3b) and a selection range of 20% and maximum precision (0% error) were chosen, giving maximum fitness $f_{max} = 200$; population sizes P of 50 individuals and evolutionary times $G = 100$ generations were used; the success rate of each experiment was evaluated over 100 independent runs; F = {+, -, *, /} and the terminal set **T** consisted only of the independent variable; and six-genic chromosomes of length 78 (head length $h = 6$) linked by addition were used. The parameters of all the five experiments are summarized in Table 12.2.

We have already seen that point mutation is by far the single most important genetic operator and that populations undergoing mutation display

Table 12.2
Success rates and parameters for a non-homogenizing system undergoing mutation (**Mut**) and homogenizing systems undergoing two-point recombination (**Rec2P**), one-point recombination (**Rec1P**), gene recombination (**RecG**), and three different kinds of recombination (**RecMix**).

	Mut	Rec2P	Rec1P	RecG	RecMix
Number of runs	100	100	100	100	100
Number of generations	100	100	100	100	100
Population size	50	50	50	50	50
Number of fitness cases	10	10	10	10	10
Head length	6	6	6	6	6
Number of genes	6	6	6	6	6
Chromosome length	78	78	78	78	78
Mutation rate	0.05	--	--	--	--
Two-point recombination rate	--	1.0	--	--	0.8
One-point recombination rate	--	--	1.0	--	0.8
Gene recombination rate	--	--	--	1.0	0.8
Selection range	20%	20%	20%	20%	20%
Precision	0%	0%	0%	0%	0%
Success rate	96%	0.04%	0.03%	0.0%	0.13%

non-homogenizing dynamics. Furthermore, mutation is the only genetic operator capable of reaching the performance peak and, for each system, this peak can be found. For this particular problem, the performance peak is shown in Figure 12.5. In this case, maximum performance is reached around a mutation rate $p_m = 0.05$. Therefore, this value will be used throughout this study. The healthy and strong non-homogenizing dynamics of Figure 12.6 was obtained for a successful run of the experiment summarized in the first column of Table 12.2.

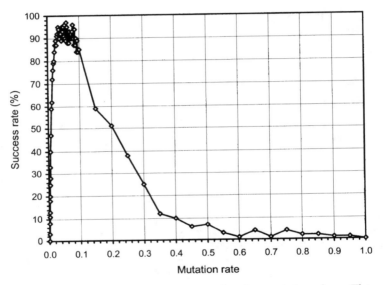

Figure 12.5. Determining the performance peak using mutation alone. The success rate was evaluated over 100 identical runs.

As for recombination, this operator is the less powerful of all GEP operators and populations undergoing recombination alone display homogenizing dynamics. Furthermore, we have also seen in the previous section that the three kinds of GEP recombination (two-point, one-point, and gene recombination) perform better at maximum rates of 1.0, and that two-point recombination is the most powerful of the three recombinational operators whereas gene recombination is the less powerful. However, for this particular problem and the particular settings chosen for this analysis, when used separately, the three kinds of recombination perform so poorly that the three recombinational operators were combined together so that the performance

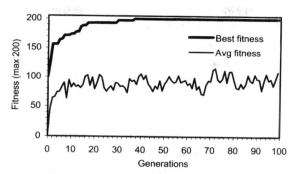

Figure 12.6. Evolutionary dynamics characteristic of non-homogenizing systems. This population evolved under a mutation rate of 0.05. Note the oscillatory pattern on average fitness and the wide gap between best and average fitness.

of the algorithm increased a little (see Table 12.2). Note that the success rate increases slightly comparatively to the individual performances obtained for the recombinational operators working separately. Thus, for this study, we will choose the general settings shown in the fifth column of Table 12.2. Notwithstanding, these populations exhibit the same kind of homogenizing dynamics characteristic of populations undergoing only one type of recombination at a time (Figure 12.7). This further reinforces the hypothesis that recombination is conservative and, therefore, plays a major role at maintaining the status quo. Note that, in this particular case, by generation 54 the plots for average and best fitness merged and all individuals became genetically identical. This might be seen as a good thing, especially if all the individuals became equal and perfect. Recall, however, that in complex real-

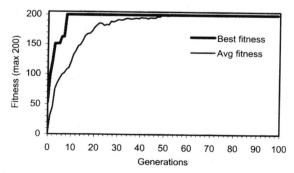

Figure 12.7. Homogenizing populations undergoing extensive recombination. This dynamics was obtained for a successful run of the experiment summarized in the fifth column of Table 12.2.

world problems, as in nature, there is no such thing as perfection and minor improvements are always possible. The disadvantages of such an evolutionary strategy, however, become evident when average fitness reaches best fitness before a perfect or good solution is found. Figure 12.8 shows such a case where the population stabilized on a mediocre solution. In this particular case, after generation 86 adaptation became impossible because all individuals share the same genetic makeup. Indeed, the small success rates typical of populations undergoing recombination alone (see, for instance, Figure 12.1 and Table 12.2) indicate that, most of the times, homogenizing populations converge before finding a good solution because they become irrevocably stuck in some local point, not necessarily optimal.

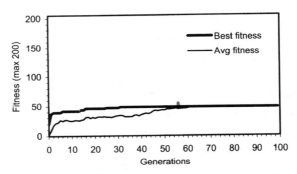

Figure 12.8. Convergence on a mediocre solution in homogenizing populations. The parameters used in this experiment are exactly the same as in Figure 12.7.

It is worth noticing that in the experiments summarized in Table 12.2, totally random initial populations were used and, therefore, the number of viable individuals in those initial populations was not controlled. In the next section I will show how the number of viable individuals in initial populations can be rigorously controlled in order to analyze the founder effect in artificial evolutionary systems.

12.2.2 The Founder Effect in Simulated Evolutionary Processes

For this analysis, a smaller, totally random "initial" population (founder population) composed of a certain number of viable individuals is created. That is, the run only starts when all the members of the founder population are viable or, in other words, have positive fitness. These founder individuals are afterwards selected and reproduced, leaving as many descendants as the actual population size $P = 50$.

As shown in Figure 12.9, for non-homogenizing populations there is no correlation between success rate and the initial diversity. Indeed, due to the constant introduction of genetic modification in the population, in non-homogenizing populations, after a certain time, the founder effect is completely erased and populations evolve, as usual, efficiently.

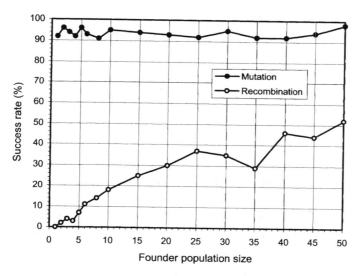

Figure 12.9. Dependence of success rate on the size of the founder population in non-homogenizing populations undergoing mutation alone (mutation rate equal to 0.05) and homogenizing populations undergoing recombination alone (two-point, one-point and gene recombination rates all equal to 0.8). The success rate was evaluated over 100 independent runs.

A very different situation happens in populations where crossover is the only source of genetic diversity and the evolutionary dynamics are homogenizing in effect. In these cases, there is a strong correlation between success rate and initial diversity. Note that populations evolve poorly under recombination, being practically incapable of adaptation in the cases where only 2-5 founder individuals are used (obviously, for cases with only one founder, homogenizing populations are altogether incapable of adaptation). It is worth emphasizing that, in these systems, even when the size of the founder population is equal to P, the success rate is significantly smaller than in populations undergoing mutation with only one viable individual in the founder population (52% and 92%, respectively).

Also worth considering is that the computational resources required to guarantee the creation of large founder populations are very expensive. Thus, systems such as GEP capable of evolving efficiently with minimal initial diversity are extremely advantageous. Furthermore, for some complex problems like, for instance, the discovery of cellular automata rules for the density-classification task of section 4.4, it is very difficult to generate randomly a viable individual, or even a mediocre one, to start the run. In those cases, systems such as GEP can use this individual as founder and continue from there, whereas systems relying on recombination alone will be stuck for a long time before they gather momentum. Indeed, in gene expression programming, due to the varied set of non-homogenizing genetic operators, there is no need for large founder populations because as long as one viable individual is randomly generated the evolutionary process can get started.

12.3 Testing the Building Block Hypothesis

We have already seen in the previous section the limits of recombination of building blocks. When systems rely exclusively on existing building blocks and are incapable of creating new ones through mutation or other mechanisms, they become severely constrained. Here, we will pursue this question further, comparing three different systems: one that constantly introduces variation in the population, and two different systems that can only recombine a particular kind of building block – GEP genes. Recall that GEP genes have defined boundaries and, through gene recombination and gene transposition, it is possible to test new combinations of these building blocks without disrupting them.

For this analysis, we are going to work with the same sequence induction problem of the previous section, using the general parameters presented in Table 12.3. For the first experiment, we are going to use a mix of all the genetic operators; for the second, we are going to use solely gene recombination at $p_{gr} = 1.0$; and for the last experiment, we are going to allow a more generalized shuffling of building blocks by combining gene recombination ($p_{gr} = 1.0$) with gene transposition ($p_{gt} = 0.5$).

In this analysis, instead of creating a founder population as was done in the previous section, we are going to study the progression of success rate

Table 12.3

Parameters for a healthy and strong system undergoing several kinds of genetic modification at the same time (**All Op**) and two other systems evolving exclusively by recombining genes (**GR** and **GR+GT**).

	All Op	GR	GR + GT
Number of runs	100	100	100
Number of generations	50	50	50
Number of fitness cases	10	10	10
Function set	+ - * /	+ - * /	+ - * /
Head length	6	6	6
Number of genes	4	4	4
Linking function	+	+	+
Chromosome length	52	52	52
Mutation rate	0.0384	--	--
One-point recombination rate	0.3	--	--
Two-point recombination rate	0.3	--	--
Gene recombination rate	0.1	1.0	1.0
IS transposition rate	0.1	--	--
IS elements length	1,2,3	--	--
RIS transposition rate	0.1	--	--
RIS elements length	1,2,3	--	--
Gene transposition rate	0.1	--	0.5
Selection range	25%	25%	25%
Precision	0%	0%	0%

with population size (Figure 12.10). As you can see, this study emphasizes further the results obtained in the previous section: systems incapable of introducing constantly new genetic material in the genetic pool evolve poorly. Furthermore, if the building blocks are only moved around and not somehow disrupted, the system is practically incapable of evolving. Only by using relatively big populations was it possible to solve this simple problem, albeit very inefficiently, when only the moving around of building blocks was available. Remember, though, that real-world problems are much more complex than the problem analyzed here and, therefore, more powerful search operators such as mutation should always be used.

In summary, the moving around of building blocks (Holland 1975) has only a limited evolutionary impact: without mutation (or other non-homogenizing operators) adaptation is so slow and requires such numbers of individuals that it becomes ineffective.

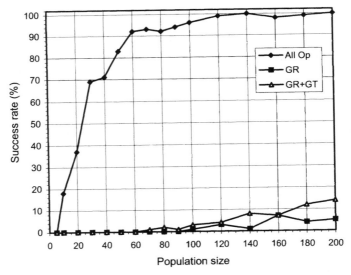

Figure 12.10. Dependence of success rate on population size in healthy and strong populations evolving under a mix of several operators (All Op) and populations evolving exclusively by recombining genes (GR: $p_{gr} = 1.0$; and GR+GT: $p_{gr} = 1.0$ and $p_{gt} = 0.5$). The success rate was evaluated over 100 identical runs.

12.4 The Role of Neutrality in Evolution

We already know that the automatic evolution of computer programs can only be done smoothly and efficiently if a genuine genotype/phenotype mapping is available. The creation of such a mapping requires some creative thinking because proteins and computer programs are very different things. Thankfully, computer programs are much easier to understand than proteins and it is not necessary to know, for instance, the rules that determine the three-dimensional structure of proteins in order to create a simple genotype/phenotype system capable of evolving computer programs. What are, then, the fundamental properties common to the DNA/protein system and an artificial system especially designed to evolve computer programs? Obviously, the first is the creation of the genome/program dyad; and second, no matter what, the genome must always produce valid programs. And how can that be accomplished?

Turning to nature for inspiration can help. How does the DNA/protein system cope with complexity? Is the information somehow fragmented in the genome? Then perhaps the fragmentation of the genome in genes could also be useful in a simple artificial evolutionary system. And what about expression in nature? Is all the information encoded in the genome always expressed? How is it possible to differentiate the information that gets to be expressed from the one that remains silent? Why is differentiation important? Might this also be of any use in artificial evolutionary systems? Although the answers to all these questions are still being sought, what is known is that, in nature, genomes are vastly redundant, with lots and lots of so called junk DNA which is never expressed: highly repetitive sequences, introns, pseudogenes, and so forth. So, most probably, the introduction of junk sequences in an artificial genome can also be useful.

The genetic representation used in gene expression programming explores both the fragmentation of the genome in genes and the existence of junk sequences or noncoding regions in the genome. As Kimura hypothesized (Kimura 1983), the accumulation of neutral mutations plays an important role in evolution. And the noncoding regions of GEP chromosomes are ideal places for the accumulation of neutral mutations. In this section, we will analyze the importance of neutral regions in the genome and, consequently, the importance of neutral mutations in evolution by using the fully functional genotype/phenotype system of gene expression programming.

For this analysis, two simple, exactly solved test problems were chosen. These problems can be solved using both unigenic and multigenic systems. On the one hand, the extent of noncoding regions in unigenic systems can be easily increased by increasing the gene length. And on the other, in multigenic systems the number of noncoding regions can be increased by increasing the number of genes.

The first problem chosen for this analysis is a function finding problem where the test function (4.1) of section 4.1.1 was used. And the second is a more difficult sequence induction problem where the test sequence (5.14) was used (this sequence was also used in sections 12.2 and 12.3).

For the function finding problem, a set of 10 random fitness cases chosen from the interval [-10, 10] was used (see Table 4.2); the fitness function was evaluated by equation (3.1b) and a selection range of 25% and a precision of 0.01% were chosen, giving maximum fitness $f_{max} = 250$; and population sizes P of 30 individuals and evolutionary times G of 50 generations were chosen.

For the sequence induction problem, as usual, the first 10 positive integers n and their corresponding a_n term were used as fitness cases (see Table 5.5); the fitness function was also evaluated by equation (3.3b) and a selection range of 25% and maximum precision (0% error) were chosen, thus giving

$f_{max} = 250$; the population size and the evolutionary time were increased, respectively, to 50 and 100 as this problem is slightly more difficult than the function finding one.

In all the experiments, the function set F = {+, -, *, /} and the terminal set T consisted only of the independent variable which was represented by a, giving T = {a}; genetic modification was introduced using a mutation rate of 0.03, an IS and RIS transposition rates of 0.1 with a set of three transposons of lengths 1, 2, and 3, and two-point and one-point recombination rates of 0.3; in multigenic systems, gene recombination and gene transposition were also used as sources of genetic modification, both at rates of 0.1 and the linking was made by addition; as usual, the selection was made by roulette-wheel sampling coupled with simple elitism and the success rate was evaluated over 100 independent runs.

12.4.1 Genetic Neutrality in Unigenic Systems

The importance of genetic neutrality in unigenic systems can be easily analyzed in GEP by increasing the gene length (Figure 12.11). After finding the most compact organization that allows the discovery of a perfect, extremely

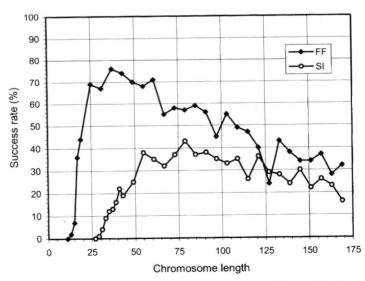

Figure 12.11. Variation of success rate with chromosome length for the function finding (FF) and sequence induction (SI) problems. The success rate was evaluated over 100 independent runs.

parsimonious solution to the problem at hand, any increase in gene length can lead to the evolution of perfect, less parsimonious solutions in which both neutral blocks and noncoding regions might appear.

For problems exactly solved by the algorithm, the most compact organization can be found in most cases. As shown in Figure 12.11, the test function (4.1) can be compactly encoded using a head length of six (corresponding to $g = 13$). One such solution composed of 13 nodes is shown below:

```
0123456789012
*++*//aaaaaaa
```

Note that, in this case, all the elements of the gene are expressed and therefore no noncoding regions exist.

The test sequence (5.14), however, requires more nodes for its correct and parsimonious expression using the chosen set of functions. As shown in Figure 12.11, an $h = 14$ (corresponding to $g = 29$) is the minimum head length necessary to solve this problem. The two perfect, parsimonious solutions shown below are expressed using, respectively, 25 and 23 nodes:

```
012345678901234567890123456789012345678
+*aa+*++**++/+aaaaaaaaaaaaaaa
```

```
012345678901234567890123456789012345678
**a+a*a++/+/+/aaaaaaaaaaaaaaa
```

Note that the first gene has a small noncoding region composed of four elements, whereas the second has a larger noncoding region with six elements.

As Figure 12.11 emphasizes, the most compact organizations are not the most efficient. For instance, in the function finding problem, the success rate obtained for the most compact organization ($g = 13$) is only 2% whereas the highest success rate, obtained for a chromosome length of 37, is 76%. In the sequence induction problem, an identical behavior is observed with a success rate of only 1% for the most compact organization ($g = 29$) and 43% for the best chromosome length ($g = 79$). Therefore, a certain amount of redundancy is fundamental for evolution to occur efficiently. Indeed, in both examples, a plateau was found where the system evolves best. Note also that highly redundant systems adapt, nonetheless, considerably better than highly compact systems, showing that evolutionary systems can cope fairly well with genetic redundancy. For instance, in the function finding experiment, the most redundant system with a gene length of 169, has a success rate of 32%, considerably higher than the 2% obtained for the most compact organization with $g = 13$. The same is observed in the sequence induction experiment, where the success rate of the most redundant organization ($g = 169$) is 16% compared to 1% for the most compact organization ($g = 29$).

The structural analysis of compact organizations and less compact ones can also be useful for understanding the role of redundancy in evolution. For instance, the following perfect solutions to the function (4.1) were discovered using, respectively, head lengths of 6, 18, and 48 (only the K-expressions are shown):

```
(a)  0123456789012
     *++*//aaaaaaa

(b)  012345678901234567890123456
     +-*+/+a*/**a*/a++-aaaaaaaaa

(c)  0123456789012345678901234567890123456789...
     /+aa++*+aa+*-a--+-*+*+/**a*a-*--+*/-/-/a...

         ...012345678901234567890123456789012345 6
         ...aa//-/*-aaaaaaaaaaaaaaaaaaaaaaaaaaaaaa
```

Note that the first solution with an $h = 6$ is expressed using 13 nodes, and therefore the entire gene was used for its expression; the second solution with an $h = 18$ is expressed using 31 nodes, and therefore has a noncoding region with six elements; and the last solution with an $h = 48$ uses for its expression only 77 of the 97 elements of the gene and therefore has a noncoding region with 20 elements. Note also that not only the length of the noncoding region increases from the most compact to the less compact (0, 6, and 20, respectively) but also increases the number of redundant or neutral motifs. For instance, two neutral motifs using a total of 14 nodes can be counted on the medium compact solution and seven neutral motifs involving 38 nodes can be counted on the less compact one. This phenomenon is known as code bloat in genetic programming and many have argued about its evolutionary function (Angeline 1994, 1996; Nordin et al. 1995). Like all kinds of genetic redundancy, neutral motifs are most probably beneficial whenever used in good measure. This can be rigorously evaluated using gene expression programming, although such an analysis would require lots of time. However, what was learned from the analysis shown in Figure 12.11 (and also from Figure 12.12 in the next section) and what is known about the evolution of proteins or technological artifacts suggest an important role for neutral regions in evolution. Their nonexistence or their excess results most probably in an inefficient evolution whereas their existence in good measure, as shown here, is extremely beneficial.

12.4.2 Genetic Neutrality in Multigenic Systems

Another way of analyzing genetic neutrality in gene expression programming, consists of increasing the number of genes. For that purpose a compact, unigenic system capable of exactly solving the problem at hand is chosen as the starting point. Thus, for the function finding problem, the starting point is a unigenic system with a head length of six (gene length equal to 13), and a head length of 14 (gene length equal to 29) for the sequence induction problem (see Figure 12.11).

As shown in Figure 12.12, the results obtained for multigenic systems further reinforce the importance of neutrality in evolution. Note that, in both experiments, the efficiency of the system increases dramatically with the introduction of a second gene and that compact unigenic systems are much less efficient than less compact, multigenic ones. For instance, in the function finding problem, the compact, unigenic system (chromosome length c equal to 13) has a success rate of 2% whereas the success rate in the two-genic system ($c = 26$) is 94%. In the sequence induction problem, the success

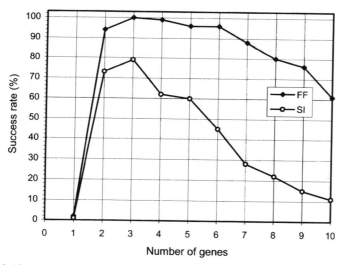

Figure 12.12. Variation of success rate with the number of genes for the function finding (FF) and sequence induction (SI) problems. The success rate was evaluated over 100 independent runs.

rate in the unigenic system ($c = 29$) is only 1%, whereas in the less compact two-genic system ($c = 58$) it is 73%. Again, a plateau is observed where systems are most efficient, showing that a certain amount of redundancy is fundamental for evolution to occur efficiently. Indeed, it is intuitively understood that a certain room to experiment, not only by forming new building blocks but also by rejecting existing ones, is essential to come up with something new and useful. If no room is there to play, only costly (if ever) one comes up with a good solution to the problem at hand. Again, note that highly redundant systems are more efficient than extremely compact ones. For instance, the 10-genic system used in the function finding problem ($c = 130$), has a success rate of 61% compared to only 2% obtained with the most compact organization ($c = 13$). The same behavior can be observed in the sequence induction problem where the highly redundant 10-genic system with a chromosome length of 290 performs slightly better than the extremely compact one ($c = 29$) (1% and 11%, respectively).

The comparison of Figures 12.11 and 12.12 also shows that multigenic systems perform considerably better than unigenic ones (see also section 12.5, The Higher Organization of Multigenic Systems). For instance, in the function finding problem, from two-genic to six-genic systems (corresponding to chromosome lengths 26 through 78) the success rates are in all cases above 94% and the best has a success rate of 100% (see Figure 12.12), whereas the best chromosome length in the unigenic system ($c = 37$ and $h = 18$) achieved only 76% (see Figure 12.11). The same phenomenon can be observed in the sequence induction problem in which from two-genic to five-genic systems (corresponding to chromosome lengths 58 through 145) the success rates are above 60% and the best has a success rate of 79% (see Figure 12.12) whereas the best chromosome length in the unigenic system ($c = 79$ and $h = 39$) achieved only 43% (see Figure 12.11).

The structural analysis of compact solutions and less compact ones can also provide some insights into the role of redundancy in evolution. For instance, the following perfect solutions to the sequence induction problem were obtained, respectively, for systems with one, three, and five genes:

```
(a)  012345678901234567890123345678
     **a+a*a++/+/+/aaaaaaaaaaaaaaaa
```

```
(b)  012345678901234567890123345678
     **a+/*+a//++//aaaaaaaaaaaaaaaa
     /*a+/*+a//++//aaaaaaaaaaaaaaaa
     -*a+/-*a//++aaaaaaaaaaaaaaaaaa
```

```
(c)  012345678901234567890123456780
     -aa+*///-+aa/*aaaaaaaaaaaaaa
     --/*a/+/**/-/-aaaaaaaaaaaaaa
     +/-*-*a-/a+*a-aaaaaaaaaaaaaa
     aa/a/*a+*+/+/+aaaaaaaaaaaaaa
     /-/*aa++**/+/+aaaaaaaaaaaaaa
```

Note that not only the total length of the noncoding regions increases from the most compact to the less compact but also increases the number of neutral motifs. Specifically, the length of the noncoding region in the unigenic solution is only six and no neutral motifs are present; for the three-genic solution, the total length of the noncoding regions is 16 and three neutral motifs involving a total of 12 nodes can be counted; for the five-genic solution, the noncoding regions encompass already 66 elements and there are six neutral motifs involving a total of 31 nodes.

Because of their clarity, the results presented in this section are extremely useful for understanding the role of genetic neutrality both in artificial and natural evolution. As shown here, there are two different kinds of neutral regions in GEP: the neutral motifs within the ORFs and the noncoding regions at the end of the ORFs. In simple replicator systems such as GP or GAs, only the former exists whereas in genotype/phenotype systems such as the DNA/protein system or GEP, both kinds exist. And the presence of noncoding regions in genotype/phenotype systems is certainly entangled with the higher efficiency observed in these systems. For instance, introns in DNA are believed to be excellent targets for crossover, allowing the recombination of different building blocks without their disruption (e.g., Maynard Smith and Szathmáry 1995). The noncoding regions of GEP genes can also be used for this purpose and, indeed, whenever the crossover points are chosen within these regions, entire ORFs are exchanged. Furthermore, the noncoding regions of GEP genes are ideal places for the accumulation of neutral mutations that can be later activated and integrated into coding regions. This is an excellent source of genetic variation and certainly contributes to the increase in performance observed in redundant systems.

But, at least in gene expression programming, the noncoding regions play another, much more fundamental role: they allow the modification of the genome by numerous high performing genetic operators. And here by "high performing" I mean genetic operators that always produce valid structures. This problem of valid structures applies only to artificial evolutionary systems for in nature there is no such thing as an invalid protein. How and why the DNA/protein system got this way is not known, but certainly there were selection pressures to get rid of imperfect genotype/phenotype mappings. The fact that the noncoding regions of GEP allow the creation of a perfect genotype/phenotype mapping, further reinforces the importance of neutrality in evolution, as a good mapping is essential for the crossing of the phenotype threshold.

On the other hand, the reason why neutral motifs within structures, be they parse trees or proteins, can boost evolution, is not so easy to understand, although I think this is another manifestation of the same phenomenon of recombining and testing smaller building blocks. In this case, the building blocks are not entire genes with clear boundaries, but smaller domains within genes. Indeed, in nature, most proteins have numerous variants in which different amino acid substitutions took place. These amino acid substitutions occur mostly outside the crucial domains of proteins such as the active sites of enzymes and, therefore, the protein variants work equally well or show slight differences in functionality. At the molecular level, these variants constitute the real genetic diversity, that is, the raw material of evolution. The neutral motifs of gene expression programming play exactly the same function, allowing the recombination and testing of different building blocks and, at the same time, allowing the creation of neutral variants that can ultimately diverge and give birth to better adapted structures.

12.5 The Higher Organization of Multigenic Systems

Gene expression programming is the only evolutionary algorithm that deals with genes as separated entities, tied up, however, in a more complex structure – the chromosome. From the analysis of the previous section, it is clear that multigenic systems are far superior than unigenic ones. Here, we are going to make a more systematic analysis by comparing multigenic and unigenic systems with exactly the same chromosome length. The problems chosen for this analysis are exactly the same of the previous section (that is, the same function finding problem and the same sequence induction problem) using also the same general settings (see Tables 12.4 and 12.5).

For this analysis, a common chromosome size of 75 was chosen for three different chromosomal organizations: a unigenic system with $h = 37$; three-genic chromosomes with $h = 12$; and five-genic chromosomes with $h = 7$. The performance of all these systems was measured in terms of success rate and is shown in Tables 12.4 and 12.5.

Table 12.4

Comparing the performance of unigenic and multigenic systems on the function finding problem.

	1 Gene	3 Genes	5 Genes
Number of runs	100	100	100
Number of generations	50	50	50
Population size	30	30	30
Number of fitness cases	10	10	10
Function set	+ - * /	+ - * /	+ - * /
Head length	37	12	7
Number of genes	1	3	5
Linking function	--	+	+
Chromosome length	75	75	75
Mutation rate	0.03	0.03	0.03
One-point recombination rate	0.3	0.3	0.3
Two-point recombination rate	0.3	0.3	0.3
Gene recombination rate	--	0.1	0.1
IS transposition rate	0.1	0.1	0.1
IS elements length	1,2,3	1,2,3	1,2,3
RIS transposition rate	0.1	0.1	0.1
RIS elements length	1,2,3	1,2,3	1,2,3
Gene transposition rate	--	0.1	0.1
Selection range	25%	25%	25%
Precision	0.01%	0.01%	0.01%
Success rate	58%	93%	98%

As expected, multigenic systems are significantly more efficient than unigenic ones and should always be our first choice. There might be problems, though, for which the fractionating of the chromosome in genes is of little advantage. For instance, when we are trying to evolve a solution to a problem best modeled by the logarithm or the square root function of some complex expression and a cellular system with automatic linking was not our first choice. But, even in those cases, the system can easily find ways of turning the unnecessary genes into neutral genes and, therefore, an efficient adaptation can still occur. But of course, in cellular and multicellular systems, the modeling of all kinds of functions can benefit from multiple genes as those systems are not constrained by a particular kind of linking function.

Table 12.5

Comparing the performance of unigenic and multigenic systems on the sequence induction problem.

	1 Gene	3 Genes	5 Genes
Number of runs	100	100	100
Number of generations	100	100	100
Population size	50	50	50
Number of fitness cases	10	10	10
Function set	+ - * /	+ - * /	+ - * /
Head length	37	12	7
Number of genes	1	3	5
Linking function	--	+	+
Chromosome length	75	75	75
Mutation rate	0.03	0.03	0.03
One-point recombination rate	0.3	0.3	0.3
Two-point recombination rate	0.3	0.3	0.3
Gene recombination rate	--	0.1	0.1
IS transposition rate	0.1	0.1	0.1
IS elements length	1,2,3	1,2,3	1,2,3
RIS transposition rate	0.1	0.1	0.1
RIS elements length	1,2,3	1,2,3	1,2,3
Gene transposition rate	--	0.1	0.1
Selection range	25%	25%	25%
Precision	0%	0%	0%
Success rate	41%	79%	96%

12.6 The Open-ended Evolution of GEP Populations

We have already seen that the populations of gene expression programming can be made to evolve efficiently because the genetic operators allow the permanent introduction of new material in the genetic pool. Here we are going to explore this question further, analyzing the variation of success rate with evolutionary time.

In this section, again was used the sequence induction problem of section 5.6.1 as it is a considerably difficult problem with an exact solution.

In the first analysis, the dependence of success rate on evolutionary time was analyzed for healthy and strong populations of 50 and 250 individuals each (Figure 12.13). In this experiment, all the genetic operators were switched on. The rates used are shown in the first column of Table 12.6.

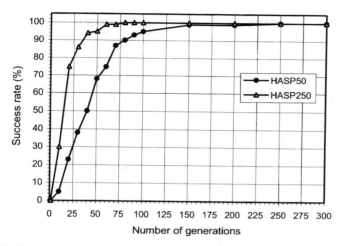

Figure 12.13. Variation of success rate with evolutionary time for healthy and strong populations of 50 (HASP50) and 250 individuals (HASP250). The success rate was evaluated over 100 identical runs.

Table 12.6
Sources of genetic variation used in healthy and strong populations (**HASP**) and homogenizing populations (**HP**).

	HASP	HP
Number of runs	100	100
Number of fitness cases	10	10
Function set	+ - * /	+ - * /
Head length	7	7
Number of genes	5	5
Linking function	+	+
Chromosome length	75	75
Mutation rate	0.044	--
One-point recombination rate	0.3	1.0
Two-point recombination rate	0.3	1.0
Gene recombination rate	0.1	1.0
IS transposition rate	0.1	--
IS elements length	1,2,3	--
RIS transposition rate	0.1	--
RIS elements length	1,2,3	--
Gene transposition rate	0.1	--
Selection range	25%	25%
Precision	0%	0%

Note that after 150 generations the success rate practically reached the maximum value for both systems. As expected, the performance, measured exclusively in terms of success rate, is superior for the system with bigger populations. However, if we compare the CPU time required for each system we will see that the 50-chromosome system is more efficient. For instance, 100 runs of 150 generations take 1'30" for the 50-chromosome system and 3'55" for populations of 250 individuals. However, if we disable the stop criterion (i.e., the system does not stop when a perfect solution is found) and let the system go through all of the 150 generations, we obtain the values of 4'14" for populations of 50 individuals and 29'42" for populations of 250 individuals. This is no idle comparison as in real-world problems perfect solutions (i.e., solutions with 0% or 0.01% of error) do not usually exist and, consequently, the system does not stop before completing the stipulated number of generations.

In summary, in gene expression programming, as long as mutation is used, it is advantageous to use small populations of 30-100 individuals for they allow an efficient evolution in record time.

These facts are further supported by the second experiment where only recombination was used as source of genetic diversity (Figure 12.14). The types of recombination used in this experiment and their rates are shown in the second column of Table 12.6.

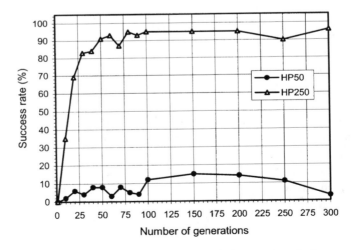

Figure 12.14. Variation of success rate with evolutionary time for homogenizing populations of 50 (HP50) and 250 individuals (HP250). The success rate was evaluated over 100 identical runs.

This analysis clearly shows the importance of the permanent introduction of genetic variation through mutation and similar non-homogenizing operators such as RIS or IS transposition. Note that populations of 50 individuals evolve, in this case, very inefficiently. And the often spoken fatalist remark about the inability of GP populations to evolve beyond 50 generations can be fully understood. As Figure 12.14 emphasizes, homogenizing systems show no appreciable increase in success rate after 50 generations. Remember that genetic programming uses almost exclusively a GP-specific recombination (sub-tree crossover) as the only source of genetic variation. Here we can see that even recombinational mechanisms more disruptive than the recombination used in genetic programming (namely, one-point and two-point recombination), are inadequate to make populations evolve efficiently. This obviously accounts for some of the superiority of GEP over GP, although, of course, the most important is the genotype/phenotype representation and all the things this representation brings with itself. The plot for the 250-members population indicates that only the use of huge population sizes will permit an efficient evolution in those systems. Note, however, that not even for such large population sizes was it possible to surpass the results obtained for populations of only 50 individuals when mutation and transposition were switched on (see Figure 12.13).

12.7 Analysis of Different Selection Schemes

It is known that the fruits of selection are better seen with time. Thus, we are going to analyze the performance of three different selection schemes using two different problems by evaluating the variation of success rate with evolutionary time.

The selection schemes I chose to study include the already known roulette-wheel selection used in all the problems of this book, a two-players tournament selection, and a deterministic selection scheme. For all the schemes, even deterministic selection, the cloning of the best individual is also done in order to allow the use of a fair amount of genetic modification without loosing the best trait.

The tournament selection with elitism works as follows: two individuals are randomly picked up and the best of them is chosen to reproduce once. If they happen to have the same fitness, then one of them is randomly chosen to reproduce. Thus, each generation, for populations of N individuals, N tournaments are made so that the population size is kept unchanged.

In the deterministic selection scheme, individuals are selected proportionately to their fitnesses, but less fit individuals are excluded from the life banquet. One can choose to exclude more or less individuals from the banquet, and I have tested several deterministic selection schemes with different exclusion levels. An exclusion factor E of 1.5 seems to be, most of the times, the best compromise.

To apply this exclusion factor, the viability (i.e., the fitness divided by average fitness) of each individual is multiplied by E. Imagine a round table where the slices begin to be occupied by rank and proportionally to viability, starting with the best individual and finishing when no more places are available on the table. The higher the exclusion factor the more individuals die without leaving offspring and only the cream of the population reproduces. Obviously, one must be careful with the degree of exclusion, otherwise the genetic diversity might be excessively reduced and evolution might be seriously compromised.

The problems chosen for this study are two test problems of sequence induction. The first is the sequence (5.14) of section 5.6.1 and the second is the more difficult sequence (12.1). The general settings chosen for both experiments are summarized in Table 12.7.

Figures 12.15 and 12.16 compare the variation of success rate obtained for the three selection schemes over a wide time span. In both experiments, the tournament selection scheme is slightly inferior to both the roulette-wheel and deterministic selection. Deciding between the roulette-wheel and

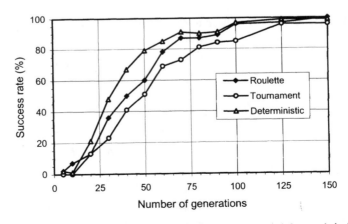

Figure 12.15. Comparison of roulette-wheel, tournament, and deterministic selection schemes on the simpler test sequence (5.14).

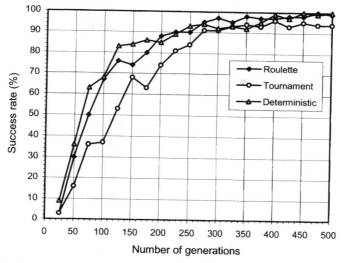

Figure 12.16. Comparison of roulette-wheel, tournament, and deterministic selection schemes on the more difficult test sequence (12.1).

Table 12.7

General settings for the sequence induction experiments used to compare different selection schemes.

	SI1	SI2
Number of runs	100	100
Population size	50	50
Number of fitness cases	10	10
Function set	+ - * /	+ - * /
Head length	7	6
Number of genes	5	7
Linking function	+	+
Chromosome length	75	91
Mutation rate	0.044	0.044
One-point recombination rate	0.3	0.3
Two-point recombination rate	0.3	0.3
Gene recombination rate	0.1	0.1
IS transposition rate	0.1	0.1
IS elements length	1,2,3	1,2,3
RIS transposition rate	0.1	0.1
RIS elements length	1,2,3	1,2,3
Gene transposition rate	0.1	0.1
Selection range	25%	25%
Precision	0%	0%

deterministic selection is more problematic as for more complex problems the plots for both schemes become more intertwined (Figure 12.16).

However, there are several reasons why one should choose the roulette-wheel selection. (1) Real-world problems are more complex than the problems analyzed here. (2) The ideal exclusion factor depends on population size and the complexity of the problem. (3) Deterministic selection requires more CPU time as individuals must be sorted by rank and, for large populations, this is a factor to take seriously into consideration. (4) Deterministic selection is not appropriate in systems undergoing recombination alone as it reduces dramatically the genetic diversity of the population. And (5), roulette-wheel selection is easy to implement and mimics nature more faithfully and therefore is much more appealing.

This book finishes here and I hope to have shown you that gene expression programming is not only easy to implement but also easy to understand. Although natural evolution seems sometimes something beyond our grasp, the simple artificial system of gene expression programming can be minutely dissected in order to reveal all its secrets.

And, most of all, I hope to have convinced you that gene expression programming can help you find very good solutions to difficult real-world problems, not easily or satisfactorily solved by conventional mathematical or statistical methods.

Bibliography

Angeline, P. J., 1994. Genetic Programming and Emergent Intelligence. In K. E. Kinnear Jr., ed., *Advances in Genetic Programming*, chapter 4, pages 75-98, MIT Press, Cambridge, MA.

Angeline, P. J., 1996. Two Self-adaptive Crossover Operators for Genetic Programming. In P. J. Angeline and K. E. Kinnear, eds., *Advances in Genetic Programming 2*, chapter 5, pages 89-110, MIT Press, Cambridge, MA.

Banzhaf, W., 1994. Genotype-Phenotype-Mapping and Neutral Variation: A Case Study in Genetic Programming. In Y. Davidor, H.-P. Schwefel, and R. Männer, eds., *Parallel Problem Solving from Nature III*, *Lecture Notes in Computer Science* 866: 322-332, Springer-Verlag.

Banzhaf, W., P. Nordin, R. E. Keller, and F. D. Francone, 1998. *Genetic Programming: An Introduction: On the Automatic Evolution of Computer Programs and its Applications*, Morgan Kaufmann.

Bonachea, D., E. Ingerman, J. Levy, and S. McPeak, 2000. An Improved Adaptive Multi-Start Approach to Finding Near-Optimal Solutions to the Euclidean TSP. In D. Whitley, D. Goldberg, E. Cantu-Paz, L. Spector, I. Parmee, and H.-G. Beyer, eds., *Proceedings of the Genetic and Evolutionary Computation Conference*, 143-150, Las Vegas, Nevada, Morgan Kaufmann.

Budihardjo, A., J. Grzymala-Busse, and L. Woolery, 1991. Program LERS_LB 2.5 as a Tool for Knowledge Acquisition in Nursing, *Proceedings of the 4th International Conference on Industrial and Engineering Applications of AI and Expert Systems*, pp. 735-740.

Cestnik, G., I. Konenenko, and I. Bratko, 1987. Assistant-86: A Knowledge-Elicitation Tool for Sophisticated Users. In I. Bratko and N. Lavrac, eds., *Progress in Machine Learning*, 31-45, Sigma Press.

Clark, P. and T. Niblett, 1987. Induction in Noisy Domains. In I. Bratko and N. Lavrac, eds., *Progress in Machine Learning*, 11-30, Sigma Press.

Cramer, N. L., 1985. A Representation for the Adaptive Generation of Simple Sequential Programs. In J. J. Grefenstette, ed., *Proceedings of the First International Conference on Genetic Algorithms and Their Applications*, 183-187, Erlbaum.

Das, R., M. Mitchell, and J. P. Crutchfield, 1994. A Genetic Algorithm Discovers Particle-based Computation in Cellular Automata. In Y. Davidor, H.-P. Schwefel, and R. Männer, eds., *Parallel Problem Solving from Nature III*, Springer-Verlag.

Dawkins, R., 1995. *River out of Eden*, Weidenfeld and Nicolson.

Eldredge, N. and S. J. Gould, 1972. Punctuated Equilibria: An Alternative to Phyletic Gradualism. In T. J. M. Schopf, ed., *Models in Paleobiology*, Freeman.

Ferreira, C., 2001. Gene Expression Programming: A New Adaptive Algorithm for Solving Problems. *Complex Systems* 13 (2): 87-129.

Ferreira, C., 2002a. Mutation, Transposition, and Recombination: An Analysis of the Evolutionary Dynamics. In H. J. Caulfield, S.-H. Chen, H.-D. Cheng, R. Duro, V. Honavar, E. E. Kerre, M. Lu, M. G. Romay, T. K. Shih, D. Ventura, P. P. Wang, Y. Yang, eds., *Proceedings of the 6th Joint Conference on Information Sciences, 4th International Workshop on Frontiers in Evolutionary Algorithms*, pages 614-617, Research Triangle Park, North Carolina, USA.

Ferreira, C., 2002b. Combinatorial Optimization by Gene Expression Programming: Inversion Revisited. In J. M. Santos and A. Zapico, eds., *Proceedings of the Argentine Symposium on Artificial Intelligence*, pages 160-174, Santa Fe, Argentina.

Ferreira, C., 2002c. Genetic Representation and Genetic Neutrality in Gene Expression Programming. *Advances in Complex Systems* 5 (4): 389-408.

Ferreira, C., 2002d. Analyzing the Founder Effect in Simulated Evolutionary Processes Using Gene Expression Programming. In A. Abraham, J. Ruiz-del-Solar, and M. Köppen, eds., *Soft Computing Systems: Design, Management and Applications*, pp. 153-162, IOS Press, Netherlands.

Ferreira, C., 2003. Function Finding and the Creation of Numerical Constants in Gene Expression Programming. In J. M. Benitez, O. Cordon, F. Hoffmann, and R. Roy, eds., *Advances in Soft Computing: Engineering Design and Manufacturing*, pages 257-266, Springer-Verlag.

Fisher, R. A., 1936. The Use of Multiple Measurements in Taxonomic Problems. *Annual Eugenics* 7 (2): 179-188. (Reprinted in *Contributions to Mathematical Statistics*, 1950. New York, John Wiley.)

Futuyma, D. J., 1998. *Evolutionary Biology*, 3rd edition, Sunderland, MA: Sinauer Associates.

Goldberg, D. E., 1989. *Genetic Algorithms in Search, Optimization, and Machine Learning*, Addison-Wesley.

Haupt, R. L. and S. E. Haupt, 1998. *Practical Genetic Algorithms*, Wiley-Interscience.

Holland, J. H., 1975. *Adaptation in Natural and Artificial Systems: An Introductory Analysis with Applications to Biology, Control, and Artificial Intelligence*, University of Michigan Press (second edition: MIT Press, 1992).

Hsu, W. W. and C.-C. Hsu, 2001. The Spontaneous Evolution Genetic Algorithm for Solving the Traveling Salesman Problem. In L. Spector, E. D. Goodman, A. Wu, W. B. Langdon, H.-M. Voigt, M. Gen, S. Sen, M. Dorigo, S. Pezeshk, M. H. Garzon, and E. Burke, eds., *Proceedings of the Genetic and Evolutionary Computation Conference*, pages 359-366, San Francisco, California, Morgan Kaufmann.

Iba, H. and T. Sato, 1992. Meta-level Strategy for Genetic Algorithms Based on Structured Representations. In *Proceedings of the Second Pacific Rim International Conference on Artificial Intelligence*, 548-554.

Iba, H., T. Sato, and H. de Garis, 1994. System Identification Approach to Genetic Programming. In *Proceedings of the First IEEE Conference on Evolutionary Computation*, Vol. I: 401-406, Piscataway, NJ, IEEE Press.

Ivakhnenko, A. G., 1971. Polynomial Theory of Complex Systems. *IEEE Transactions on Systems, Man, and Cybernetics* 1 (4): 364-378.

Johnson, D. S. and L. A. McGeoch, 1997. The Traveling Salesman Problem: A Case Study. In E. H. L. Aarts and J. K. Lenstra, eds., *Local Search in Combinatorial Optimization*, 215-310, Wiley & Sons, New York.

Juillé, H. and J. B. Pollack, 1998. Coevolving the "Ideal" Trainer: Application to the Discovery of Cellular Automata Rules. In J. R. Koza, W. Banzhaf, K. Chellapilla, M. Dorigo, D. B. Fogel, M. H. Garzon, D. E. Goldberg, H. Iba, and R. L. Riolo, eds., *Genetic Programming 1998: Proceedings of the Third Annual Conference*, Morgan Kaufmann, San Francisco, CA.

Kargupta, H. and R. E. Smith, 1991. System Identification with Evolving Polynomial Networks. In R. K. Belew and L. B. Booker, eds., *Proceedings of the Fourth International Conference on Genetic Algorithms*, 370-376, San Mateo, California, Morgan Kaufmann.

Katayama, K. and H. Narihisa, 1999. Iterated Local Search Approach Using Gene Transformation to the Traveling Salesman Problem. In W. Banzhaf, ed., *Proceedings of the Genetic and Evolutionary Computation Conference*, 321-328, Morgan Kaufmann.

Kimura, M., 1983. *The Neutral Theory of Molecular Evolution*, Cambridge University Press, Cambridge, UK.

Keith, M. J. and M. C. Martin, 1994. Genetic Programming in C++: Implementation Issues. In K. E. Kinnear, ed., *Advances in Genetic Programming*, MIT Press.

Keller, R. E. and W. Banzhaf, 1996. Genetic Programming Using Genotype-Phenotype Mapping from Linear Genomes into Linear Phenotypes. In J. R. Koza, D. E. Goldberg, D. B. Fogel, and R. L. Riolo, eds., *Genetic Programming 1996: Proceedings of the First Annual Conference*, MIT Press.

Koza, J. R., F. H. Bennett III, D. Andre, and M. A. Keane, 1999. *Genetic Programming III: Darwinian Invention and Problem Solving*. San Francisco: Morgan Kaufmann.

Koza, J. R., 1992. *Genetic Programming: On the Programming of Computers by Means of Natural Selection*, Cambridge, MA: MIT Press.

Koza, J. R., 1994. *Genetic Programming II: Automatic Discovery of Reusable Programs*, Cambridge, MA: MIT Press.

Langley, P., H. A. Simon, G. L. Bradshaw, and J. M. Zytkow, 1987. *Scientific Discovery: Computational Explorations of the Creative Process*, MIT Press.

Larrañaga, P., C. M. H. Kuijpers, and R. H. Murga, Tackling the Traveling Salesman Problem: Representations and Operators, *Technical Report*, 1998.

Lynch, M. and J. S. Conery, 2000. The Evolutionary Fate and Consequences of Duplicated Genes, *Science* 290: 1151-1155.

Margulis, L., 1970. *Origin of Eukaryotic Cells*, Yale University Press.

Margulis, L. and D. Sagan, 1986. *Origins of Sex: Three Billion Years of Recombination*, Yale University Press.

Margulis, L. and D. Sagan, 1997. *What is Sex?*, Simon and Schuster, New York.

Mathews, C. K., K. E. van Holde, and K. G. Ahern, 1999. *Biochemistry*, 3rd ed., Benjamin/Cummings.

Maynard Smith, J. and E. Szathmáry, 1995. *The Major Transitions in Evolution*, W. H. Freeman.

Mayr, E., 1954. Change of Genetic Environment and Evolution. In J. Huxley, A. C. Hardy, and E. B. Ford, eds., *Evolution as a Process*, 157-180, Allen and Unwin, London.

Mayr, E., 1963. *Animal Species and Evolution*, Harvard University Press, Cambridge, Massachusetts.

Merz, P. and B. Freisleben, 1997. Genetic Local Search for the TSP: New Results. In T. Bäck, Z. Michalewicz, and X. Yao, eds., *Proceedings of the 1997 IEEE International Conference on Evolutionary Computation*, 159-164, Piscataway, NJ, IEEE Press.

Michalewicz, Z., 1996. *Genetic algorithms + Data Structures = Evolution Programs*, 3rd ed., Springer-Verlag, Berlin, Heidelberg.

Michalski, R., I. Mozetic, J. Hong, and N. Lavrac, 1986. The Multi-Purpose Incremental Learning System AQ15 and its Testing Applications to Three Medical Domains. *Proceedings of the Fifth National Conference on Artificial Intelligence*, 1041-1045, Philadelphia, PA, Morgan Kaufmann.

Mitchell, M., 1996. *An Introduction to Genetic Algorithms*, MIT Press.

Mitchell, M., P. T. Hraber, and J. P. Crutchfield, 1993. Revisiting the Edge of Chaos: Evolving Cellular Automata to Perform Computations. *Complex Systems* 7: 89-130.

Mitchell, M., J. P. Crutchfield, and P. T. Hraber, 1994. Evolving Cellular Automata to Perform Computations: Mechanisms and Impediments. *Physica D* 75: 361-391.

Nikolaev, N. I. and H. Iba, 2001. Accelerated Genetic Programming of Polynomials. *Genetic Programming and Evolvable Machines* 2: 231-257.

Nordin, P., F. Francone, and W. Banzhaf, 1995. Explicitly Defined Introns and Destructive Crossover in Genetic Programming. In J. P. Rosca, ed., *Proceedings of the Workshop on Genetic Programming: From Theory to Real-World Applications*, pages 6-22, Tahoe City, California.

Papadimitriou, C. H. and K. Steiglitz, 1982. *Combinatorial Optimization*, Prentice Hall.

Prechelt, L., 1994. PROBEN1 – A Set of Neural Network Benchmark Problems and Benchmarking Rules. *Technical Report* 21/94, Karlsruhe University, Germany.

Quinlan, J. R., 1986. Induction of Decision Trees. *Machine Learning* 1 (1): 81-106.

Reinelt, G., 1994. The Traveling Salesman: Computational Solutions for TSP Applications. *Lecture Notes in Computer Science* 496: 188-192, Springer-Verlag, Berlin, Germany.

Ryan, C., J. J. Collins, and M. O'Neill, 1998. Grammatical Evolution: Evolving Programs for an Arbitrary Language. In *Proceedings of the First European Workshop on Genetic Programming*, *Lecture Notes in Computer Science* 1391: 83-95, Springer-Verlag.

Tank, D. W. and J. J. Hopfield, 1987. Collective Computation in Neuronlike Circuits. *Scientific American* 257 (6): 104-114.

Toffoli, T. and N. Margolus, 1987. *Cellular Automata Machines: A New Environment for Modeling*, MIT Press.

Urey, H. C., 1952. *The Planets: Their Origin and Development*, Yale University Press.

Whitley, L. D. and J. D. Schaffer, eds., 1992. *COGANN-92: International Workshop on Combinations of Genetic Algorithms and Neural Networks*. IEEE Computer Society Press.

Wolfram, S., 1983. Statistical Mechanics of Cellular Automata. *Reviews of Modern Physics* 55: 601-644. (Reprinted in S. Wolfram, 1994. *Cellular Automata and Complexity: Collected Papers*, Addison-Wesley.)

Wolfram, S., 1986. *Theory and Applications of Cellular Automata*, World Scientific.

Zielinski, L. and J. Rutkowski, 2004. Design Tolerancing with Utilization of Gene Expression Programming and Genetic Algorithm. In *Proceedings of the International Conference on Signals and Electronic Systems*, Poznan, Poland.

Index